Lecture Notes in Computational Vision and Biomechanics

Volume 16

For further volumes:
http://www.springer.com/series/8910

The research related to the analysis of living structures (Biomechanics) has been a source of recent research in several distinct areas of science, for example, Mathematics, Mechanical Engineering, Physics, Informatics, Medicine and Sport. However, for its successful achievement, numerous research topics should be considered, such as image processing and analysis, geometric and numerical modelling, biomechanics, experimental analysis, mechanobiology and enhanced visualization, and their application to real cases must be developed and more investigation is needed. Additionally, enhanced hardware solutions and less invasive devices are demanded.

On the other hand, Image Analysis (Computational Vision) is used for the extraction of high level information from static images or dynamic image sequences. Examples of applications involving image analysis can be the study of motion of structures from image sequences, shape reconstruction from images and medical diagnosis. As a multidisciplinary area, Computational Vision considers techniques and methods from other disciplines, such as Artificial Intelligence, Signal Processing, Mathematics, Physics and Informatics. Despite the many research projects in this area, more robust and efficient methods of Computational Imaging are still demanded in many application domains in Medicine, and their validation in real scenarios is matter of urgency.

These two important and predominant branches of Science are increasingly considered to be strongly connected and related. Hence, the main goal of the LNCV&B book series consists of the provision of a comprehensive forum for discussion on the current state-of-the-art in these fields by emphasizing their connection. The book series covers (but is not limited to):

- Applications of Computational Vision and Biomechanics
- Biometrics and Biomedical Pattern Analysis
- Cellular Imaging and Cellular Mechanics
- Clinical Biomechanics
- Computational Bioimaging and Visualization
- Computational Biology in Biomedical Imaging
- Development of Biomechanical Devices
- Device and Technique Development for Biomedical Imaging
- Digital Geometry Algorithms for Computational Vision and Visualization
- Experimental Biomechanics
- Gait & Posture Mechanics
- Multiscale Analysis in Biomechanics
- Neuromuscular Biomechanics
- Numerical Methods for Living Tissues
- Numerical Simulation
- Software Development on Computational Vision and Biomechanics
- Grid and High Performance Computing for Computational Vision and Biomechanics
- Image-based Geometric Modeling and Mesh Generation
- Image Processing and Analysis
- Image Processing and Visualization in Biofluids
- Image Understanding
- Material Models
- Mechanobiology
- Medical Image Analysis
- Molecular Mechanics
- Multi-Modal Image Systems
- Multiscale Biosensors in Biomedical Imaging
- Multiscale Devices and Biomems for Biomedical Imaging
- Musculoskeletal Biomechanics
- Sport Biomechanics
- Virtual Reality in Biomechanics
- Vision Systems

Jorge Belinha

Meshless Methods in Biomechanics

Bone Tissue Remodelling Analysis

Jorge Belinha
Mechanical Engineering Department
Universidade do Porto
Porto
Portugal

ISSN 2212-9391 ISSN 2212-9413
ISBN 978-3-319-06399-7 ISBN 978-3-319-06400-0 (eBook)
DOI 10.1007/978-3-319-06400-0
Springer Cham Heidelberg New York Dordrecht London

Library of Congress Control Number: 2014937281

Printed on acid-free paper

Springer is part of Springer Science+Business Media (www.springer.com)

*To my Family and to my Love, for all the support and encouragement****

To my mentors, Professor Lúcia Dinis and Professor Natal Jorge, for the guidance and friendship

To the unbreakable power of the human spirit and the transcendent force of friendship, love and faith

Thank you

Preface

Mechanical forces act on tissues and organs inducing their movement and/or deformation. This area is a well-known subject of research being called as biomechanics. Sometimes the consequences of mechanical forces are evident, such as when a bone fractures or a ligament is injured by stretching. In the last decades, innovative research has been developed to accurately determine the mechanical behaviour of tissues and organs. Knowledge of how tissues deform and fail is important, but it is even more important to know how mechanical forces act on tissues to maintain health and to regulate biological processes. Therefore, the scientific community is now modelling the actions and their associated responses, ranging from the organ scale, passing through biological tissue, down to individual cells and molecules.

Nowadays, numerical simulation plays a fundamental role in many branches of science, in particular in mechanics. Computational biomechanics is one of the areas in which the numerical simulation of very complex processes takes place. Computational biomechanics is a relatively recent and emergent discipline. Since the appearance of the finite element method in the 1950s it has been undoubtedly the most extended tool to perform such simulation in that field. However, the high complexity of the geometries involved (bones, soft tissues, organs, etc.) and frequently, the large deformations that usually appear make the aspects related to the mesh creation and control an important fact to be considered. Although the finite element method appeared more than five decades ago, other numerical methods such as meshless methods have been successfully used to solve problems in engineering and applied sciences. One can say that the term 'meshless method' refers to a broad class of numerical techniques for solving a growing number of science and engineering applications without the dependence on an underlying computational mesh. Currently, the diversity of problems analysed by these methods is very large, and ranges from fracture mechanics, fluid mechanics, laminated composites, multiscale problems and biomechanics, among others.

This book is a significant contribution to the state of the art in the field of computational biomechanics, from the application of meshless methods in biomechanics to the evaluation of stresses in hip prosthesis replacement. The Natural Neighbour Radial Point Interpolation Method (NNRPIM), a recent truly meshless method is presented and developed with special focus on biomechanics. The theoretical fundaments of NNRPIM is presented and it is extended to several engineering fields such as solid mechanics static and dynamic linear analysis and structural nonlinear analysis. A special focus is on its application to achieve biomechanical analysis of bone remodelling.

January 2014 Renato Natal Jorge

Contents

About the Author

The author, Jorge Belinha, joined our research group, the 'Unit of Design and Experimental Validation', at Institute of Mechanical Engineering (IDMEC) in 2002. Since then he has been an active member in our research team and has dedicated his activities in the area of Meshless Methods and Applications. He has produced an M.Sc. and a Ph.D. thesis and has produced a significant number of research papers in relevant international journals. Additionally, he has participated in a number of international and national conferences where he presented papers and discussed his results with other researchers.

He is actually our main expert in meshless methods and is keen on developing applications in various fields, namely in Biomechanics, where we in the group have a number of members working. He is easy to communicate with and has a friendly approach towards all members of the research unit and other researchers in Faculty of Engineering of the University of Porto (FEUP) and outside institutions.

He has lectured a number of courses in the area of Meshless Methods and supervised several M.Sc. theses in the referred field. He has also lectured in other fields, namely Solid Mechanics and Finite Element Method. He is very keen on participating in activities organised by him and/or other researchers, and is always an active member participating in or creating research proposals either for internal or external financing.

For me it has been a pleasure working with him since the very beginning. We have had long conversations about his work and he is always very keen on proposing new ideas that will allow him to exploit the meshless methods, besides the fact of being a hard and conscientious worker who can be trusted and makes everyone to be a good friend. With this short note I expect to give a good impression about the qualities of the author.

FEUP, January 2014

Lúcia Maria de Jesus Simas Dinis
(Associate Professor in Faculty of Engineering
of the University of Porto)

Chapter 1
Introduction

Abstract The bioengineering design process can be divided in many specific phases. The present work lean over three of these phases: the Modulation, the Simulation and the Analysis. All this process is recurrent by nature, always looking for the numerical approach which better reproduces the studied phenomenon. At the present time, there are many numerical methods available and capable to successfully handle the referred bioengineering design process phases. This work presents, develop and extends a new advance discretization meshless technique. From the simple solid mechanical problems to the complex nonlinear bone tissue remodelling analysis in biomechanics, this work shows that the proposed numerical method is flexible and accurate.

1.1 Meshless Methods

In the last few years meshless methods for numerically solving partial differential equations came into focus of interest, especially in the engineering community. In the meshless methods [1–3] the nodes can be arbitrary distributed, once the field functions are approximated within an influence-domain rather than an element. In opposition to the no-overlap rule between elements in the Finite Element Method (FEM) [4, 5], in meshless methods the influence-domains may and must overlap each other. It is possible to define and classify a numerical method by three fundamental modules: the field approximation (or interpolation) function, the used formulation and the integration.

1.1.1 Approximation or Interpolation Functions

There are many approximation (or interpolation) functions available. The most relevant are the Taylor approximation, the moving least-square approximation, the

J. Belinha, *Meshless Methods in Biomechanics*, Lecture Notes in Computational Vision and Biomechanics 16, DOI: 10.1007/978-3-319-06400-0_1,

reproducing kernel approximation, the hp-cloud approximation function, the polynomial interpolation, the parametric interpolation, the radial interpolation and the Sibson interpolation. The approximation, or interpolation, function requires a domain of applicability, outside this domain the function assumes zero values. In the FEM this domain is the 'element' and in the meshless methods this domain is called 'influence-domain'. In meshless methods it is necessary to determine the influence-domain for each node within the nodal distribution discretizing the problem domain, as a consequence the shape and the size of the influence-domain vary with the considered node. The technique used to defined the influence-domains varies with the used meshless method.

1.1.2 Formulation

Additionally, meshless methods can be classified in two categories, a first category that pursues the strong form solution and another that seeks the weak form solution. The strong formulation uses directly the partial differential equations governing the studied physical phenomenon in order to obtain the solution. The weak formulation uses a variational principle to minimize the residual weight of the differential equations ruling the phenomenon. The residual is obtained by substituting the exact solution by an approximated function affected by a test function. The existent distinct weak form solution methods are dependent on the used test function. Surprisingly a differential equation may have solutions which are not exactly differentiable and the weak formulation allows to find such solutions. Weak form solutions are very important since many differential equations governing the real world phenomena do not admit sufficiently smooth solutions, then the only way of solving such equations is using the weak formulation.

1.1.3 Integration

In order to obtain the integral of the residual weight of the differential equations it is necessary to select an integration scheme. The integration can be made using a background mesh, covering the entire problem domain, composed by integration points. These integration points must have a defined influence area and weight (which corresponds to the theoretical infinitesimal mass portion defined in the integral expression) and should not overlap each other. This background mesh for integration proposes generally is nodal independent, which jeopardises the 'meshless' denomination of such numerical methods. There are other integration schemes often used, the point collocation and the nodal integration. In these integration schemes the node represents the integration point, the influence area is the node influence-domain and the integration weight is the node influence volume. In this case the integration mesh is the nodal distribution itself. Some authors have the opinion that such integration schemes are less accurate, which is not true. Meshless methods using these integrations schemes are considered 'truly' meshless methods. In this work a hybrid integration scheme, totally nodal dependent, is presented and applied, ensuring that the proposed meshless method is 'truly' meshless.

1.1.4 Relevant Meshless Methods

The initially created meshless methods used approximation functions, since it produces smoother solutions, the implementation of the influence-domain concept was easier and the background integration scheme was nodal independent. The first meshless method using the Moving Least Square approximants (MLS) in the construction of the approximation function was the Diffuse Element Method (DEM) [6]. The MLS was proposed by Lancaster and Salkauskas [7] for surface fitting. Belytschko evolved the DEM and developed one of the most popular meshless methods, the Element Free Galerkin Method (EFGM) [8], which uses a nodal independent background integration mesh. One of the oldest meshless methods is the Smooth Particle Hydrodynamics Method (SPH) [9], which is in the origin of the Reproducing Kernel Particle Method (RKPM) [10].

Another very popular approximant meshless method is the meshless local Petrov-Galerkin method (MLPG) [11], initially created to solve linear and nonlinear potential problems, which later evolved towards the Method of the Finite Spheres (MFS) [12]. The Finite Point Method (FPM) [13–15] uses for integration proposes a stabilization technique in the collocation point method. Another approximation method, distinct from the previous, is the Radial Basis Function Method (RBFM) [16, 17]. It uses the radial basis functions, respecting a Euclidean norm, to approximate the variable fields within the entire domain or in small domains. It does not require an integration mesh and, in opposition to the previous referred meshless methods, uses the strong form formulation. Initially used to approximate multidimensional data [18] only latter it was applied by others [19, 20] to the analysis of solid mechanics differential equations.

Although approximants meshless methods have been successfully applied in computational mechanics there were several problems not completely solved. One of these problems, and perhaps the most important unsolved issue, was the lack of the Kronecker delta property on the approximation functions, which difficult the imposition of essential and natural boundary conditions.

To address the above problem, several interpolant meshless methods were developed in the last few years. The most relevant are the Point Interpolation Method (PIM) [21], the Point Assembly Method [22], the Radial Point Interpolation Method (RPIM) [23, 24], Meshless Finite Element Method (MFEM) [25]. The Natural Neighbour Finite Element Method (NNFEM) [26, 27] or the Natural Element Method (NEM) [28–30] are meshless methods that use the Sibson interpolation functions. The combination between the NEM and the RPIM originated the Natural Neighbour Radial Point Interpolation Method (NNRPIM) [31], which is the object of this work. More recently, the Natural Radial Element Method (NREM) [32–34] was developed. The NREM is an efficient and accurate truly meshless method, which presents a low order nodal connectivity.

1.2 Natural Neighbour Radial Point Interpolation Method

All the numerical examples presented in this book are analysed considering the Natural Neighbour Radial Point Interpolation Method (NNRPIM). The NNRPIM is the product of the combination of the Radial Point Interpolators (RPI) with the Natural Neighbours geometric concept. The RPI started with the Point Interpolation Method (PIM) [21]. The technique consisted in constructing polynomial interpolants, possessing the Kronecker delta property, based only on a group of arbitrarily distributed points. However this technique has too many numerical problems, for instance the perfect alignment of the nodes produces singular solutions in the interpolation function construction process. As so, this technique evolved and the Radial Point Interpolation Method (RPIM) [23] was created. Within this meshless method the Radial Basis Function (RBF) was added in the construction process of the interpolation function, stabilizing the procedure. The RBF used in these early works were the Gaussian and the multiquadric RBF. Initially the RBF was developed for data surface fitting, and later, with the work developed by Kansa [16, 17], the RBF was used for solving partial differential equations. However the RPIM uses, unlike Kansa's algorithm, the concept of "influence-domain" instead of "global-domain", generating sparse and banded stiffness matrices, more adequate to complex geometry problems.

The NNRPIM is the next step in the RPI. In order to impose the nodal connectivity, the 'influence-domain' is substituted by the 'influence-cell' concept. In order to obtain the influence-cells the NNRPIM relies on geometrical and mathematical constructions such as the Voronoï diagrams [35] and the Delaunay tessellation [36]. Thus, resorting to Voronoï cells, a set of influence-cells are created departing from an unstructured set of nodes. The Delaunay triangles, which are the dual of the Voronoï cells, are applied to create a node-depending background mesh used in the numerical integration of the NNRPIM interpolation functions. Due to the integration mesh total dependency on the nodal distribution, the NNRPIM can be considered a truly meshless method. Unlike the FEM, where geometrical restrictions on elements are imposed for the convergence of the method, in the NNRPIM there are no such restrictions, which permits a total random node distribution for the discretized problem. The NNRPIM interpolation functions, used in the Galerkin weak form, are constructed in a similar process to the RPIM, with some differences that modify the method performance.

Although the NNRPIM is a recent developed meshless method [31] it has been extended to many fields of the computational mechanics, such as the static analysis of isotropic and orthotropic plates [37] and the functionally graded material plate analysis [38], the 3D shell-like approach [39] for laminated plates and shells [40].

The dynamic analysis of several solid-mechanic problems was also studied [41–44]. The NNRPIM was also tested in more demanding applications such as the material nonlinearity [45] and the large deformation analysis [46].

1.3 Bone Tissue Remodelling Analysis

The process where bone tissue progressively modifies its morphology in order to adapt to any new external load is known as bone remodelling and it was firstly empirically noticed and reported by Wolff in 1892 [47]. Since then many increasingly sophisticated theoretical and numerical models have been developed.

Many diverse stimuli have been defined as a function of strain, stress or strain energy. A first mathematical formulation for the 'Wolff's law' was presented by Pauwels [48], in which it was assumed the existence of an optimal mechanical stimulus balancing the bone tissue resorption and deposition [49]. The 'self-optimization' concept was introduced by Carter and co-workers [50–52], disclosing the functional adaptation mathematical law for the bone trabecular structure. The algorithm proposed by Carter assumes that the mechanical stimulus is proportional to the effective stress field and permits to consider several mechanical cases.

The fabric tensor was included in the trabecular bone material model proposed by Cowin and co-workers [53]. The fabric tensor, a symmetric second order tensor, permits to correlate the trabecular microstructural arrangement and the material elasticity tensor [54, 55]. Considering the fabric tensor concept, the algorithm assumes that the trabecular arrangement seeks to adapt until an equilibrium strain state is achieved [53]. Since the number of bone remodelling parameters required by Cowin algorithm represents a disadvantage, Huiskes and co-workers [56] simplified successfully the algorithm by considering the Strain Energy Density (SED) as the mechanical stimulus for the bone tissue remodelling. This proposed 'adaptive elasticity' model [56] permitted to predict a numerical trabecular arrangement very similar with the one observed in clinical X-ray images. Along with a detailed state of the art review on bone tissue remodelling analysis using the SED optimization criterion, Pettermann and co-workers [49] suggested an approach that considers the bone spatial distribution adaptation combined with the reorientation of the material axis and the stiffness parameters. Other efficient bone tissue remodelling algorithm approaches can be found in the literature, such the cell biology based remodelling algorithm [57], the accumulated damage model [58] or the continuum damage-repair algorithm [59].

During the remodelling process, principal stress variable values and directions change locally, leading to a global anisotropic behaviour. Even so the bone clearly shows an anisotropic behaviour, early bone models were assumed isotropic. In addition to the numerical difficulties encountered when anisotropic material behaviour is considered [60, 61], these early remodelling works had to deal with the lack of a comprehensive data bank incorporating the mechanical properties of bone as a function of the material direction.

Therefore, based in experimental studies, several anisotropic bone material laws were proposed and developed [62]. Although many initial anisotropic bone material laws suggested distinct mathematical laws for the cortical and the trabecular bone, this work considers the recent experimental study of Zioupos and co-workers [63], in which it is shown that the law governing the cortical and trabecular bone mechanical behaviour is in fact the same. This idea is being corroborated by the analysis of high-resolution three-dimensional images, from which some authors were able to estimate the homogenized anisotropic mechanical properties [64].

1.3.1 Bone Tissue Remodelling Due to Femoral Implants

Generally, the hip replacement orthopaedic surgery is performed to answer to one of the following needs: to relieve arthritis pain; to repair a physically damaged joint; to replace the joint functionality after a hip fracture. This surgery, in which the hip joint is replaced by a prosthetic implant, is presently the most common orthopaedic surgical procedure. The prosthetic implant used in hip replacement orthopaedic surgery is generally composed by three parts: the acetabular cup, the articular interface and the femoral stem. The selection process of the femoral implant is a very important step, since the success of the medical operation depends on choosing the right implant for the right patient. Currently, there are a numerous variety of femoral implants available in the specialized market, with distinct shapes, materials and functionalities. The hip replacement therapy, with cemented or cementless femoral implants, induces in the femur bone the adaptive remodelling effect [65, 66]. The main purpose of using cementless femoral implants is to avoid stress-shielding, by ensuring a smooth physiological transfer of loads from the prosthetic head to the femur diaphysis. However, it has been verified that distinct cementless femoral implants models lead to stress-shielding [66]. In a large number of research works studding the quantification of the mass changes of the bone tissue after a hip replacement, it was observed a significant degree of atrophy in the proximal region due to an effective lack of loads or due to a shift of the load magnitudes [66, 67].

The Finite Element Method (FEM) [68] is the most common numerical tool used to analyse femoral bone implants, allowing to estimate the stress field changes on the bone domain produced by the insertion of the femoral implant system [69]. Using the FEM analysis combined with a bone tissue remodelling algorithm, it is also possible to predict the long-term influence of the prosthetic replacement in the bone mass of the femur [66, 70]. In the literature it is possible to find several research works studying the influence of: the prosthesis geometry; the material properties of the implant stem; the type of fixation [71–73]. Other authors have compared distinct stem shapes performance for hip prosthesis using a FEM static and dynamic analysis [74, 75].

1.3.2 Bone Tissue Remodelling Due to Dental Implants

Dental implants are an efficient therapy usually prescribed to partial edentulous or total edentulous patients. These biocompatible medical devices are fixed prostheses subjected to recurring loads and should transfer efficiently and smoothly, to the bone tissue, the loads applied during masticatory activity. The dental implant is surgically placed into the mandibular or maxillary bone to support a prosthetic tooth crown. The implant insertion changes the natural mechanic scenario in the mandible or maxillary bone, triggering the adaptation and remodelling of the trabecular structure of the bone tissue in the implant surroundings.

In the numerical analysis of dental implants, the FEM is also the discrete numerical method most frequently used to obtain the variable fields required by the numerous remodelling algorithms described in the literature [76–78]. In the work of Lin and co-workers [79] it is presented an extensive review on mandible and maxillary bone remodelling as a result of dental implant insertion. Using the FEM, in the work of Mellal and co-workers [80], the bone remodelling predictions due to the implant insertion were assessed using three different models, and in the work of Li and co-workers [81] it was presented an innovate bone remodelling algorithm capable of simulating both underload and overload bone resorptions in dental implant treatments. More recently other authors [82, 83] were able to predict a more accurate the mandible bone trabecular architecture, achieving solutions more close with clinical results [84].

1.3.3 Meshless Methods in Biomechanics

When compared with the FEM, meshless methods possess several advantages such as the re-meshing efficiency, which permits to deal with the large distortions of soft materials (muscles, internal organs, skin, etc.) or to explicitly simulate the fluid flow (the hemodynamics, the swallow, the respiration, etc.). The accuracy and smoothness of the stress fields obtained with meshless methods are also very useful to predict the remodelling process of biological tissues and the rupture or damage of such biomaterials. Additionally, recent works [73, 85] shown that combined with scanning techniques, such as CT and MRI, meshless methods are more efficient than the FEM.

Regarding the study of biomechanical problems considering large strains, the NEM was used to study the human lateral collateral ligament and the human knee joint, showing clear advantages over the FEM [86]. More recently, the EFGM was extended to the nonlinear explicit dynamic analysis to simulate the brain tissue response [87]. The results confirmed the accuracy of the EFGM to deal with highly demanding nonlinear hyperelastic biomaterials.

The SPH is frequently used to simulate hemodynamics. Researchers were able to simulate the motion of a deformable red blood cell in flowing blood plasma [88] and to study numerically the effect of red blood cells on the primary thrombus formation [89].

Meshless method gradually begin to enter the bone remodelling application field [90]. Liew and co-workers [91] presented one of the first works dealing with bone structures and using meshless methods. A simple stress analysis of a femoral bone model was performed and some meshless methods limitations are identified. All the meshless limitations indicated [91] do not represent a difficulty for the NNRPIM, which can easily deal with the non-convex boundaries and the material discontinuities in the bone structure. Other authors applied the meshless methods to the bone tissue analysis [92, 93] and more recently Belinha and co-workers [94, 95] presented a new bone tissue remodelling algorithm relying on the meshless method accuracy.

1.4 Book Purpose

The present book results from the author's MSc and PhD monographs [96, 97] and from the support texts of the unit course: "Meshless Methods: A Introduction Course" directed and lectured by the author. Therefore, the main purpose of this work is to provide an explanatory academic text book on meshless methods (with a particular emphasis on the NNRPIM), which can be used and understood by university students and researchers interested in meshless methods for computational mechanics.

Regarding the bone tissue remodelling analysis, in this book a gradient remodelling algorithm is used, in which the bone tissue anisotropic material properties gradually vary through the model domain, in accordance with the proposed phenomenological anisotropic material law and the lower SED regions [94, 95].

Thus, the present work expects to contribute with new efficient numerical tools and strategies, such is the use of the proposed bone tissue gradient remodelling algorithm combined with the NNRPIM, in order to improve the trabecular bone tissue remodelling analysis field.

1.5 Meshless Method Software

During the last 10 years the author developed an original meshless code, containing more than 50.000 code lines, written in FORTRAN©. The meshless code permits to analyse several solid mechanical problems using the EFGM, the RPIM, the NNRPIM, the NREM and the FEM (for comparison purposes).

The code is capable of solving various engineering problems, such as:

- Static and dynamic linear problems;
- Nonlinear large deformation static problems;
- Elasto-plastic static problems;
- Structural topology optimization problems;
- Crack opening path problems;
- Bone tissue remodelling analysis;

Additionally, the software permits to analyse the problem considering the three-dimensional deformation theory, the plane stress and the plane strain two-dimensional deformation theory, several plate shear deformation theories, such as the Reissner-Mindlin deformation theory, the third-order shear deformation theory and the unconstrained third-order shear deformation theory. The developed meshless software is being continuously improved with new application fields and analysis options.

The program allows the user to choose the material disposition along the solid domain, as well as the principal material orientation for the case of an anisotropic analysis. The program reads external mesh files generated in the CAD softwares. Nonetheless it also possesses an algorithm which permits the user to create the nodal distribution discretizing the problem domain without using the external CAD softwares. In order to visualize the obtained results the program produces data files which can be open in external CAD softwares, permitting to explicitly analyse and understand the deformation field and the stress distribution along the solid domain. Besides the pre-processing phase and the pos-processing phase, the global NNR-PIM analysis program is independent from any other external software. All the examples shown in this book were analysed using the author's meshless software.

1.6 Book Arrangement

This book is divided in seven chapters. In the first chapter it is introduced the book purpose. A general overview on the meshless method state of the art is shown and the used meshless method is briefly presented. Regarding the bone tissue remodelling numerical analysis topic, the most relevant research works are also mentioned.

In the second chapter the solid mechanics fundamentals are briefly presented. The continuum formulation is presented, along with the used weak formulation and the discrete system equations obtained for the elastostatic and elastodynamic solid mechanical problem.

The meshless method procedure is fully presented in the third chapter. The main differences and similarities between the meshless methods and the FEM are shown and explained. The enforcement of the nodal connectivity in meshless methods is described in detail, along with the introduction of meshless concepts, such as "influence-domains" and the "influence-cells". Additionally, numerical tools to construct background integration meshes are presented. The presented integration schemes permit to obtain nodal dependent and nodal independent background integration meshes. In the end of the chapter the numerical implementation using meshless methods is fully addressed.

The fourth chapter shows how to construct shape functions for meshless methods. The "support-domain" concept is firstly presented. Then, two of the most popular shape functions used in meshless methods are presented in detail: approximation functions constructed using the moving least square (MLS) approximation and interpolation functions obtained with the radial point interpolator (RPI) technique. The construction of the meshless shape functions is explicitly presented and the most important properties are demonstrated.

The fifth chapter is dedicated to the two-dimensional and the three-dimensional linear analysis. First, the RPIM and NNRPIM patch test are presented. From the patch test results, the optimal values of the most important parameters of the RPI shape functions are obtained, allowing to improve and stabilize the meshless method. Afterwards, static and dynamic benchmark examples are solved with the RPIM and the NNRPIM. The obtained results validated the meshless method and prompted advances and improvements in the RPI formulation.

The mechanobiolgy analysis can be found in the sixth chapter, first the basic concepts of the bone biology are introduced. Then, bone tissue phenomenological laws are presented, which permit to correlate the bone tissue local apparent density with the bone tissue local mechanical properties. The mathematical law proposed by the author is presented in detail. This chapter ends with an extensive presentation of the most relevant numerical approaches for the prediction of the bone tissue remodelling. Additionally, the bone tissue remodelling algorithm used in this book is presented explicitly.

In the seventh chapter it is possible to find mechanobiologic applications, where the proposed phenomenological bone tissue material law and the used remodelling algorithm are applied. Two-dimensional and three-dimensional benchmark examples are studied to validate the bone trabecular remodelling algorithm. The maxillary central incisor, the calcaneus bone and the femoral bone are analysed. In all bone examples, the obtained trabecular bone architecture is in very good agreement with the real bone X-ray images. This chapter ends with bone tissue remodelling studies in which the remodelling is triggered by the insertion of an implant. The bone tissue remodelling analysis of the mandibular bone due to the insertion of a dental implant is analysed. Trabecular distributions are obtained and the osseointegration process is studied. A similar study is performed to study the bone tissue remodelling of the femur bone when a stem is inserted in the femur head along the diaphysis.

References

1. Belytschko T, Krongauz Y, Organ D, Fleming M, Krysl P (1996) Meshless methods: an overview and recent developments. Comput Methods Appl Mech Eng 139(1):3–47
2. Gu YT (2005) Meshfree methods and their comparisons. Int J Comput Methods 2(4):477–515
3. Nguyen VP, Rabczuk T, Bordas S, Duflot M (2008) Meshless methods: A review and computer implementation aspects. Math Comput Simul 79(3):763–813
4. Zienkiewicz OC, Taylor RL (1994) The Finite Element Method, 4th edn. McGraw-Hill, London
5. Bathe KJ (1996) Finite element procedures. Prentice-Hall, Englewood Cliffs
6. Nayroles B, Touzot G, Villon P (1992) Generalizing the finite element method: diffuse approximation and diffuse elements. Comput Mech 10:307–318
7. Lancaster P, Salkauskas K (1981) Surfaces generation by moving least squares methods. Math Comput 37:141–158

8. Belytschko T, Lu YY, Gu L (1994) Element-free galerkin method. Int J Numer Meth Eng 37:229–256
9. Gingold RA, Monaghan JJ (1977) Smoothed particle hydrodynamics: theory and applications to non-spherical stars. Mon Not Astron Soc 181:375–389
10. Liu WK, Jun S, Zhang YF (1995) Reproducing kernel particle methods. Int J Numer Meth Fluids 20(6):1081–1106
11. Atluri SN, Zhu T (1998) A new meshless local Petrov-Galerkin (MLPG) approach in computational mechanics. Comput Mech 22(2):117–127
12. De S, Bathe KJ (2000) The method of finite spheres. Comput Mech 25(4):329–345
13. Oñate E, Idelsohn S, Zienkiewicz OC, Taylor RL, Sacco CA (1996) A stabilized finite point method for analysis of fluid mechanics problems. Comput Methods Appl Mech Eng 139(1–4):315–346
14. Oñate E, Idelsohn S, Zienkiewicz OC, Taylor RL (1996) A finite point method in computational mechanics—applications to convective transport and fluid flow. Int J Numer Meth Eng 39:3839–3866
15. Oñate E, Perazzo F, Miquel J (2001) A finite point method for elasticity problems. Comput Struct 79(22–25):2151–2163
16. Kansa EJ (1990) Multiquadrics-a scattered data approximation scheme with applications to computational fluid-dynamics-I surface approximations and partial derivative estimates. Comput Math Appl 19(8–9):127–145
17. Kansa EJ (1990) Multiquadrics-a scattered data approximation scheme with applications to computational fluid-dynamics-II solutions to parabolic, hyperbolic and elliptic partial differential equations. Comput Math Appl 19(8–9):147–161
18. Hardy RL (1990) Theory and applications of the multiquadrics - Biharmonic method (20 years of discovery 1968-1988). Comput Math Appl 19(8-9): 163–208
19. Ferreira AJM, Roque CMC, Martins PALS (2004) Radial basis functions and higher-order shear deformation theories in the analysis of laminated composite beams and plates. Compos Struct 66(1–4):287–293
20. Tiago CM, Leitão VMA (2006) Application of radial basis functions to linear and nonlinear structural analysis problems. Comput Math Appl 51(8):1311–1334
21. Liu GR, Gu YT (2001) A point interpolation method for two-dimensional solids. Int J Numer Meth Eng 50:937–951
22. Liu GR (2002) A point assembly method for stress analysis for two-dimensional solids. Int J Solid Struct 39:261–276
23. Wang JG, Liu GR (2002) A point interpolation meshless method based on radial basis functions. Int J Numer Meth Eng 54:1623–1648
24. Wang JG, Liu GR (2002) On the optimal shape parameters of radial basis functions used for 2-D meshless methods. Comput Methods Appl Mech Eng 191:2611–2630
25. Sergio R, Idelsohn S, Oñate E, Calvo N, Del Pin F (2003) The meshless finite element method. Int J Numer Meth Eng 58(6):893–912
26. Traversoni L (1994) Natural neighbour finite elements. In: International conference on hydraulic engineering software. Hydrosoft proceedings of computational mechanics publications, vol 2, pp 291–297
27. Sukumar N, Moran B, Semenov AY, Belikov VV (2001) Natural neighbour Galerkin methods. Int J Numer Meth Eng 50(1):1–27
28. Braun J, Sambridge M (1995) A numerical method for solving partial differential equations on highly irregular evolving grids. Nature 376:655–660
29. Sukumar N, Moran B, Belytschko T (1998) The natural element method in solid mechanics. Int J Numer Meth Eng 43(5):839–887
30. Cueto E, Doblaré M, Gracia L (2000) Imposing essential boundary conditions in the natural element method by means of density-scaled -shapes. Int J Numer Meth Eng 49(4):519–546
31. Dinis LMJS, Jorge RMN, Belinha J (2007) Analysis of 3D solids using the natural neighbour radial point interpolation method. Comput Methods Appl Mech Eng 196(13–16):2009–2028

32. Belinha J, Jorge RMN, Dinis LMJS (2013) The natural radial element method. Int J Numer Meth Eng 93(12):1286–1313
33. Belinha J, Jorge RMN, Dinis LMJS (2013) Composite laminated plate analysis using the natural radial element method. Compos Struct 103(1):50–67
34. Belinha J, Jorge RMN, Dinis LMJS (2013) Analysis of thick plates by the natural radial element method. Int J Mech Sci 76(1):33–48
35. Voronoï GM (1908) Nouvelles applications des paramètres continus à la théorie des formes quadratiques. Deuxième Mémoire: Recherches sur les parallélloèdres primitifs. Journal für die reine und angewandte Mathematik 134:198–287
36. Delaunay B (1934) Sur la sphére vide. A la memoire de Georges Voronoï. Izv. Akad. Nauk SSSR, Otdelenie Matematicheskih i Estestvennyh Nauk 7:793–800
37. Dinis LMJS, Jorge RMN, Belinha J (2008) Analysis of plates and laminates using the natural neighbour radial point interpolation method. Eng Anal Boundary Elem 32(3):267–279
38. Dinis LMJS, Jorge RMN, Belinha J (2010) An unconstrained third-order plate theory applied to functionally graded plates using a meshless method. Mech Adv Mater Struct 17:1–26
39. Dinis LMJS, Jorge RMN, Belinha J (2010) Composite laminated plates: a 3d natural neighbour radial point interpolation method approach. J Sandwich Struct Mater 12(2): 119–138
40. Dinis LMJS, Jorge RMN, Belinha J (2010) A 3D shell-like approach using a natural neighbour meshless method: isotropic and orthotropic thin structures. Compos Struct 92(5):1132–1142
41. Dinis LMJS, Jorge RMN, Belinha J (2009) The natural neighbour radial point interpolation method: dynamic applications. Eng Comput 26(8):911–949
42. Dinis LMJS, Jorge RMN, Belinha J (2011) The dynamic analysis of thin structures using a radial interpolator meshless method. In: Vasques CMA, Dias Rodrigues J (eds). Vibration and Strucutural Acoustics Analysis. Springer, Netherlands, p 1–20
43. Dinis LMJS, Jorge RMN, Belinha J (2011) Static and dynamic analysis of laminated plates based on an unconstrained third order theory and using a radial point interpolator meshless method. Comput Struct 89(19–20):1771–1784
44. Dinis LMJS, Jorge RMN, Belinha J (2011) A natural neighbour meshless method with a 3d shell-like approach in the dynamic analysis of thin 3d structures. Thin-Walled Struct 49(1):185–196
45. Dinis LMJS, Jorge RMN, Belinha J (2008) The radial natural neighbour interpolators extended to elastoplasticity. In: Ferreira AJM, Kansa EJ, Fasshauer GE, Leitao VMA (eds) Progress on meshless methods. Springer, Netherlands, pp 175–198
46. Dinis LMJS, Jorge RMN, Belinha J (2009) Large deformation applications with the radial natural neighbours interpolators. Comput Modell Eng Sci. 44(1):1–34
47. Wolff J (1986) The law of bone remodeling (Das Gesetzder Transformationder Knochen, Hirschwald, 1892). Springer, Berlin
48. Pauwels F (1956) Gesammelte abhandlungen zur funktionellen anatomie des bewegungsapparates. Springer, Berlin
49. Pettermann H, Reiter T, Rammerstorfer FG (1997) Computational simulation of internal bone remodeling. Arch Comput Methods Eng 4(4):295–323
50. Carter DR, Fyhrie DP, Whalen RT (1987) Trabecular bone density and loading history: regulation of connective tissue biology by mechanical energy. J Biomech 20(8):785–794
51. Whalen RT, Carter DR, Steele CR (1988) Influence of physical activity on the regulation of bone density. J Biomech 21(10):825–837
52. Carter DR, Orr TE, Fyhrie DP (1989) Relationship between loading history and femoral cancellous bone architecture. J Biomech 22(3):231–244
53. Cowin SC, Sadegh AM, Luo GM (1992) An evolutionary Wolff's law for trabecular architecture. J Biomech Eng 114(1):129–136
54. Cowin SC (1985) The relationship between the elasticity tensor and the fabric tensor. Mech Mater 4(2):137–147

55. Turner CH, Cowin SC, Rho JY, Ashman RB, Rice JC (1990) The fabric dependence of the orthotropic elastic constants of cancellous bone. J Biomech 23(6):549–561
56. Huiskes R, Weinans H, Grootenboer HJ, Dalstra M, Fudala B, Slooff TJ (1987) Adaptive bone-remodelling theory applied to prosthetic-design analysis. J Biomech 20(11–12):1135–1150
57. Hart RT, Davy DT (1989) Theories of bone modeling and remodeling. In: Cowin SC (ed) Bone mechanics. CRC Press, Boca Raton, pp 253–277
58. Prendergast PJ, Taylor D (1994) Prediction of bone adaptation using damage accumulation. J Biomech 27(8):1067–1076
59. Doblaré M, García JM (2002) Anisotropic bone remodelling model based on a continuum damage-repair theory. J Biomech 35(1):1–17
60. Beaupré GS, Orr TE, Carter DR (1990) An approach for time dependent bone modelling and remodelling. A preliminary remodelling simulation. J Orthop Res 8(5):662–670
61. Jacobs CR, Simo JC, Beaupré GS, Carter DR (1997) Adaptive bone remodeling incorporating simultaneous density and anisotropy considerations. J Biomech 30(6):603–613
62. Wirtz DC, Schiffers N, Pandorf T, Radermacher K, Weichert D, Forst R (2000) Critical evaluation of known bone material properties to realize anisotropic FE-simulation of the proximal femur. J Biomech 33(10):1325–1330
63. Zioupos P, Cook RB, Hutchinsonc JR (2008) Some basic relationships between density values in cancellous and cortical bone. J Biomech 41:1961–1968
64. Sansalone V, Naili S, Bousson V, Bergot C, Peyrin F, Zarka J, Laredo JD, Haiat G (2010) Determination of the heterogeneous anisotropic elastic properties of human femoral bone: from nanoscopic to organ scale. J Biomech 43(10):1857–1863
65. Huiskes R, Weinans H, Dalstra M (1989) Adaptive bone remodeling and biomechanical design considerations for noncemented total hip arthroplasty. Orthopedics 12:1255–1267
66. Herrera A, Panisello JJ, Ibarz E, Cegoñino J, Puértolas JÁ, Gracia L (2007) Long-term study of bone remodelling after femoral stem: a comparison between dexa and finite element simulation. J Biomech 40(16):3615–3625
67. Herrera A, Canales V, Anderson J, García-Araujo C, Murcia-Mazón A, Tonino AJ (2004) Seven to ten years follow-up of an anatomic hip prosthesis: an international study. Clin Orthop 423:129–137
68. Zienkiewicz OC, Taylor RL (1994) The Finite Element Method, 4th edn. McGraw-Hill, London
69. Yosibash Z, Alon Katz A, Milgrom C (2013) Toward verified and validated FE simulations of a femur with a cemented hip prosthesis. Med Eng Phys 35(7):978–987
70. Gong H, Kong L, Zhang R, Fang J, Zhao M (2013) A femur-implant model for the prediction of bone remodeling behaviour induced by cement less stem. J Bionic Eng 10:350–358
71. Weinans H, Huiskes R, Grootenboer HJ (1994) Effects of fit and bonding characteristics of femoral stems on adaptive bone remodelling. J Biomech Eng 116(4):393–400
72. van Rietbergen B, Huiskes R (2001) Load transfer and stress shielding of the hydroxyapatite-ABG hip: a study of stem length and proximal fixation. J Arthroplasty 16(8):55–63
73. Wong K, Wang L, Zhang H, Liu H, Shi P (2010) Meshfree implementation of individualized active cardiac dynamics. Comput Med Imaging Graph 34: 91–103
74. Kayabasi O, Erzincanli F (2006) Finite element modelling and analysis of a new cemented hip prosthesis. Adv Eng Softw 37:477–483
75. Senalp AZ, Kayabasi O, Kurtaran H (2007) Static, dynamic and fatigue behaviour of newly designed stem shapes for hip prosthesis using finite element analysis. Mater Des 28:1577–1583
76. Cook SD, Weinstein AM, Klawitter JJ (1982) A 3-dimensional finite-element analysis of a porous rooted Co-Cr-Mo alloy dental implant. J Dent Res 61(1):25–29
77. Geng JP, Tan KBC, Liu GR (2001) Application of finite element analysis in implant dentistry: a review of the literature. J Prosthet Dent 85(6):585–598

78. Choi AH, Conway RC, Ben Nissan B (2004) Finite element analysis of ceramic dental implants incorporated into the human mandible. Key Eng Mater-Bioceramics 16 254–256:707–712
79. Lin D, Li Q, Li W, Swain M (2009) Dental implant induced bone remodeling and associated algorithms. J Mech Behav Biomed Mater 2:410–432
80. Mellal A, Wiskott HWA, Botsis J, Scherrer SS, Belser UC (2004) Stimulating effect of implant loading on surrounding bone—comparison of three numerical models and validation by in vivo data. Clin Oral Implant Res 15:239–248
81. Li J, Li H, Shi L, Fok ASL, Ucer C, Devlin H, Horner K, Silikas N (2007) A mathematical model for simulating the bone remodeling process under mechanical stimulus. Dent Mater 23:1073–1078
82. Chou HY, Jagodnik JJ, Muftu S (2008) Predictions of bone remodeling around dental implant systems. J Biomech 41:1365–1373
83. Lian Z, Guan H, Ivanovski S, Loo YC, Johnson NW, Zhang H (2010) Effect of bone to implant contact percentage on bone remodelling surrounding a dental implant. Int J Oral Maxillofac Surg 39:690–698
84. Watzak G, Zechner W, Ulm C, Tangl S, Tepper G, Watzek G (2005) Histologic and histomorphometric analysis of three types of dental implants following 18 months of occlusal loading: a preliminary study in baboons. Clin Oral Implant Res 16:408–416
85. Chen T, Kim S, Goyal S, Jabbour S, Zhou J, Rajagopal G, Haffty B, Yue N (2010) Object-constrained meshless deformable algorithm for high speed 3D nonrigid registration between CT and CBCT. Med Phys 37:197–210
86. Doweidar MH, Calvo B, Alfaro I, Groenenboom P, Doblaré M (2010) A comparison of implicit and explicit natural element methods in large strains problems: application to soft biological tissues modeling. Comput Methods Appl Mech Eng 199(25–28):1691–1700
87. Zhang GY, Wittek A, Joldes GR, Jin X, Miller K (2013) A three-dimensional nonlinear meshfree algorithm for simulating mechanical responses of soft tissue. Eng Anal Boundary Elem. ISSN 0955-7997. http://dx.doi.org/10.1016/j.enganabound.2013.08.014
88. Tsubota K-i, Wada S, Yamaguchi T (2006) Particle method for computer simulation of red blood cell motion in blood flow. Comput Methods Programs Biomed 83(2):139–146
89. Mori D, Yano K, Tsubota K-i, Ishikawa T, Wada S, Yamaguchi T (2008) Computational study on effect of red blood cells on primary thrombus formation. Thromb Res 123(1):114–121
90. Doblaré M, Cueto E, Calvo B, Martínez MA, Garcia JM, Cegoñino J (2005) On the employ of meshless methods in biomechanics. Comput Methods Appl Mech Eng 194:801–821
91. Liew KM, Wu HY, Ng TY (2002) Meshless method for modeling of human proximal femur: treatment of nonconvex boundaries and stress analysis. Comput Mech 28:390–400
92. Lee JD, Chen Y, Zeng X, Eskandarian A, Oskard M (2007) Modeling and simulation of osteoporosis and fracture of trabecular bone by meshless method. Int J Eng Sci 45:329–338
93. Taddei F, Pani M, Zovatto L, Tonti E, Viceconti M (2008) A new meshless approach for subject-specific strain prediction in long bones: evaluation of accuracy. Clin Biomech 23(9):1192–1199
94. Belinha J, Jorge RMN, Dinis LMJS (2013) A meshless microscale bone tissue trabecular remodelling analysis considering a new anisotropic bone tissue material law. Comput Methods Biomech Biomed Eng 16(11):1170–1184
95. Belinha J, Jorge RMN, Dinis LMJS (2012) Bone tissue remodelling analysis considering a radial point interpolator meshless method. Eng Anal Boundary Elem 36(11):1660–1670
96. Belinha J (2004) Elasto-plastic analysis considering the element free galerkin method. MSc Dissertation, Faculty of Engineering of University of Porto, p 259
97. Belinha J (2010) The natural neighbour radial point interpolation method—solid mechanics and mechanobiology applications. PhD Dissertation, Faculty of Engineering of University of Porto, p 282

Chapter 2
Solid Mechanics Fundamentals

Abstract In this chapter the mechanical fundamentals behind the numerical applications presented in this work are developed. Firstly it is present a brief exposition of the used continuum formulation, where the solid kinematics and constitutive equations are shown. Following it is presented the used weak form and the consequent generated discrete equation system. Next, the dynamic analysis equations are presented and transient analysis basic concepts are introduced.

2.1 Continuum Formulation

The continuum mechanics is the foundation of the nonlinear numerical analysis. Solids and structures subjected to loads or forces become stressed. The stresses lead to strains, which can be interpreted as deformations or relative displacements. Solid Mechanics and Structural Mechanics deals, for a given solid and boundary condition (external forces and displacements constrains), with the relationship between stress and strain and the relationship between strain and displacements [1–3]. Solids can show different behaviours, depending on the solid material stress-strain curve. In this work only linear elastic materials are considered. In elastic materials the deformation in the solid caused by loading disappears fully with the unloading, in contrast, plastic materials show a residual deformation (which cannot be naturally recovered) that remains after the total unload process. The material properties on the solid can also be anisotropic, i.e., the material property varies with the direction [4]. On an anisotropic material the deformation caused by a load applied in a given direction causes a different deformation if the same load is applied in a distinct direction. Composite laminates are generally constituted by layers of anisotropic material. There are many material constants to be considered and defined in order to fully describe an anisotropic material, which is the reason why so often the engineering problems reduce the analysis to an isotropic material analysis. Isotropic materials are a special case of anisotropic materials, where only two independent material properties need to be known, the

J. Belinha, *Meshless Methods in Biomechanics*, Lecture Notes in Computational Vision and Biomechanics 16, DOI: 10.1007/978-3-319-06400-0_2,

Young modulus and the Poisson ratio. In this chapter the rigid solid motion and deformation are described, with emphasis on rotation, which plays an important role in nonlinear continuum mechanics. Also the concepts of strain and stress in nonlinear mechanics are introduced. The equilibrium and the constitutive equations are presented afterwards.

2.1.1 Kinematics

The general motion of a deformable body is represented in Fig. 2.1. The body, in the initial position $t = 0$, is considered to be an assemblage of material particles, labelled by the coordinates $\boldsymbol{X}$, with respect to the Cartesian basis e. The current position of a particle is defined at time t by the coordinates $\boldsymbol{x}$.

The motion can be mathematically described by a mapping function ϕ between initial and current particle positions,

$$\boldsymbol{x} = \phi(\boldsymbol{X}, t) \tag{2.1}$$

It is considered the material description, the Lagragian description, since the variation of the solid deformation is described with respect to the initial coordinates $\boldsymbol{X}$, at time t.

2.1.1.1 Deformation Gradient

The deformation gradient $\boldsymbol{F}$, is a key quantity in finite deformation analysis, since it is involved in all equations relating quantities before deformation (initial configuration) with the correspondent quantities after the finite deformation (current configuration). The deformation gradient tensor $\boldsymbol{F}$ can be defined as,

$$\boldsymbol{F} = \frac{\partial \phi}{\partial \boldsymbol{X}} = \nabla \phi \tag{2.2}$$

Alternatively to Eq. (2.1) the motion can be expressed by,

$$\boldsymbol{x} = \boldsymbol{x}(\boldsymbol{X}, t) \tag{2.3}$$

which permits the deformation gradient to be written as,

$$\boldsymbol{F} = \frac{\partial \boldsymbol{x}}{\partial \boldsymbol{X}} \tag{2.4}$$

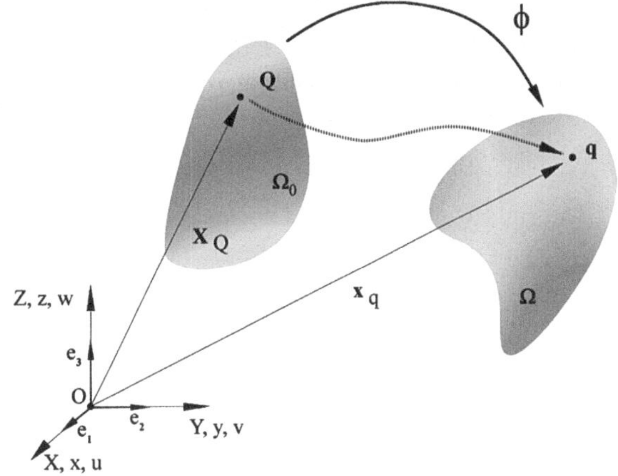

Fig. 2.1 General motion of a deformable body

As so, for a three-dimensional deformation problem the deformation gradient tensor of an initial material position $\boldsymbol{X} = \{X \quad Y \quad Z\}$ in respect to a current material position $\boldsymbol{x} = \{x \quad y \quad z\}$ can be presented as,

$$\boldsymbol{F} = \begin{bmatrix} \frac{\partial x}{\partial X} & \frac{\partial x}{\partial Y} & \frac{\partial x}{\partial Z} \\ \frac{\partial y}{\partial X} & \frac{\partial y}{\partial Y} & \frac{\partial y}{\partial Z} \\ \frac{\partial z}{\partial X} & \frac{\partial z}{\partial Y} & \frac{\partial z}{\partial Z} \end{bmatrix} \tag{2.5}$$

The determinant of the $\boldsymbol{F}$ is denoted by J and is called the 'Jacobian determinant'.

$$J = \det(\boldsymbol{F}) \tag{2.6}$$

The Jacobian determinant can be used to relate the integral of a given functional f in the current and in the initial configuration by,

$$\int\limits_{\Omega} f(\boldsymbol{x}, t)\, d\Omega = \int\limits_{\Omega_0} f(\phi(\boldsymbol{X}, t), t)\, J\, d\Omega_0. \tag{2.7}$$

2.1.1.2 Strain

Consider the change of the scalar product of the two elemental vectors from $d\boldsymbol{X}_1 = \boldsymbol{Q}_1 - \boldsymbol{P}$ and $d\boldsymbol{X}_2 = \boldsymbol{Q}_2 - \boldsymbol{P}$, initial configuration, to $d\boldsymbol{x}_1 = \boldsymbol{q}_1 - \boldsymbol{p}$ and $d\boldsymbol{x}_2 = \boldsymbol{q}_2 - \boldsymbol{p}$, current configuration, as a general measure of deformation. Where $\boldsymbol{Q}_1$ and $\boldsymbol{Q}_2$ are two material particles in the neighbourhood of a material particle $\boldsymbol{P}$ for the initial configuration and $\boldsymbol{q}_1$ and $\boldsymbol{q}_2$ and $\boldsymbol{p}$ the same respective material particles in the current configuration. Equation (2.4) permits the following relations $d\boldsymbol{x}_1 = \boldsymbol{F}\,d\boldsymbol{X}_1$ and $d\boldsymbol{x}_2 = \boldsymbol{F}\,d\boldsymbol{X}_2$, and the spatial scalar product $d\boldsymbol{x}_1 \cdot d\boldsymbol{x}_2$ can be found in terms of the material vectors $d\boldsymbol{X}_1$ and $d\boldsymbol{X}_2$ as,

$$d\boldsymbol{x}_1 \cdot d\boldsymbol{x}_2 = d\boldsymbol{X}_1 \cdot \boldsymbol{F}^T \boldsymbol{F} d\boldsymbol{X}_2 \tag{2.8}$$

The right Cauchy-Green deformation tensor is defined by

$$\boldsymbol{C} = \boldsymbol{F}^T \boldsymbol{F} \tag{2.9}$$

Which operates directly on the material vectors $d\boldsymbol{X}_1$ and $d\boldsymbol{X}_2$. Alternatively the initial scalar product $d\boldsymbol{X}_1 \cdot d\boldsymbol{X}_2$ can be obtained in terms of spatial vectors $d\boldsymbol{x}_1$ and $d\boldsymbol{x}_1$ using the left Cauchy-Green deformation tensor $\boldsymbol{b}$,

$$\boldsymbol{b} = \boldsymbol{F}\boldsymbol{F}^T \tag{2.10}$$

The change in scalar product can now be found in terms of the material vectors $d\boldsymbol{X}_1$ and $d\boldsymbol{X}_2$ and the Lagrange or Green strain tensor $\boldsymbol{E}$ can be defined as,

$$\frac{1}{2}(d\boldsymbol{x}_1 \cdot d\boldsymbol{x}_2 - d\boldsymbol{X}_1 \cdot d\boldsymbol{X}_2) = d\boldsymbol{X}_1 \cdot \boldsymbol{E}\,d\boldsymbol{X}_2 \tag{2.11}$$

where the strain tensor $\boldsymbol{E}$ is expressed as

$$\boldsymbol{E} = \frac{1}{2}(\boldsymbol{C} - \boldsymbol{I}). \tag{2.12}$$

2.1.1.3 Polar Decomposition

The tensor $\boldsymbol{F}$ can be expressed as the product of the orthogonal rotation tensor $\boldsymbol{R}$ by the symmetric stretch tensor $\boldsymbol{U}$,

$$\boldsymbol{F} = \boldsymbol{R}\boldsymbol{U} \tag{2.13}$$

where

$$\boldsymbol{R}^T\boldsymbol{R} = \boldsymbol{I} \quad \text{and} \quad \boldsymbol{U} = \boldsymbol{U}^T \tag{2.14}$$

Such decomposition is called, polar decomposition, and Eq. (2.9) can be expressed as,

$$\boldsymbol{C} = \boldsymbol{F}^T\boldsymbol{F} = \boldsymbol{U}^T\boldsymbol{R}^T\boldsymbol{R}\boldsymbol{U} = \boldsymbol{U}^T\boldsymbol{I}\boldsymbol{U} = \boldsymbol{U}\boldsymbol{U} \tag{2.15}$$

In order to actually obtain $\boldsymbol{U}$ from Eq. (2.15) it is first necessary to evaluate the principal directions of $\boldsymbol{C}$, represented by the eigenvectors set $\boldsymbol{W}_i$ and the correspondent eigenvalues λ_i, with $i = \{1, 2, 3\}$ for the three-dimensional case. In this manner $\boldsymbol{C}$ can be defined as,

$$\boldsymbol{C} = \sum_{i=1}^{3} \lambda_i^2 \boldsymbol{W}_i \otimes \boldsymbol{W}_i \tag{2.16}$$

since the eigenvectors $\boldsymbol{W}_i$ are in fact orthogonal unit vectors, because $\boldsymbol{C} = \boldsymbol{C}^T$. As so, with Eqs. (2.15) and (2.16) it is possible to write the material stretch tensor $\boldsymbol{U}$ as,

$$\boldsymbol{U} = \sum_{i=1}^{3} \lambda_i \boldsymbol{W}_i \otimes \boldsymbol{W}_i \tag{2.17}$$

Once the stretch tensor $\boldsymbol{U}$ is known, the rotation tensor $\boldsymbol{R}$ can be obtained without difficulty from Eq. (2.13).

$$\boldsymbol{R} = \boldsymbol{F}\,\boldsymbol{U}^{-1}. \tag{2.18}$$

2.1.1.4 Stress

In a large deformation analysis a body can experience a large rotation and/or a large strain. The defined stress terms together with the obtained strain terms enables to express the virtual work as an integral over the known body volume, expressing in this manner the change in the body configuration. Both strain tensor and stress tensor are referred to the same deformed state. The Cauchy stress tensor, here defined as $\boldsymbol{\Lambda}$, is a symmetric tensor and it represents the stresses of the current configuration. For the three-dimensional case it can be defined as,

$$\boldsymbol{\Lambda} = \begin{bmatrix} \sigma_{xx} & \sigma_{xy} & \sigma_{xz} \\ \sigma_{yx} & \sigma_{yy} & \sigma_{yz} \\ \sigma_{zx} & \sigma_{zy} & \sigma_{zz} \end{bmatrix} \tag{2.19}$$

In this work it is used the Voigt notation, since the development of fourth order tensors is less practical. In Voigt notation the tensors are expressed in column vectors, so the stress tensor $\boldsymbol{\Lambda}$ is reduced to the stress vector $\boldsymbol{\sigma}$,

$$\boldsymbol{\sigma} = \{\, \sigma_{xx} \quad \sigma_{yy} \quad \sigma_{zz} \quad \sigma_{xy} \quad \sigma_{yz} \quad \sigma_{zx} \,\}^{\mathrm{T}} \tag{2.20}$$

and the strain tensor $\boldsymbol{E}$ to the strain vector $\boldsymbol{\varepsilon}$,

$$\boldsymbol{\varepsilon} = \{\, \varepsilon_{xx} \quad \varepsilon_{yy} \quad \varepsilon_{zz} \quad \varepsilon_{xy} \quad \varepsilon_{yz} \quad \varepsilon_{zx} \,\}^{\mathrm{T}} \tag{2.21}$$

The use of vectors is more practical in the programming process.

2.1.1.5 Principal Stress

Another way of describing the Cauchy stress tensor, which completely defines the stress state in an interest point, is through,

$$\Lambda = \begin{bmatrix} t^{(\hat{e}_1)} \\ t^{(\hat{e}_2)} \\ t^{(\hat{e}_3)} \end{bmatrix} = \begin{bmatrix} \sigma_{xx} & \sigma_{xy} & \sigma_{xz} \\ \sigma_{yx} & \sigma_{yy} & \sigma_{yz} \\ \sigma_{zx} & \sigma_{zy} & \sigma_{zz} \end{bmatrix} \tag{2.22}$$

where $\hat{e}_1$, $\hat{e}_2$ and $\hat{e}_3$ are the versors of the coordinate system and $t^{(\hat{e}_i)}$ is the stress vector on a plane normal to $\hat{e}_1$ passing through the interest point, Fig. 2.2(a). Following Cauchy's stress theorem, if the stress vectors of three orthogonal planes, with a common point, are known, then the stress vector on any other plane passing through that point can be found through the coordinate transformation equations [5]. Thus, the stress vector $t^{(n)}$ in a point belonging to an inclined plane, Fig. 2.2(b), can be defined by,

$$t^{(n)} = \boldsymbol{n} \cdot \sigma_{ij} = \begin{bmatrix} n_1 & n_2 & n_3 \end{bmatrix} \cdot \begin{bmatrix} \sigma_{xx} & \sigma_{xy} & \sigma_{xz} \\ \sigma_{yx} & \sigma_{yy} & \sigma_{yz} \\ \sigma_{zx} & \sigma_{zy} & \sigma_{zz} \end{bmatrix} \tag{2.23}$$

where $\boldsymbol{n}$ is the inclined plane normal vector. The relation in Eq. (2.23) leads to the transformation rule of the stress tensor. The initial stress tensor σ_{ij}, defined in the $\boldsymbol{x}_i$ coordinate system, can be transformed in a new stress tensor σ'_{ij}, defined in another $\boldsymbol{x}'_i$ coordinate system by the relation,

$$\Lambda' = \boldsymbol{A}\,\Lambda \boldsymbol{A}^T \tag{2.24}$$

being $\boldsymbol{A}$ the rotation matrix. Developing Eq. (2.24),

$$\begin{bmatrix} \sigma'_{xx} & \sigma'_{xy} & \sigma'_{xz} \\ \sigma'_{yx} & \sigma'_{yy} & \sigma'_{yz} \\ \sigma'_{zx} & \sigma'_{zy} & \sigma'_{zz} \end{bmatrix} = \begin{bmatrix} a_{11} & a_{12} & a_{13} \\ a_{21} & a_{22} & a_{23} \\ a_{31} & a_{32} & a_{33} \end{bmatrix} \begin{bmatrix} \sigma_{xx} & \sigma_{xy} & \sigma_{xz} \\ \sigma_{yx} & \sigma_{yy} & \sigma_{yz} \\ \sigma_{zx} & \sigma_{zy} & \sigma_{zz} \end{bmatrix} \begin{bmatrix} a_{11} & a_{21} & a_{31} \\ a_{12} & a_{22} & a_{32} \\ a_{13} & a_{23} & a_{33} \end{bmatrix} \tag{2.25}$$

The a_{ij} coefficients can be understood as the projection of the $\boldsymbol{x}'_i$ coordinate system versors in the $\boldsymbol{x}_i$ coordinate system versors. Therefore, the angle between the versors of each coordinate system can be defined as,

$$\gamma_{ij} = \cos^{-1}(a_{ij}) \tag{2.26}$$

Through Eq. (2.26) and Fig. 2.3 it is possible to comprehend better the physical meaning of the a_{ij} coefficients and the respective angles.

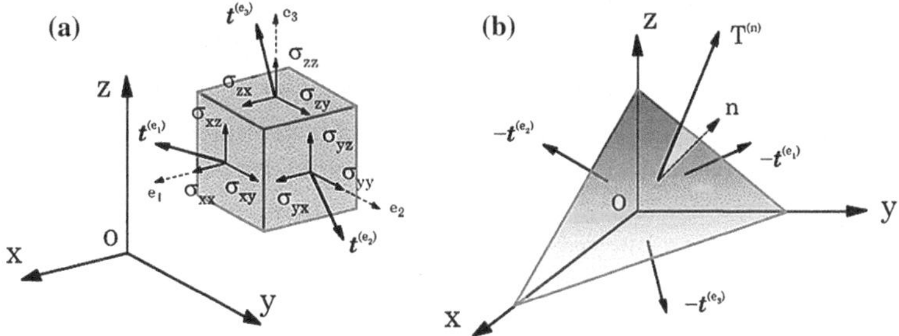

Fig. 2.2 **a** Three-dimensional stress components. **b** Stress vector acting on a plane with normal vector ***n***

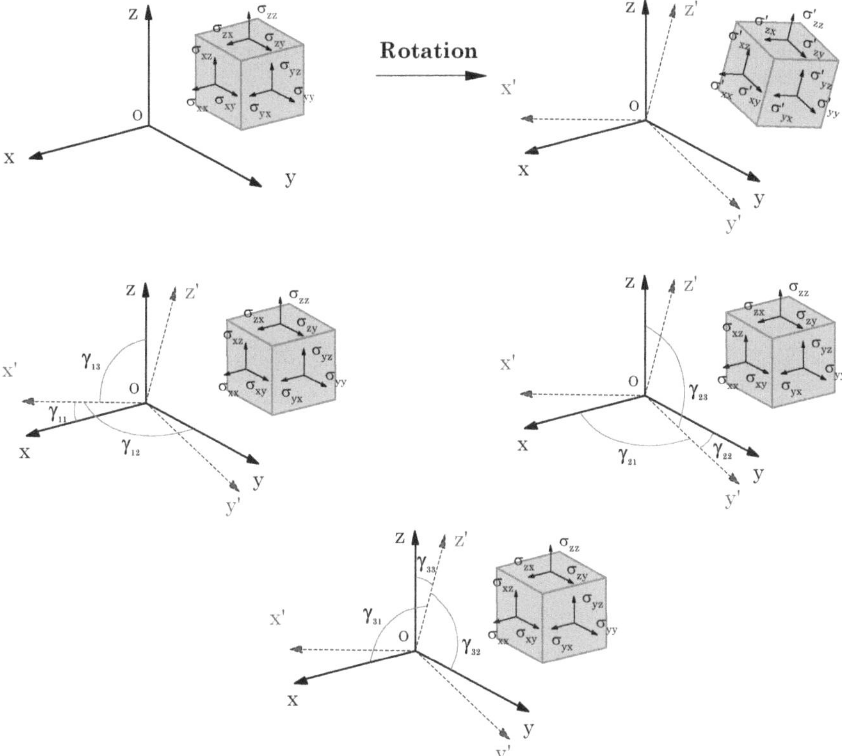

Fig. 2.3 Stress tensor transformation and respective angles

Let ***P*** be an interest point of a considered stressed body. There are at least three planes, orthogonal with each other, crossing ***P*** where the corresponding stress vector is normal to the plane. These planes are called principal planes and the normal vectors of each plane are called principal directions. The stress vectors are parallel to the plane normal vectors and are called principal stresses.

The stress tensor is a physical quantity, independent of the coordinate system chosen to represent it. Therefore, there are certain invariants associated with it which are also independent of the coordinate system. Being a second order tensor, the stress tensor has associated three independent invariant quantities. One set of such invariants are the principal stresses of the stress tensor, which are just the eigenvalues of the stress tensor. Their direction vectors are the principal directions or eigenvectors. A stress vector parallel to the normal vector $\boldsymbol{n}$ is given by,

$$\boldsymbol{t}^{(\boldsymbol{n})} = \lambda \boldsymbol{n} = \sigma_n \boldsymbol{n} \tag{2.27}$$

being λ a constant of proportionality, and in this particular case the magnitude of σ_n, the principal stress in $\boldsymbol{n}$ direction. Knowing $t_i^{(n)} = \sigma_{ij} n_j$ and $n_i = \delta_{ij} n_j$, where δ_{ij} is the Kronecker delta, the following development can be performed,

$$t_i^{(n)} = \lambda n_i \Rightarrow \sigma_{ij} n_j = \lambda \delta_{ij} n_j \Rightarrow \left(\sigma_{ij} - \lambda \delta_{ij}\right) n_j = 0 \tag{2.28}$$

which is a homogeneous system, three linear equations for three n_j unknowns. To obtain the n_j non-zero solution, the matrix determinant must be equal to zero,

$$\left|\sigma_{ij} - \lambda \delta_{ij}\right| = \begin{vmatrix} \sigma_{xx} - \lambda & \sigma_{xy} & \sigma_{xz} \\ \sigma_{yx} & \sigma_{yy} - \lambda & \sigma_{yz} \\ \sigma_{zx} & \sigma_{zy} & \sigma_{zz} - \lambda \end{vmatrix} = 0 \tag{2.29}$$

which leads to the following cubic equation,

$$\left|\sigma_{ij} - \lambda \delta_{ij}\right| = -\lambda^3 + I_1 \lambda^2 - I_2 \lambda + I_3 = 0 \tag{2.30}$$

being I_1, I_2 and I_3 the stress invariants,

$$I_1 = \sigma_{kk} \tag{2.31}$$

$$I_2 = \frac{1}{2}\left(\sigma_{ii}\sigma_{jj} - \sigma_{ij}\sigma_{ji}\right) \tag{2.32}$$

$$I_3 = \det\left(\sigma_{ij}\right) \tag{2.33}$$

The three roots $\lambda_1 = \sigma_1$, $\lambda_2 = \sigma_2$ and $\lambda_3 = \sigma_3$ of Eq. (2.30) are the eigenvalues or principal stresses, which are unique. Therefore the stress invariants have always the same value regardless of the orientation of the chosen coordinate system. For each eigenvalue λ, exists a non-trivial solution $\boldsymbol{n}$ on Eq. (2.28). These n_j solutions, called eigenvectors, are the principal directions, which defines the plane where the respective stress acts. Applying Eq. (2.25),

$$\begin{bmatrix} \sigma_1 & 0 & 0 \\ 0 & \sigma_2 & 0 \\ 0 & 0 & \sigma_3 \end{bmatrix} = \begin{bmatrix} n_{11} & n_{12} & n_{13} \\ n_{21} & n_{22} & n_{23} \\ n_{31} & n_{32} & n_{33} \end{bmatrix} \begin{bmatrix} \sigma_{xx} & \sigma_{xy} & \sigma_{xz} \\ \sigma_{yx} & \sigma_{yy} & \sigma_{yz} \\ \sigma_{zx} & \sigma_{zy} & \sigma_{zz} \end{bmatrix} \begin{bmatrix} n_{11} & n_{21} & n_{31} \\ n_{12} & n_{22} & n_{32} \\ n_{13} & n_{23} & n_{33} \end{bmatrix} \tag{2.34}$$

The principal stresses and principal directions characterize the stress in $\boldsymbol{P}$ and are independent of the orientation of the coordinate system.

2.1.2 Constitutive Equations

The following relation between the stress rate and the strain rate is assumed,

$$d\boldsymbol{\sigma} = \boldsymbol{c}\ d\boldsymbol{\varepsilon} \tag{2.35}$$

The material constitutive matrix is defined by $\boldsymbol{c}$ and if material nonlinear relations exists between $\boldsymbol{\sigma}$ and $\boldsymbol{\varepsilon}$, then $\boldsymbol{c} = \boldsymbol{c}^{ep}$. With Eq. (2.35) the following relation can be established,

$$d\boldsymbol{\varepsilon} = \boldsymbol{c}^{-1}\ d\boldsymbol{\sigma} \tag{2.36}$$

being $\boldsymbol{s} = \boldsymbol{c}^{-1}$ and defined for the three-dimensional case as,

$$\boldsymbol{s} = \begin{bmatrix} \frac{1}{E_{xx}} & -\frac{\upsilon_{yx}}{E_{yy}} & -\frac{\upsilon_{zx}}{E_{zz}} & 0 & 0 & 0 \\ -\frac{\upsilon_{xy}}{E_{xx}} & \frac{1}{E_{yy}} & -\frac{\upsilon_{zy}}{E_{zz}} & 0 & 0 & 0 \\ -\frac{\upsilon_{xz}}{E_{xx}} & -\frac{\upsilon_{yz}}{E_{yy}} & \frac{1}{E_{zz}} & 0 & 0 & 0 \\ 0 & 0 & 0 & \frac{1}{G_{xy}} & 0 & 0 \\ 0 & 0 & 0 & 0 & \frac{1}{G_{yz}} & 0 \\ 0 & 0 & 0 & 0 & 0 & \frac{1}{G_{zx}} \end{bmatrix} \tag{2.37}$$

The material constitutive matrix $\boldsymbol{c}$ is obtained by inverting the material compliance matrix $\boldsymbol{s}$, which is here defined for an three-dimensional anisotropic material. The elements on matrix $\boldsymbol{s}$ are obtained experimentally. E_{ii} is the Young modulus in direction i, υ_{ij} is the Poisson ratio which characterizes the deformation rate in direction j when a force is applied in direction i, G_{ij} is the shear modulus which characterizes the variation angle between directions i and j. Due to symmetry the following relation can be established,

$$E_i\, \upsilon_{ji} = E_j\, \upsilon_{ij} \tag{2.38}$$

For the two-dimensional case the plane stress and plane strain [5] deformation theory assumptions can be presumed. Considering the plane stress assumptions,

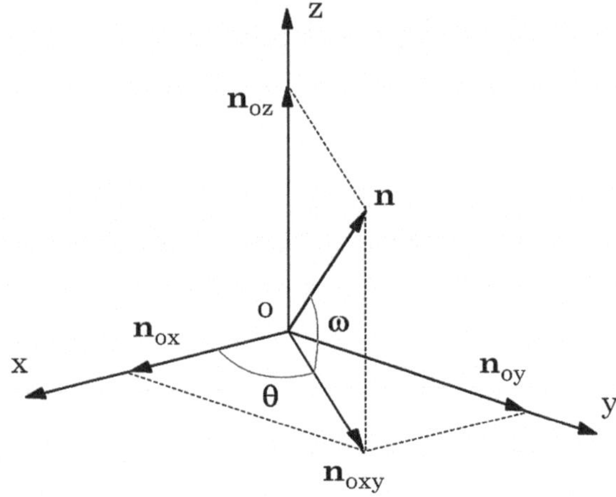

Fig. 2.4 Projection of vector $\boldsymbol{n}$ in the coordinate axis and in the oxy plane

$\sigma_{zx} = \sigma_{zy} = \sigma_{zz} = 0$, the material compliance matrix $\boldsymbol{s}$ is obtained directly from the three-dimensional compliance matrix $\boldsymbol{s}$,

$$\boldsymbol{s} = \begin{bmatrix} \frac{1}{E_{xx}} & -\frac{\upsilon_{yx}}{E_{yy}} & 0 \\ -\frac{\upsilon_{xy}}{E_{xx}} & \frac{1}{E_{yy}} & 0 \\ 0 & 0 & \frac{1}{G_{xy}} \end{bmatrix} \tag{2.39}$$

For the plane strain deformation theory it is considered $\varepsilon_{zx} = \varepsilon_{zy} = \varepsilon_{zz} = 0$ and the material compliance matrix $\boldsymbol{s}$ is defined as,

$$\boldsymbol{s} = \begin{bmatrix} \frac{1}{E_{xx}} - \frac{\upsilon_{zx}\upsilon_{xz}}{E_{xx}} & -\frac{\upsilon_{yx}}{E_{yy}} - \frac{\upsilon_{zx}\upsilon_{yz}}{E_{yy}} & 0 \\ -\frac{\upsilon_{xy}}{E_{xx}} - \frac{\upsilon_{zy}\upsilon_{xz}}{E_{xx}} & \frac{1}{E_{yy}} - \frac{\upsilon_{zy}\upsilon_{yz}}{E_{yy}} & 0 \\ 0 & 0 & \frac{1}{G_{xy}} \end{bmatrix} \tag{2.40}$$

In the case of an anisotropic material, it is possible to rotate the material constitutive matrix $\boldsymbol{c}$ and orientate the material directions with a vector. Consider a known vector $\boldsymbol{n}$ in the Euclidean space $\mathbb{R}^3$, Fig. 2.4, and the respective projections on the coordinate axis and in the oxy plane. As it is known,

$$\theta = \cos^{-1}\left(\frac{\boldsymbol{n}_{oxy} \cdot \boldsymbol{n}_{ox}}{\left\|\boldsymbol{n}_{oxy}\right\| \cdot \left\|\boldsymbol{n}_{ox}\right\|}\right) \quad \text{and} \quad \omega = \cos^{-1}\left(\frac{\boldsymbol{n}_{oxy} \cdot \boldsymbol{n}}{\left\|\boldsymbol{n}_{oxy}\right\| \cdot \left\|\boldsymbol{n}\right\|}\right) \tag{2.41}$$

With the obtained angle information it is now possible to rotate the material matrix using the rotational transformation matrix and therefore align the material ox axis with the known vector $\boldsymbol{n}$. The rotational transformation matrix that permits an anticlockwise rotation along the ox axis of a known angle β can be defined as,

$$\boldsymbol{T}_{ox} = \begin{bmatrix} 1 & 0 & 0 & 0 & 0 & 0 \\ 0 & \cos^2\beta & \sin^2\beta & 0 & -\sin 2\beta & 0 \\ 0 & \sin^2\beta & \cos^2\beta & 0 & \sin 2\beta & 0 \\ 0 & 0 & 0 & \cos\beta & 0 & \sin\beta \\ 0 & \sin\beta\cdot\cos\beta & -\sin\beta\cdot\cos\beta & 0 & \cos^2\beta-\sin^2\beta & 0 \\ 0 & 0 & 0 & -\sin\beta & 0 & \cos\beta \end{bmatrix} \tag{2.42}$$

Along the oy axis,

$$\boldsymbol{T}_{oy} = \begin{bmatrix} \cos^2\beta & 0 & \sin^2\beta & 0 & 0 & \sin 2\beta \\ 0 & 1 & 0 & 0 & 0 & 0 \\ \sin^2\beta & 0 & \cos^2\beta & 0 & 0 & -\sin 2\beta \\ 0 & 0 & 0 & \cos\beta & -\sin\beta & 0 \\ 0 & 0 & 0 & \sin\beta & \cos\beta & 0 \\ -\sin\beta\cdot\cos\beta & 0 & \sin\beta\cdot\cos\beta & 0 & 0 & \cos^2\beta-\sin^2\beta \end{bmatrix} \tag{2.43}$$

Along the oz axis,

$$\boldsymbol{T}_{oz} = \begin{bmatrix} \cos^2\beta & \sin^2\beta & 0 & -\sin 2\beta & 0 & 0 \\ \sin^2\beta & \cos^2\beta & 0 & \sin 2\beta & 0 & 0 \\ 0 & 0 & 1 & 0 & 0 & 0 \\ \sin\beta\cdot\cos\beta & -\sin\beta\cdot\cos\beta & 0 & \cos^2\beta-\sin^2\beta & 0 & 0 \\ 0 & 0 & 0 & 0 & \cos\beta & -\sin\beta \\ 0 & 0 & 0 & 0 & \sin\beta & \cos\beta \end{bmatrix} \tag{2.44}$$

The material matrix after rotation can be defined as,

$$\boldsymbol{c}^{current} = [\boldsymbol{T}_{oz}]_0^{\mathrm{T}}\left[[\boldsymbol{T}_{oy}]_{\omega}^{\mathrm{T}}\boldsymbol{c}^{initial}[\boldsymbol{T}_{oy}]_{\omega}\right][\boldsymbol{T}_{oz}]_0. \tag{2.45}$$

2.2 Weak Form

The strong form system equations are the partial differential system equations governing the studied physic phenomenon. In contrast, the weak form requires a weaker consistency on the adopted approximation (or interpolation) functions. The ideal would be obtaining the exact solution from strong form system equations, however this is usually an extremely difficult task in complex practical engineering problems. Formulations based on weak forms are able to produce stable algebraic system equations and to give a discretized system of equations which leads to more accurate results. These are the reasons why so many prefer the weak form to obtain the approximated solution.

In this work the discrete equation system is obtained using the Galerkin weak form, which is a variational method [6]. For meshless methods used in this book the discrete system of equations is obtained similarly with the FEM, with some differences inherent to the meshless approach. The discrete equations for the static and the dynamic approach are developed and shown for the basic three-dimensional deformation theory.

2.2.1 Weak Form of Galerkin

Consider the solid with a domain Ω bounded by Γ, Fig. 2.5. The continuous solid surface on which the external forces $\bar{\boldsymbol{t}}$ are applied is denoted as Γ_t (natural boundary) and the surface where the displacements are constrained is denoted as Γ_u (essential boundary).

The Galerkin weak form is a variational principle based on the energy principle. Of all possible displacement configurations satisfying the compatibility conditions, the essential boundary conditions (kinematical and displacement) and the initial and final time conditions, the real solution correspondent configuration is the one which minimizes the Lagrangian functional L,

$$L = T - U + W_f \tag{2.46}$$

being T the kinetic energy, U is the strain energy and W_f is the work produced by the external forces. The kinetic energy is defined by,

$$T = \frac{1}{2} \int_{\Omega} \rho \dot{\boldsymbol{u}}^{\mathrm{T}} \dot{\boldsymbol{u}} \, \mathrm{d}\Omega \tag{2.47}$$

where the solid volume is defined by Ω and $\dot{\boldsymbol{u}}$ is the displacement first derivative with respect to time, i.e., the velocity. The solid mass density is defined by ρ. The strain energy, for elastic materials, is defined as,

$$U = \frac{1}{2} \int_{\Omega} \boldsymbol{\varepsilon}^{\mathrm{T}} \boldsymbol{\sigma} \, \mathrm{d}\Omega \tag{2.48}$$

being $\boldsymbol{\varepsilon}$ the strain vector and $\boldsymbol{\sigma}$ the stress vector. The work produced by the external forces can be expressed as,

$$W_f = \int_{\Omega} \boldsymbol{u}^{\mathrm{T}} \boldsymbol{b} \, \mathrm{d}\Omega + \int_{\Gamma_t} \boldsymbol{u}^{\mathrm{T}} \bar{\boldsymbol{t}} \, \mathrm{d}\Gamma \tag{2.49}$$

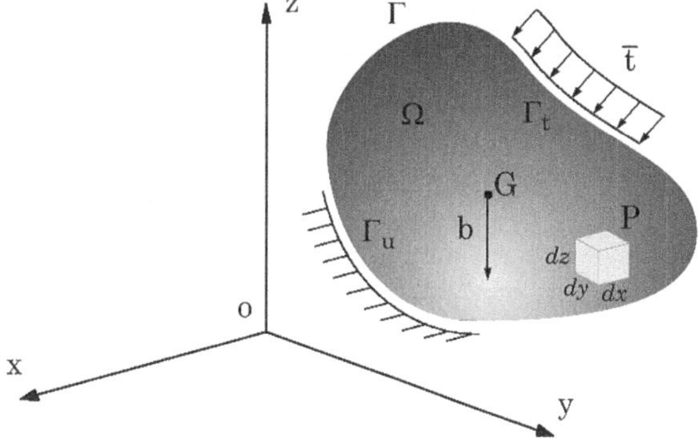

Fig. 2.5 Continuous solid subject to volume forces and external forces

where $\boldsymbol{u}$ represents the displacement, $\boldsymbol{b}$ the body forces and Γ_t the traction boundary where the external forces $\bar{\boldsymbol{t}}$ are applied. By substitution the Lagrangian functional L can be rewritten as,

$$L = \frac{1}{2}\int_{\Omega} \rho \dot{\boldsymbol{u}}^{\mathrm{T}} \dot{\boldsymbol{u}}\, \mathrm{d}\Omega - \frac{1}{2}\int_{\Omega} \boldsymbol{\varepsilon}^{T} \boldsymbol{\sigma}\, \mathrm{d}\Omega + \int_{\Omega} \boldsymbol{u}^{T} \boldsymbol{b}\, \mathrm{d}\Omega + \int_{\Gamma_t} \boldsymbol{u}^{T} \bar{\boldsymbol{t}}\, \mathrm{d}\Gamma \tag{2.50}$$

and then minimized,

$$\delta \int_{t_1}^{t_2} \left[\frac{1}{2}\int_{\Omega} \rho \dot{\boldsymbol{u}}^{\mathrm{T}} \dot{\boldsymbol{u}}\, \mathrm{d}\Omega - \frac{1}{2}\int_{\Omega} \boldsymbol{\varepsilon}^{\mathrm{T}} \boldsymbol{\sigma}\, \mathrm{d}\Omega + \int_{\Omega} \boldsymbol{u}^{T} \boldsymbol{b}\, \mathrm{d}\Omega + \int_{\Gamma_t} \boldsymbol{u}^{T} \bar{\boldsymbol{t}}\, \mathrm{d}\Gamma \right] \mathrm{d}t = 0 \tag{2.51}$$

Moving the variation operator δ inside the integrals,

$$\int_{t_1}^{t_2} \left[\frac{1}{2}\int_{\Omega} \delta\left(\rho \dot{\boldsymbol{u}}^{\mathrm{T}} \dot{\boldsymbol{u}}\right) \mathrm{d}\Omega - \frac{1}{2}\int_{\Omega} \delta\left(\boldsymbol{\varepsilon}^{T} \boldsymbol{\sigma}\right) \mathrm{d}\Omega + \int_{\Omega} \delta \boldsymbol{u}^{T} \boldsymbol{b}\, \mathrm{d}\Omega + \int_{\Gamma_t} \delta \boldsymbol{u}^{T} \bar{\boldsymbol{t}}\, \mathrm{d}\Gamma \right] \mathrm{d}t = 0 \tag{2.52}$$

Since all operations are linear, changing the order of operation does not affect the result. In the first term of Eq. (2.52) the time integral can be moved inside the spatial integral,

$$\int_{t_1}^{t_2} \left[\frac{1}{2}\int_{\Omega} \delta\left(\rho \dot{\boldsymbol{u}}^{\mathrm{T}} \dot{\boldsymbol{u}}\right) \mathrm{d}\Omega \right] \mathrm{d}t = \frac{1}{2}\int_{\Omega} \left[\int_{t_1}^{t_2} \delta\left(\rho \dot{\boldsymbol{u}}^{\mathrm{T}} \dot{\boldsymbol{u}}\right) \mathrm{d}t \right] \mathrm{d}\Omega \tag{2.53}$$

Using the chain rule of variation and then the scalar property, the integral can be rewritten as,

$$\int_{t_1}^{t_2} \delta(\rho\dot{\boldsymbol{u}}^{\mathrm{T}}\dot{\boldsymbol{u}})\mathrm{d}t = \rho\int_{t_1}^{t_2} (\delta\dot{\boldsymbol{u}}^{\mathrm{T}}\dot{\boldsymbol{u}} + \dot{\boldsymbol{u}}^{\mathrm{T}}\delta\dot{\boldsymbol{u}})\mathrm{d}t = 2\rho\int_{t_1}^{t_2} (\delta\dot{\boldsymbol{u}}^{\mathrm{T}}\dot{\boldsymbol{u}})\mathrm{d}t \tag{2.54}$$

And knowing that $\dot{\boldsymbol{u}}^{\mathrm{T}}\dot{\boldsymbol{u}}$ is a scalar and $\dot{\boldsymbol{u}} = \partial\boldsymbol{u}/\partial t$,

$$\int_{t_1}^{t_2} (\delta\dot{\boldsymbol{u}}^{\mathrm{T}}\dot{\boldsymbol{u}})\mathrm{d}t = \int_{t_1}^{t_2} \left(\frac{\partial\delta\boldsymbol{u}^{\mathrm{T}}}{\partial t}\frac{\partial\boldsymbol{u}}{\partial t}\right)\mathrm{d}t \tag{2.55}$$

Then integrating by parts, with respect to time,

$$\int_{t_1}^{t_2} \left(\frac{\partial\delta\boldsymbol{u}^{\mathrm{T}}}{\partial t}\;\frac{\partial\boldsymbol{u}}{\partial t}\right) dt = \int_{t_1}^{t_2} \left(-\delta\boldsymbol{u}^{\mathrm{T}}\;\frac{\partial^2\boldsymbol{u}}{\partial t^2}\right)dt + \left(\delta\boldsymbol{u}^{\mathrm{T}}\frac{\partial\boldsymbol{u}}{\partial t}\right)\Bigg|_{t_1}^{t_2} \tag{2.56}$$

Notice that $\boldsymbol{u}$ satisfies, by imposition, the conditions at the initial time, t_1, and final time, t_2, leading to a null $\delta\boldsymbol{u}$ at t_1 and t_2. Therefore the last term in Eq. (2.56) vanishes. Considering the last development and switching the integration order again, Eq. (2.53) becomes,

$$\int_{t_1}^{t_2} \left[\frac{1}{2}\int_{\Omega} \delta(\rho\dot{u}^{\mathrm{T}}\dot{u})\mathrm{d}\Omega\right]\mathrm{d}t = -\int_{t_1}^{t_2} \left[\rho\int_{\Omega} (\delta\boldsymbol{u}^{\mathrm{T}}\ddot{u})\mathrm{d}t\right] \tag{2.57}$$

being $\ddot{u} = \partial^2\boldsymbol{u}/\partial t^2$ the acceleration. The second term on Eq. (2.52) can also be developed. The integrand function in the second integral term can be written as follows,

$$\delta(\boldsymbol{\varepsilon}^{\mathrm{T}}\boldsymbol{\sigma}) = \delta\boldsymbol{\varepsilon}^{\mathrm{T}}\boldsymbol{\sigma} + \boldsymbol{\varepsilon}^{\mathrm{T}}\delta\boldsymbol{\sigma} \tag{2.58}$$

as the two terms in Eq. (2.58) are in fact scalars, the transpose does not affect the result, as so,

$$\boldsymbol{\varepsilon}^{\mathrm{T}}\delta\sigma = (\boldsymbol{\varepsilon}^{\mathrm{T}}\delta\boldsymbol{\sigma})^{\mathrm{T}} = \delta\boldsymbol{\sigma}^{\mathrm{T}}\boldsymbol{\varepsilon} \tag{2.59}$$

Using the constitutive equation $\boldsymbol{\sigma} = \boldsymbol{c}\boldsymbol{\varepsilon}$ and the symmetric property of the material matrix, $\boldsymbol{c}^{\mathrm{T}} = \boldsymbol{c}$, it is possible to write,

$$\delta\boldsymbol{\sigma}^{\mathrm{T}}\varepsilon = \delta(\boldsymbol{c}\boldsymbol{\varepsilon})^{\mathrm{T}}\varepsilon = \delta\boldsymbol{\varepsilon}^{\mathrm{T}}\boldsymbol{c}^{\mathrm{T}}\varepsilon = \delta\boldsymbol{\varepsilon}^{\mathrm{T}}\boldsymbol{c}\varepsilon = \delta\boldsymbol{\varepsilon}^{\mathrm{T}}\boldsymbol{\sigma} \tag{2.60}$$

Therefore Eq. (2.58) becomes,

$$\delta(\boldsymbol{\varepsilon}^{\mathrm{T}}\boldsymbol{\sigma}) = 2\delta\boldsymbol{\varepsilon}^{\mathrm{T}}\boldsymbol{\sigma} \tag{2.61}$$

simplifying the second term in Eq. (2.52),

$$\int_{t_1}^{t_2} \left[\frac{1}{2} \int_{\Omega} \delta(\boldsymbol{\varepsilon}^{\mathrm{T}}\boldsymbol{\sigma}) \mathrm{d}\Omega \right] \mathrm{d}t = \int_{t_1}^{t_2} \left[\int_{\Omega} \delta\boldsymbol{\varepsilon}^{\mathrm{T}}\boldsymbol{\sigma} \mathrm{d}\Omega \right] \mathrm{d}t \tag{2.62}$$

Equation (2.52) now becomes,

$$\int_{t_1}^{t_2} \left[-\rho \int_{\Omega} (\delta\boldsymbol{u}^{\mathrm{T}}\, \ddot{\boldsymbol{u}})\, \mathrm{d}\Omega - \int_{\Omega} \delta\boldsymbol{\varepsilon}^{T}\boldsymbol{\sigma}\, \mathrm{d}\Omega + \int_{\Omega} \delta\boldsymbol{u}^{T}\boldsymbol{b}\, \mathrm{d}\Omega + \int_{\Gamma_t} \delta\boldsymbol{u}^{T}\bar{\boldsymbol{t}}\, \mathrm{d}\Gamma \right] \mathrm{d}t = 0 \tag{2.63}$$

To satisfy Eq. (2.63) for all possible choices of the integrand of the time integration has to be null, leading to the following expression,

$$-\rho \int_{\Omega} (\delta\boldsymbol{u}^{\mathrm{T}}\, \ddot{\boldsymbol{u}}) \mathrm{d}\Omega - \int_{\Omega} \delta\boldsymbol{\varepsilon}^{T}\boldsymbol{\sigma}\, \mathrm{d}\Omega + \int_{\Omega} \delta\boldsymbol{u}^{T}\boldsymbol{b}\, \mathrm{d}\Omega + \int_{\Gamma_t} \delta\boldsymbol{u}^{T}\bar{\boldsymbol{t}}\, \mathrm{d}\Gamma = 0 \tag{2.64}$$

This last equation is known as the 'Galerkin weak form', which can also be viewed as the principle of virtual work. The principle of virtual work states that if a solid body is in equilibrium, the virtual work produced by the body inner stresses and the body applied external forces should vanish when the body experiments a virtual displacement. Considering the stress-strain relation, $\boldsymbol{\sigma} = \boldsymbol{c}\,\boldsymbol{\varepsilon}$, and the strain-displacement relation, $\boldsymbol{\varepsilon} = \boldsymbol{L}\boldsymbol{u}$, Eq. (2.64) can be rearranged in the following expression,

$$\int_{\Omega} \delta(\boldsymbol{L}\boldsymbol{u})^{T}\boldsymbol{c}(\boldsymbol{L}\boldsymbol{u})\mathrm{d}\Omega - \int_{\Omega} \delta\boldsymbol{u}^{T}\boldsymbol{b}\, \mathrm{d}\Omega - \int_{\Gamma_t} \delta\boldsymbol{u}^{T}\bar{\boldsymbol{t}}\, \mathrm{d}\Gamma + \int_{\Omega} \rho(\delta\boldsymbol{u}^{\mathrm{T}}\, \ddot{\mathbf{u}})\, \mathrm{d}\Omega = 0 \tag{2.65}$$

which is the generic Galerkin weak form written in terms of displacement, very useful in solid mechanical problems. In static problems the fourth term of Eq. (2.65) disappears.

2.3 Discrete System of Equations

The discrete equations for meshless methods are obtained from the principle of virtual work by using the meshless shape functions as trial and test functions. The domain Ω is discretized in a nodal distribution, and each node possesses an "influence-domain", which imposes the nodal connectivity between the neighbouring nodes. The meshless trial function $\boldsymbol{u}(\boldsymbol{x}_I)$ is given by,

$$\boldsymbol{u}(\boldsymbol{x}_I) = \sum_{i=1}^{n} \varphi_i(\boldsymbol{x}_I) u_i \tag{2.66}$$

being $\varphi_i(\boldsymbol{x}_I)$ the meshless approximation or interpolation function and u_i are the nodal displacements of the n nodes belonging to the influence-domain of interest node $\boldsymbol{x}_I$. Considering the NNRPIM, it is known that the NNRPIM interpolation function satisfies the condition,

$$\varphi_i(\boldsymbol{x}_j) = \delta_{ij} \tag{2.67}$$

where δ_{ij} is the Kronecker delta, $\delta_{ij} = 1$ if $i = j$ and $\delta_{ij} = 0$ if $i \neq j$. Following Eq. (2.66), the test functions (or virtual displacements) are defined as,

$$d\boldsymbol{u}(\boldsymbol{x}_I) = \sum_{i=1}^{n} \varphi_i(\boldsymbol{x}_I) du_i \tag{2.68}$$

where du_i are the nodal values for the test function.

2.3.1 Weak Form of Galerkin

The meshless formulation can be established in terms of a weak form of the differential equation under consideration, Eq. (2.64). In the solid mechanics context this implies the use of the virtual work equation.

$$L = \int_{\Omega} \sigma \, d\varepsilon \, \mathrm{d}\Omega - \int_{\Omega} \boldsymbol{b} \cdot d\boldsymbol{u} \, \mathrm{d}\Omega - \int_{\Gamma} \bar{\boldsymbol{t}} \cdot d\boldsymbol{u} \, \mathrm{d}\Gamma + \rho \int_{\Omega} \left(d\boldsymbol{u}^{\mathrm{T}} \; \ddot{\mathbf{u}}\right) \mathrm{d}\Omega = 0 \tag{2.69}$$

The virtual deformation $d\varepsilon$ is defined by,

$$d\boldsymbol{\varepsilon} = \overline{\boldsymbol{B}} d\boldsymbol{u} \tag{2.70}$$

where $\overline{\boldsymbol{B}}$ is the deformation matrix. Thus, the virtual work of the first term in Eq. (2.69), using Eq. (2.70), can be expressed as,

$$L_1 = \int_{\Omega} d\boldsymbol{u}^{\mathrm{T}} \overline{\boldsymbol{B}}^{\mathrm{T}} \boldsymbol{\sigma} \, d\Omega \tag{2.71}$$

The strain vector can be divided in two parts, the linear part and the nonlinear part,

$$\boldsymbol{\varepsilon} = \boldsymbol{\varepsilon}_0 + \boldsymbol{\varepsilon}_{NL} \tag{2.72}$$

which can also be presented as,

$$\boldsymbol{\varepsilon} = \underbrace{\boldsymbol{L}\boldsymbol{\theta}}_{\varepsilon_0} + \underbrace{\frac{1}{2}\boldsymbol{A}\,\boldsymbol{\theta}}_{\varepsilon_{NL}} = \left(\boldsymbol{L} + \frac{1}{2}\boldsymbol{A}\right)\boldsymbol{\theta} \tag{2.73}$$

Matrix $\boldsymbol{L}$ is defined as,

$$\boldsymbol{L} = \begin{bmatrix} \boldsymbol{e}_1^{\mathrm{T}} & \underset{[1\times3]}{\mathbf{0}} & \underset{[1\times3]}{\mathbf{0}} \\ \underset{[1\times3]}{\mathbf{0}} & \boldsymbol{e}_2^{\mathrm{T}} & \underset{[1\times3]}{\mathbf{0}} \\ \underset{[1\times3]}{\mathbf{0}} & \underset{[1\times3]}{\mathbf{0}} & \boldsymbol{e}_3^{\mathrm{T}} \\ \boldsymbol{e}_2^{\mathrm{T}} & \boldsymbol{e}_1^{\mathrm{T}} & \underset{[1\times3]}{\mathbf{0}} \\ \underset{[1\times3]}{\mathbf{0}} & \boldsymbol{e}_3^{\mathrm{T}} & \boldsymbol{e}_2^{\mathrm{T}} \\ \boldsymbol{e}_3^{\mathrm{T}} & \underset{[1\times3]}{\mathbf{0}} & \boldsymbol{e}_1^{\mathrm{T}} \end{bmatrix} = \begin{bmatrix} 1 & 0 & 0 & 0 & 0 & 0 & 0 & 0 & 0 \\ 0 & 0 & 0 & 0 & 1 & 0 & 0 & 0 & 0 \\ 0 & 0 & 0 & 0 & 0 & 0 & 0 & 0 & 1 \\ 0 & 1 & 0 & 1 & 0 & 0 & 0 & 0 & 0 \\ 0 & 0 & 0 & 0 & 0 & 1 & 0 & 1 & 0 \\ 0 & 0 & 1 & 0 & 0 & 0 & 1 & 0 & 0 \end{bmatrix} \tag{2.74}$$

Being $\boldsymbol{e}_i$ the coordinate i director column vector,

$$\boldsymbol{I} = \begin{bmatrix} \boldsymbol{e}_1 & \boldsymbol{e}_2 & \boldsymbol{e}_3 \end{bmatrix} \tag{2.75}$$

The column vector $\boldsymbol{\theta}$ is defined by,

$$\boldsymbol{\theta} = \boldsymbol{G}\boldsymbol{u} \tag{2.76}$$

The geometric matrix $\boldsymbol{G}$ is defined by,

$$\boldsymbol{G}^{\mathrm{T}} = \begin{bmatrix} \frac{\partial\varphi}{\partial x} & 0 & 0 & \frac{\partial\varphi}{\partial y} & 0 & 0 & \frac{\partial\varphi}{\partial z} & 0 & 0 \\ 0 & \frac{\partial\varphi}{\partial x} & 0 & 0 & \frac{\partial\varphi}{\partial y} & 0 & 0 & \frac{\partial\varphi}{\partial z} & 0 \\ 0 & 0 & \frac{\partial\varphi}{\partial x} & 0 & 0 & \frac{\partial\varphi}{\partial y} & 0 & 0 & \frac{\partial\varphi}{\partial z} \end{bmatrix} \tag{2.77}$$

which produces the following column vector $\boldsymbol{\theta}$,

$$\boldsymbol{\theta} = \begin{bmatrix} \boldsymbol{\theta}_x \\ \boldsymbol{\theta}_y \\ \boldsymbol{\theta}_z \end{bmatrix} \quad \text{being} \quad \boldsymbol{\theta}_\xi = \begin{bmatrix} \frac{\partial u}{\partial \xi} \\ \frac{\partial v}{\partial \xi} \\ \frac{\partial w}{\partial \xi} \end{bmatrix} \tag{2.78}$$

The current configuration displacement is considered in matrix $\boldsymbol{A}$, which corresponds in Eq. (2.73) to the actualized component.

$$\boldsymbol{A} = \begin{bmatrix} \boldsymbol{\theta}_x^T & \underset{[1x3]}{\boldsymbol{0}} & \underset{[1x3]}{\boldsymbol{0}} \\ \underset{[1x3]}{\boldsymbol{0}} & \boldsymbol{\theta}_y^T & \underset{[1x3]}{\boldsymbol{0}} \\ \underset{[1x3]}{\boldsymbol{0}} & \underset{[1x3]}{\boldsymbol{0}} & \boldsymbol{\theta}_z^T \\ \boldsymbol{\theta}_y^T & \boldsymbol{\theta}_x^T & \underset{[1x3]}{\boldsymbol{0}} \\ \underset{[1x3]}{\boldsymbol{0}} & \boldsymbol{\theta}_z^T & \boldsymbol{\theta}_y^T \\ \boldsymbol{\theta}_z^T & \underset{[1x3]}{\boldsymbol{0}} & \boldsymbol{\theta}_x^T \end{bmatrix} = \begin{bmatrix} \frac{\partial u}{\partial x} & \frac{\partial v}{\partial x} & \frac{\partial w}{\partial x} & 0 & 0 & 0 & 0 & 0 & 0 \\ 0 & 0 & 0 & \frac{\partial u}{\partial y} & \frac{\partial v}{\partial y} & \frac{\partial w}{\partial y} & 0 & 0 & 0 \\ 0 & 0 & 0 & 0 & 0 & 0 & \frac{\partial u}{\partial z} & \frac{\partial v}{\partial z} & \frac{\partial w}{\partial z} \\ \frac{\partial u}{\partial y} & \frac{\partial v}{\partial y} & \frac{\partial w}{\partial y} & \frac{\partial u}{\partial x} & \frac{\partial v}{\partial x} & \frac{\partial w}{\partial x} & 0 & 0 & 0 \\ 0 & 0 & 0 & \frac{\partial u}{\partial z} & \frac{\partial v}{\partial z} & \frac{\partial w}{\partial z} & \frac{\partial u}{\partial y} & \frac{\partial v}{\partial y} & \frac{\partial w}{\partial y} \\ \frac{\partial u}{\partial z} & \frac{\partial v}{\partial z} & \frac{\partial w}{\partial z} & 0 & 0 & 0 & \frac{\partial u}{\partial x} & \frac{\partial v}{\partial x} & \frac{\partial w}{\partial x} \end{bmatrix} \tag{2.79}$$

The deformation matrix $\overline{\boldsymbol{B}}$, dependent of $\boldsymbol{u}$, can be defined as,

$$\overline{\boldsymbol{B}} = \boldsymbol{B}_0 + \boldsymbol{B}_{\mathrm{NL}}(\boldsymbol{u}) \tag{2.80}$$

since it varies with the deformation of the solid. The linear part of the deformation matrix is represented by $\boldsymbol{B}_0$ and the nonlinear contribution by $\boldsymbol{B}_{NL}$. For the three-dimensional case,

$$\boldsymbol{B}_0^{\mathrm{T}} = \begin{bmatrix} \frac{\partial \varphi}{\partial x} & 0 & 0 & \frac{\partial \varphi}{\partial y} & 0 & \frac{\partial \varphi}{\partial z} \\ 0 & \frac{\partial \varphi}{\partial y} & 0 & \frac{\partial \varphi}{\partial x} & \frac{\partial \varphi}{\partial z} & 0 \\ 0 & 0 & \frac{\partial \varphi}{\partial z} & 0 & \frac{\partial \varphi}{\partial y} & \frac{\partial \varphi}{\partial x} \end{bmatrix} \tag{2.81}$$

and

$$\boldsymbol{B}_{\mathrm{NL}} = \boldsymbol{A}\,\boldsymbol{G} \tag{2.82}$$

The nonlinear deformation is actualized through matrix $\boldsymbol{A}$, which contains the displacement current configuration.

2.3.2 *Stiffness Matrix*

The tangential stiffness matrix $\boldsymbol{K}_T$ can be determined considering the variation of the virtual work of Eq. (2.71), in order to the generalized displacements $d\boldsymbol{u}$,

$$dL_1 = d\left[\int_{\Omega} \overline{\boldsymbol{B}}^{\mathrm{T}} \boldsymbol{\sigma} \mathrm{d}\Omega\right] \tag{2.83}$$

which can be developed as,

$$dL_1 = \int_{\Omega} d\overline{\boldsymbol{B}}^{\mathrm{T}} \boldsymbol{\sigma} \mathrm{d}\Omega + \int_{\Omega} \overline{\boldsymbol{B}}^{\mathrm{T}} d\boldsymbol{\sigma} \mathrm{d}\Omega = \boldsymbol{K}_{\mathrm{T}} d\boldsymbol{u} \tag{2.84}$$

Using Eqs. (2.35) and (2.70) the following relation is obtained,

$$d\boldsymbol{\sigma} = \boldsymbol{c}\overline{\boldsymbol{B}} d\boldsymbol{u} \tag{2.85}$$

In the deformation matrix $\overline{\boldsymbol{B}}$ only the nonlinear part $d\boldsymbol{B}_{\mathrm{NL}}$ is dependent of $\boldsymbol{u}$, Eq. (2.80), thus $d\overline{\boldsymbol{B}} = d\boldsymbol{B}_{\mathrm{NL}}$ and therefore,

$$dL_1 = \int_{\Omega} d\boldsymbol{B}_{\mathrm{NL}}^{\mathrm{T}} \boldsymbol{\sigma} \, \mathrm{d}\Omega + \int_{\Omega} \overline{\boldsymbol{B}}^{\mathrm{T}} \boldsymbol{c} \, \overline{\boldsymbol{B}} \, \mathrm{d}\Omega \tag{2.86}$$

Where the stiffness matrix can be presented as,

$$\boldsymbol{K}_T = \boldsymbol{K}_{\sigma} + \boldsymbol{K}_0 + \boldsymbol{K}_{\mathrm{NL}} \tag{2.87}$$

Being,

$$\boldsymbol{K}_{\sigma} = \int_{\Omega} d\boldsymbol{B}_{\mathrm{NL}}^{\mathrm{T}} \, \boldsymbol{\sigma} \, d\Omega \tag{2.88}$$

$$\boldsymbol{K}_0 = \int_{\Omega} \boldsymbol{B}_0^T \, \boldsymbol{c} \, \boldsymbol{B}_0 \, d\Omega \tag{2.89}$$

$$\boldsymbol{K}_{\mathrm{NL}} = \int_{\Omega} \left(\boldsymbol{B}_0^T \, \boldsymbol{c} \, \boldsymbol{B}_{\mathrm{NL}} + \boldsymbol{B}_{\mathrm{NL}}^T \, \boldsymbol{c} \, \boldsymbol{B}_{\mathrm{NL}} + \boldsymbol{B}_{\mathrm{NL}}^T \, \boldsymbol{c} \, \boldsymbol{B}_0 \right) d\Omega \tag{2.90}$$

The initial stress matrix or geometric matrix $\boldsymbol{K}_{\sigma}$ is defined as,

$$\boldsymbol{K}_{\sigma} d\boldsymbol{u} = \int_{\Omega} \boldsymbol{G}^T d\boldsymbol{A}^T \boldsymbol{\sigma} \, d\Omega \tag{2.91}$$

The variation of matrix $\boldsymbol{A}$ in order to $\boldsymbol{u}$ can be defined as,

$$d\boldsymbol{A}^{\mathrm{T}} = \begin{bmatrix} d\boldsymbol{\theta}_x & \underset{[3\times1]}{\boldsymbol{0}} & \underset{[3\times1]}{\boldsymbol{0}} & d\boldsymbol{\theta}_y & \underset{[3\times1]}{\boldsymbol{0}} & d\boldsymbol{\theta}_z \\ \underset{[3\times1]}{\boldsymbol{0}} & d\boldsymbol{\theta}_y & \underset{[3\times1]}{\boldsymbol{0}} & d\boldsymbol{\theta}_x & d\boldsymbol{\theta}_z & \underset{[3\times1]}{\boldsymbol{0}} \\ \underset{[3\times1]}{\boldsymbol{0}} & \underset{[3\times1]}{\boldsymbol{0}} & d\boldsymbol{\theta}_z & \underset{[3\times1]}{\boldsymbol{0}} & d\boldsymbol{\theta}_y & d\boldsymbol{\theta}_x \end{bmatrix} \tag{2.92}$$

As so, the term $d\boldsymbol{A}^T\boldsymbol{\sigma}$ can be represented as,

$$d\boldsymbol{A}^{\mathrm{T}}\boldsymbol{\sigma} = \begin{bmatrix} \sigma_{xx} \underset{[3\times3]}{\boldsymbol{I}} & \tau_{xy} \underset{[3\times3]}{\boldsymbol{I}} & \tau_{xz} \underset{[3\times3]}{\boldsymbol{I}} \\ \tau_{yx} \underset{[3\times3]}{\boldsymbol{I}} & \sigma_{yy} \underset{[3\times3]}{\boldsymbol{I}} & \tau_{yz} \underset{[3\times3]}{\boldsymbol{I}} \\ \tau_{zx} \underset{[3\times3]}{\boldsymbol{I}} & \tau_{zy} \underset{[3\times3]}{\boldsymbol{I}} & \sigma_{zz} \underset{[3\times3]}{\boldsymbol{I}} \end{bmatrix} d\boldsymbol{\theta} = \boldsymbol{Z}\,d\boldsymbol{\theta} = \boldsymbol{Z}\boldsymbol{G}\,d\boldsymbol{u} \tag{2.93}$$

and therefore Eq. (2.88) can be presented as,

$$\boldsymbol{K}_{\sigma} = \int_{\Omega} \boldsymbol{G}^{T}\boldsymbol{Z}\boldsymbol{G}\,d\Omega \tag{2.94}$$

Therefore, the initial stress matrix $\boldsymbol{K}_{\sigma}$ takes into consideration the actualized stress field.

2.3.3 Mass Matrix

The virtual work of the last term in Eq. (2.69) can be expressed and developed as,

$$dL_4 = d\left[\rho \int_{\Omega} \left(d\boldsymbol{u}^{\mathrm{T}}\ \ddot{\boldsymbol{u}}\right) d\Omega\right] = \boldsymbol{M}\,d\ddot{\boldsymbol{u}} \tag{2.95}$$

where the mass matrix $\boldsymbol{M}$ can be defined as,

$$\boldsymbol{M} = \int_{\Omega} \boldsymbol{H}^{\mathrm{T}}\boldsymbol{\rho}\boldsymbol{H}\,\mathrm{d}\Omega \tag{2.96}$$

being $\boldsymbol{H}$ the interpolation function matrix for the interest point i defined as,

$$\boldsymbol{H}_i = \varphi_i\boldsymbol{I} \tag{2.97}$$

Where φ_i is the interpolation function for interest node i and $\boldsymbol{I}$ is the identity matrix defined in Eq. (2.75). The density diagonal matrix can be defined as,

$$\boldsymbol{\rho} = \rho\boldsymbol{I} \tag{2.98}$$

being ρ the solid material density.

2.3.4 Force Vector

The virtual work of the middle terms in Eq. (2.69) can be expressed and developed as,

$$dL_2 = d\left[\int_{\Omega} \boldsymbol{b} \cdot d\boldsymbol{u} \, \mathrm{d}\Omega\right] = \boldsymbol{f}_b \tag{2.99}$$

and

$$dL_3 = d\left[\int_{\Gamma} \bar{\boldsymbol{t}} \cdot d\boldsymbol{u} \, \mathrm{d}\Gamma\right] = \boldsymbol{f}_{\bar{t}} \tag{2.100}$$

being the total force vector $\boldsymbol{f}$ defined as,

$$\boldsymbol{f}_b + \boldsymbol{f}_{\bar{t}} = \boldsymbol{f} \tag{2.101}$$

Thus, the total force vector $\boldsymbol{f}$ can be developed in a matrix form,

$$\boldsymbol{f} = \int_{\Omega} \boldsymbol{H}^{\mathrm{T}} \boldsymbol{b} \, \mathrm{d}\Omega + \int_{\Gamma} \boldsymbol{H}^{\mathrm{T}} \bar{\boldsymbol{t}} \, \mathrm{d}\Gamma. \tag{2.102}$$

2.3.5 Essential Boundary Conditions Imposition

If the shape functions of the meshless method possess the Kronecker delta property, then the boundary conditions can be imposed directly as in the FEM. The continuum analysis involves two types of boundary conditions, the essential boundary conditions (displacement related) and the natural boundary conditions (force related). Neglecting dumping effects and assuming that the matricial form of the equilibrium equations resulting from virtual work expression, Eq. (2.69), can be presented as,

$$\boldsymbol{K}\boldsymbol{u} + \boldsymbol{M}\ddot{\boldsymbol{u}} = \boldsymbol{f} \tag{2.103}$$

Such equation can be rewritten as,

$$\begin{bmatrix} \boldsymbol{K}_{cc} & \boldsymbol{K}_{cd} \\ \boldsymbol{K}_{dc} & \boldsymbol{K}_{dd} \end{bmatrix} \begin{bmatrix} \boldsymbol{u}_c \\ \boldsymbol{u}_d \end{bmatrix} + \begin{bmatrix} \boldsymbol{M}_{cc} & \boldsymbol{M}_{cd} \\ \boldsymbol{M}_{dc} & \boldsymbol{M}_{dd} \end{bmatrix} \begin{bmatrix} \ddot{u}_c \\ \ddot{u}_d \end{bmatrix} = \begin{bmatrix} \boldsymbol{f}_c \\ \boldsymbol{f}_d \end{bmatrix} \tag{2.104}$$

where $\boldsymbol{u}_c$ are the unknown displacements and $\boldsymbol{u}_d$ the known, or prescribed, displacements. The vectors $\boldsymbol{f}_c$ and $\boldsymbol{f}_d$ correspond respectively to the known applied

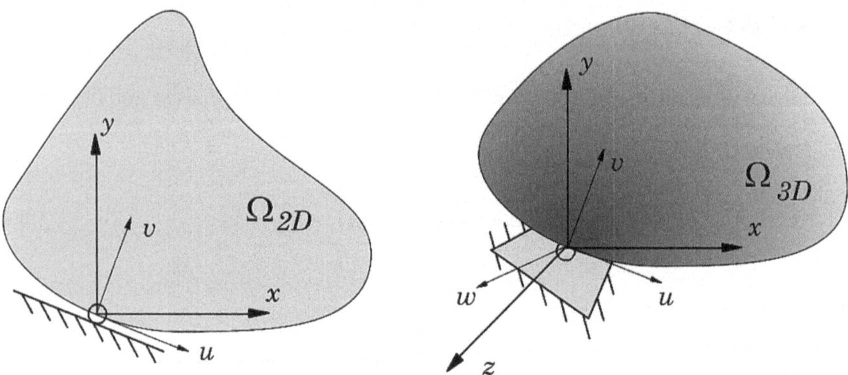

Fig. 2.6 Essential boundary condition nonaligned with the global axis

loads (external and body forces) and to the unknown reactions due the imposed displacement constrains. With the Eq. (2.104) it is assumed that the displacement components considered are axial aligned with the prescribed displacements. If this is not the case it is required the identification of all prescribed displacement orientations and transform locally the discrete equilibrium equations to correspond to the global axis. Thus,

$$\boldsymbol{u} = \boldsymbol{T}\overline{\boldsymbol{u}} \tag{2.105}$$

where $\overline{\boldsymbol{u}}$ is the vector of nodal point degrees of freedom in the required directions. The transformation matrix $\boldsymbol{T}$ is defined by Eq. (2.106) and Fig. 2.6, which is a typical representation of the constrained displacements in 2D and 3D analysis.

$$\boldsymbol{T}_{2D} = \begin{bmatrix} u_x & v_x \\ u_y & v_y \end{bmatrix} \quad \text{and} \quad \boldsymbol{T}_{3D} = \begin{bmatrix} u_x & v_x & w_x \\ u_y & v_y & w_y \\ u_z & v_z & w_z \end{bmatrix} \tag{2.106}$$

Using Eqs. (2.105) and (2.106) it is possible to write,

$$\overline{\boldsymbol{K}}\,\overline{\boldsymbol{u}} + \overline{\boldsymbol{M}}\,\ddot{\overline{\boldsymbol{u}}} = \overline{\boldsymbol{f}} \tag{2.107}$$

where,

$$\overline{\boldsymbol{M}} = T^{\mathrm{T}}\boldsymbol{M}\,\boldsymbol{T} \tag{2.108}$$

$$\overline{\boldsymbol{K}} = T^{\mathrm{T}}\boldsymbol{K}\,\boldsymbol{T} \tag{2.109}$$

$$\overline{\boldsymbol{f}} = T^{\mathrm{T}}\boldsymbol{f} \tag{2.110}$$

Notice that the matrix multiplications in Eqs. (2.108), (2.109) and (2.110) involve changes only in those columns and rows of $\boldsymbol{M}$, $\boldsymbol{K}$ and $\boldsymbol{f}$ that are actually affected by the prescribed displacement. In practice, the transformation can be effectively carried out on the local level just prior to adding the local matrices to the global assembled matrices.

2.3.6 Dynamic Equations

The equilibrium equations governing the linear dynamic response can be represented as in Eq. (2.103). The fundamental mathematical method used to solve Eq. (2.103) is the separation of variables. In order to change the equilibrium equations to the modal generalized displacements [7] it is proposed the following transformation:

$$\boldsymbol{u}(t) = \boldsymbol{\Phi}\boldsymbol{x}(t) \tag{2.111}$$

where $\boldsymbol{\Phi}$ is a $m \times m$ square matrix containing m spatial vectors independent of the time variable t, $\boldsymbol{x}(t)$ is a time dependent vector and $m = 2N$ for the 2D case and $m = 3N$ for the 3D case, being N the total number of nodes in the problem domain. From Eq. (2.111) also follows that $\dot{\boldsymbol{u}}(t) = \boldsymbol{\Phi}\dot{\boldsymbol{x}}(t)$ and $\ddot{\boldsymbol{u}}(t) = \boldsymbol{\Phi}\ddot{\boldsymbol{x}}(t)$. The components of $\boldsymbol{u}(t)$ are called generalized displacements. For which the solution can be presented in the form,

$$\boldsymbol{u}(t) = \boldsymbol{\phi}\ \sin(\omega\,(t - t_0)) \tag{2.112}$$

being $\boldsymbol{\phi}$ the vector of order m, t the time variable, the constant initial time is defined by t_0 and ω is the vibration frequency vector. Substituting Eqs. (2.112) into (2.103) the generalized eigenproblem is obtained, from which $\boldsymbol{\phi}$ and ω must be determined,

$$\boldsymbol{K}\,\boldsymbol{\phi} = \omega^2\,\boldsymbol{M}\,\boldsymbol{\phi} \tag{2.113}$$

Equation (2.113) yields the m eigensolutions,

$$\begin{bmatrix} \boldsymbol{K}\,\boldsymbol{\phi}_1 = \omega_1^2\,\boldsymbol{M}\,\boldsymbol{\phi}_1 \\ \boldsymbol{K}\,\boldsymbol{\phi}_2 = \omega_2^2\,\boldsymbol{M}\,\boldsymbol{\phi}_2 \\ \vdots \\ \boldsymbol{K}\,\boldsymbol{\phi}_m = \omega_m^2\,\boldsymbol{M}\,\boldsymbol{\phi}_m \end{bmatrix} \tag{2.114}$$

The vector $\boldsymbol{\phi}_i$ is called the *ith* mode shape vector and ω_i is the corresponding frequency of vibration. Defining a matrix $\boldsymbol{\Phi}$ whose columns are the eigenvectors $\boldsymbol{\phi}_i$,

$$\boldsymbol{\Phi} = [\boldsymbol{\phi}_1 \quad \boldsymbol{\phi}_2 \quad \ldots \quad \boldsymbol{\phi}_m] \tag{2.115}$$

and a diagonal matrix $\boldsymbol{\Omega}$ which stores the eigenvalues ω_i,

$$\boldsymbol{\Omega} = \begin{bmatrix} \omega_1^2 & 0 & \cdots & 0 \\ 0 & \omega_2^2 & \cdots & 0 \\ \vdots & \vdots & \ddots & \vdots \\ 0 & 0 & \cdots & \omega_m^2 \end{bmatrix} \tag{2.116}$$

the m solutions can be written as,

$$\boldsymbol{K}\boldsymbol{\Phi} = \boldsymbol{M}\,\boldsymbol{\Phi}\,\boldsymbol{\Omega} \tag{2.117}$$

It is required that the space functions satisfy the following stiffness and mass orthogonality conditions,

$$\boldsymbol{\Phi}^{\mathrm{T}}\boldsymbol{K}\,\boldsymbol{\Phi} = \boldsymbol{\Omega} \tag{2.118}$$

and

$$\boldsymbol{\Phi}^{\mathrm{T}}\boldsymbol{M}\,\boldsymbol{\Phi} = \boldsymbol{I} \tag{2.119}$$

After substituting Eq. (2.111) and its time derivatives into Eq. (2.103) and pre-multiplying it by $\boldsymbol{\Phi}^{\mathrm{T}}$, the equilibrium equation that corresponds to the modal generalized displacement is obtained,

$$\ddot{\boldsymbol{x}}(t) + \boldsymbol{\Omega}\boldsymbol{x}(t) = \boldsymbol{\Phi}^{\mathrm{T}}\boldsymbol{F}(t) \tag{2.120}$$

The initial conditions on $\boldsymbol{x}(t)$ are obtained using Eq. (2.111) and considering the the $\boldsymbol{M}$-orthonormality of $\boldsymbol{\Phi}^{\mathrm{T}}$ at time $t = 0$,

$$\left[\begin{array}{l} \boldsymbol{x}_0 = \boldsymbol{\Phi}^{\mathrm{T}}\boldsymbol{M}\boldsymbol{u}_0 \\ \dot{\boldsymbol{x}}_0 = \boldsymbol{\Phi}^{\mathrm{T}}\boldsymbol{M}\dot{\boldsymbol{u}}_0 \end{array}\right. \tag{2.121}$$

Equation (2.120) can be represented as m individual equations of the form,

$$\left[\begin{array}{l} \ddot{x}_i(t) + \omega_i^2 x_i(t) = f_i(t) \\ f_i(t) = \boldsymbol{\phi}_i^T \mathbf{F}(t) \end{array}\right. \tag{2.122}$$

with the initial conditions,

$$\left[\begin{array}{l} x_i^{t=0} = \boldsymbol{\phi}_i^T \boldsymbol{M}\boldsymbol{u}_0 \\ \dot{x}_i^{t=0} = \boldsymbol{\phi}_i^T \boldsymbol{M}\dot{\boldsymbol{u}}_0 \end{array}\right. \tag{2.123}$$

For the complete response, the solution to all m equations in Eq. (2.122) must be calculated and then the modal point displacements are obtained by superposition of the response in each mode.

$$\boldsymbol{u}(t) = \sum_{i=1}^{m} \boldsymbol{\phi}_i x_i(t) \tag{2.124}$$

Therefore the response analysis requires, first, the solution of the eigenvalues and eigenvectors of the problem, Eq. (2.113), then the solution of the decoupled equilibrium equations in Eq. (2.122) and, finally, the superposition of the response in each eigenvector as expressed in Eq. (2.124).

2.3.7 Forced Vibrations

In this book when forced vibrations are imposed only three different time-dependent loading conditions are considered, $\boldsymbol{f}(t) = \boldsymbol{f} \times g(t)$. A time constant load—load case A,

$$g_A(t) = 1 \tag{2.125}$$

A transient load—load case B,

$$\begin{cases} g_B(t) = 1 & \text{if} \quad t \leq t_i \\ g_B(t) = 0 & \text{if} \quad t > t_i \end{cases} \tag{2.126}$$

And a harmonic load—load case C,

$$g_C(t) = \sin(\gamma t) \tag{2.127}$$

The solution of each equation in Eq. (2.123) can be calculated using the Duhamel integral,

$$x_i(t) = \frac{1}{\omega_i} \int_0^t f_i(\tau) \sin(\omega_i(t - \tau)) d\tau + \alpha_i \sin(\omega_i t) + \beta_i \cos(\omega_i t) \tag{2.128}$$

where α_i and β_i are determined from the initial conditions: Eq. (2.123) and $f_i(t) = \boldsymbol{\phi}_i^T \boldsymbol{f}(t)$. For load case A and load case B the obtained solution is defined as,

$$x_i(t) = \frac{f_i(t)}{\omega_i^2}(1 - \cos(\omega_i t)) + \frac{\dot{x}_i^{t=0}}{\omega_i} \sin(\omega_i t) + x_i^{t=0} \cos(\omega_i t) \tag{2.129}$$

For load case C the obtained solution is,

$$x_i(t) = \frac{f_i(t)}{\omega_i^2 - \gamma^2} \left(\sin(\gamma t) - \frac{\gamma}{\omega_i} \sin(\omega_i t) \right) \tag{2.130}$$

References

1. Fung YC (1965) Foundations of solid mechanics. Englwood Cliffs, Prentice-Hall, New Jersey, USA
2. Malvern LE (1969) Introduction of the Mechanics of a Continuous Medium. Englwood Cliffs, Prentice-Hall, New Jersey, USA
3. Hodge PG (1970) Continuum mechanics. Mc Graw-Hill, New York
4. Lekhnitskii SG (1968) Anisotropic Plates. Gordon and Breach Science Publishers, New York-London-Paris
5. Timoshenko S, Goodier JN (1970) Theory of Elasticity. 3rd ed. Singapore, McGraw Hill
6. Reddy JN (1986) Applied functional analysis and variational methods in engineering. McGraw-Hill, Singapore
7. Bathe KJ (1996) Finite element procedures. Prentice-Hall, Englewood Cliffs

Chapter 3
Meshless Methods Introduction

Abstract In this chapter the most important meshless method concepts are detailed introduced. The chapter stars with a generic description on the meshless procedure. Additionally, it is presented a brief comparison between procedures of the finite element method (FEM) and the meshless method. Afterwards the meshless method nodal connectivity is addressed. Techniques to enforce the nodal connectivity in meshless methods are presented, such as the classic "influence-domain" concept and the recently developed "influence-cell" methodology. Then, it is presented a broad description of the integration schemes used in the numerical examples shown in this book: the Gauss-Legendre quadrature scheme and a flexible nodal based integration scheme. The final section of this chapter presents explicitly the generic numerical implementation of approximation and interpolation meshless methods based on the Galerkin weak formulation.

3.1 Meshless Generic Procedure

Like the majority of other nodal dependent discretization numerical methods, most of the meshless methods approaches respect the following outline. First the problem geometry is studied and the solid domain and contour is established, then the essential and natural boundary conditions are identified, Fig. 3.1a. Afterwards the problem domain and boundary is numerically discretized by a nodal set following a regular or irregularly distribution, Fig. 3.1b and c.

This nodal distribution do not form a mesh, since it is not required by the meshless method any kind of previous information about the relation between each node in order to construct the approximation (or interpolation) functions of the unknown variable field functions. The only information required by truly meshless methods is the spatial location of each node discretizing the problem domain.

J. Belinha, *Meshless Methods in Biomechanics*, Lecture Notes in Computational Vision and Biomechanics 16, DOI: 10.1007/978-3-319-06400-0_3,

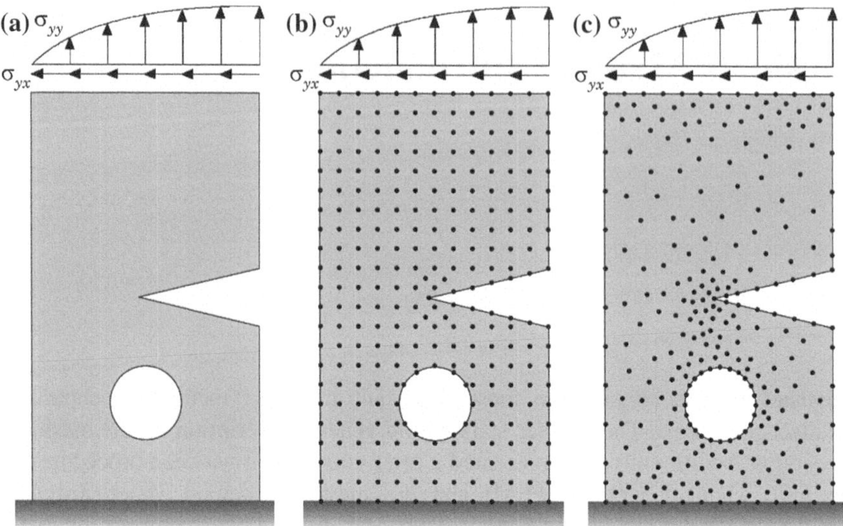

Fig. 3.1 **a** Solid domain. **b** Regular nodal discretization example. **c** Irregular nodal discretization example

Similarly with mesh dependent discretization numerical methods, in meshless methods the nodal density of the discretization, as well as the nodal spatial distribution, affects the method performance. A fine nodal distribution leads generally to more accurate results, however the computational cost grows with the increase of the total number of nodes. The unbalanced distribution of the nodes discretizing the problem domain can lead to a lower accuracy. Generally locations with predictable stress concentrations, such as: domain discontinuities; convex boundaries; crack tips; essential boundaries; natural boundaries; etc., should present a higher nodal density when compared with locations in which smooth stress distributions are expected [1, 2], Fig. 3.1c.

After the nodal discretization a background integration mesh is constructed, nodal dependent or nodal independent many numerical methods require it in order to numerically integrate the weak form equations governing the physic phenomenon. The integration mesh can have the size of the problem domain or even a larger one, without affecting too much the final results [3].

It is common to use Gaussian integration meshes, as in the FEM, fitted to the problem domain, Fig. 3.2a, however other approaches in meshless methods are also valid, Fig. 3.2b. Another way to integrate the weak form equations is using the nodal integration, which resorts to the Voronoï diagrams in order to obtain the integration weight on each node, Fig. 3.2c.

As it is perceptible, using the nodal integration, the nodal distribution will additionally serve as an integration mesh. Nonetheless, the nodal integration generally leads to the decrease of accuracy, being necessary to implement a stabilization method which increases the computational cost [4–6].

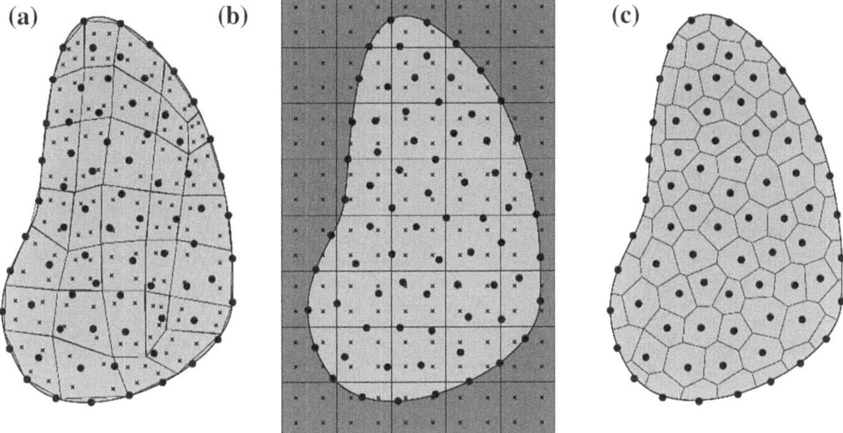

Fig. 3.2 **a** Fitted Gaussian integration mesh. **b** General Gaussian integration mesh. **c** Voronoï diagram for nodal integration

With the nodal distribution defined and the integration mesh constructed the nodal connectivity can be imposed. In the FEM the nodal connectivity is predefined by the used of 'elements', however in the meshless methods there are no elements. Thus, for each interest point $\boldsymbol{x}_I$ of the problem domain concentric areas or volumes are defined, and the nodes inside these areas or volumes belong to the influence-domain of node $\boldsymbol{x}_i$. In the majority of the meshless methods the interest points are equivalent to the integration points from the background integration mesh, however there are meshless methods using the collocation point methodology or the nodal integration scheme, in which the interest points are the nodes of the nodal discretization. The shape and size of the influence-domain, which depends on the relative position of the interest point, affects the quality of the results, Fig. 3.3a. It is recommend that all influence-domains possess approximately the same number of nodes inside, as so the size of the influence-domain should be dependent on the nodal density around the interest point. In Fig. 3.3b it is shown a bad choice in the influence-domain strategy.

Afterwards the field variables can be obtained with the approximation or interpolation function. Take for example the displacement field $\boldsymbol{u}$ as the field variable under consideration. In meshless methods the displacement components $\boldsymbol{u}_I = (u_x u_y u_z)$ at any interest point $\boldsymbol{x}_I$ within the problem domain are approximated or interpolated using the nodal displacement of the nodes inside the influence-domain of such interest point $\boldsymbol{x}_I$,

$$\boldsymbol{u}(\boldsymbol{x}_I) = \sum_{i=1}^{n} \varphi_i(\boldsymbol{x}_I)\, \boldsymbol{u}(\boldsymbol{x}_i) \tag{3.1}$$

being n the number of nodes inside the influence-domain of the interest point $\boldsymbol{x}_I$, $\boldsymbol{u}(\boldsymbol{x}_i)$ are the displacement components of each node within the influence-domain

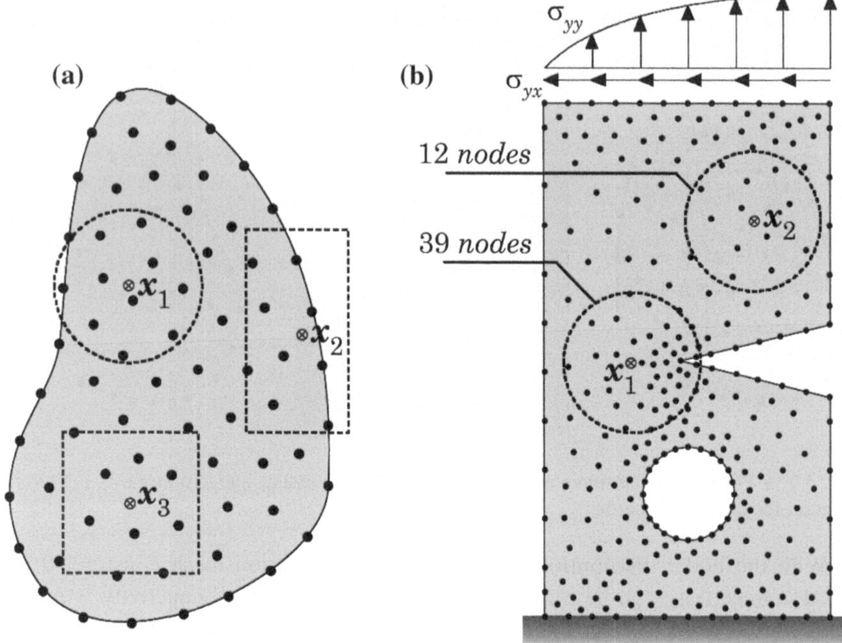

Fig. 3.3 **a** Influence-domains with different sizes and shapes. **b** Example of a bad choice in the size of the influence-domain, interest point $\boldsymbol{x}_1$ has much more nodes inside the influence-domain when compared with interest point $\boldsymbol{x}_2$

and $\varphi_i(\boldsymbol{x}_I)$ is the approximation or interpolation function value of the ith node obtained using only the n nodes inside the influence-domain.

Establishing the equation system is the next step. In meshless methods, the discrete equations can be formulated using the approximation or interpolation functions applied to the strong or weak form formulation. In the case of the meshless methods using the weak form of Galerkin, the discrete equations can be obtained applying to the differential equation governing the physic phenomenon the weighted residual method of Galerkin. The produced equations are then arranged in a local nodal matrix form and assembled into a global equation system matrix. In the case of a static problem the global equation system matrix is a set of algebraic equations, for the case of the free vibration analysis or buckling analysis is a set of eigenvalue equations and in the case of a dynamic (time dependent) analysis is a set of differential equations.

In order to obtain the distinct solutions for the distinct types of analyses, one must choose the appropriate solver. For static problems with the global equation system matrix the displacement field is obtained. To obtain the solution field, a linear algebraic equation solver is required. In this work for small systems it is used a Gauss elimination method and for larger equation systems a LU decomposition method. For the free-vibration and buckling analysis it is required the use

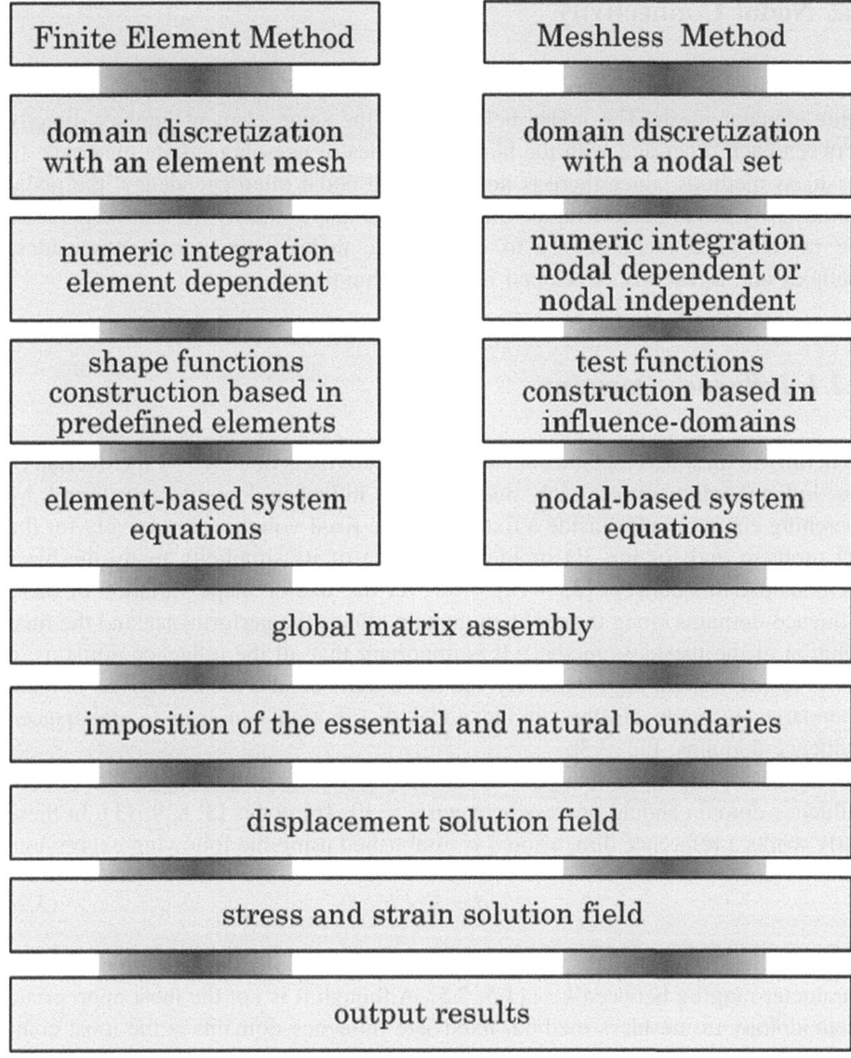

Fig. 3.4 Comparison between the flow-charts for a finite element method formulation and generic meshless method formulation

of an eigenvalue equation solver to obtain the eigenvalue and the eigenvector solution. In this work it was used the Jacobi's method. In the dynamic analysis the variation through time of the displacement, velocity and acceleration are the variable fields to be obtained. In this book the solution of the dynamic equations is obtained using the modal superposition method.

In order to clarify the main differences between the finite element method and a standard meshless method, in Fig. 3.4 it is presented a comparative flow-chart of both numerical approaches.

3.2 Nodal Connectivity

In the Finite Element Method the nodal connectivity is assure with the predefined finite element mesh. The nodes belonging to the same element interact directly between each other and with the boundary nodes of neighbour finite elements. In meshless methods, since there is no predefined nodal interdependency, the nodal connectivity has to be determined after the nodal distribution. This section presents the mostly used methodology to enforce the nodal connectivity in meshless methods and a recently developed flexible technique.

3.2.1 Influence-Domains

Generally in meshless methods the nodal connectivity is obtained by the overlap of the influence-domain of each node. These influence-domains are found by searching enough nodes inside a fixed area or a fixed volume, respectively for the 2D problem and for the 3D problem. Because of its simplicity many meshless methods use this concept [3, 7–10]. However the size or shape variation of these influence-domains along the problem domain affects the performance and the final solution of the meshless method. It is important that all the influence-domains in the problem contain approximately the same number of nodes. Irregular domain boundaries or node clusters in the nodal distribution can lead to unbalanced influence-domains, Fig. 3.3b.

Regardless the used meshless technique, previous works suggest that each influence-domain should possess between $n = [9, 16]$ nodes [3, 8, 9, 11]. In these early works a reference dimension d is established using the following expression,

$$d = k \times h \tag{3.2}$$

Being h the average nodal spacing in the surroundings of $\boldsymbol{x}_i$ and k a dimensionless parameter ranging between $k = [1.5, 2.5]$. Although it is not the most appropriate methodology in meshless method, fixed size influence-domains is the most common technique to establish the nodal connectivity. For a two dimensional space, in Fig. 3.5 two types of fixed size domains are suggested: the rectangular shape influence-domain and the circular shape influence-domain, Fig. 3.5a and b.

In the case of the fixed rectangular influence-domains, initially the dimensions d_x and d_y are established, then, for each interest point $\boldsymbol{x}_I$, the n nodes inside the $d_x \times d_y$ rectangle centred in $\boldsymbol{x}_I$ are identified. In the case of the fixed circular influence-domain it is settled the dimension d_R and, once again, for each interest point $\boldsymbol{x}_I$, the n nodes inside the circle with centre in $\boldsymbol{x}_I$ and radius d_R are identified. Notice that using this process an interest point $\boldsymbol{x}_i$ near the domain boundary possesses less nodes inside the respective influence-domains in comparison with an inner interest point $\boldsymbol{x}_j$. Using an analogy with the FEM, the use of fixed domains is the

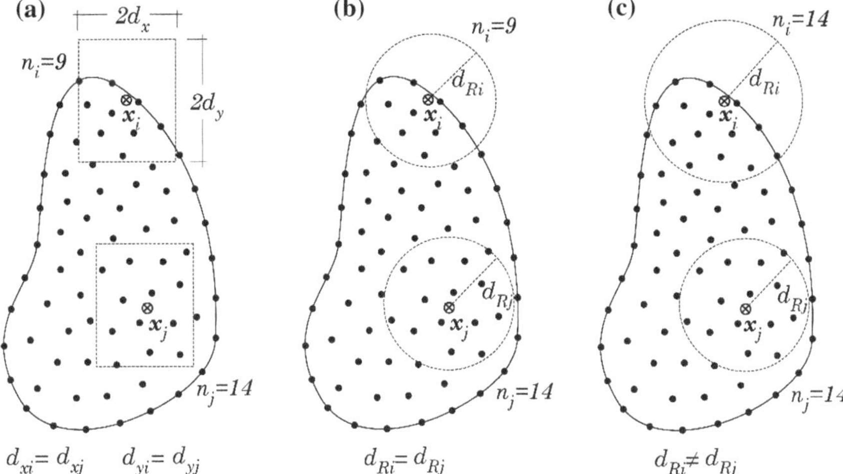

Fig. 3.5 **a** Fixed *rectangular* shaped influence-domain. **b** Fixed *circular* shaped influence-domain. **c** Flexible *circular* shaped influence-domain

same as using triangular or quadrilateral elements in the boundary and high-order elements in the centre of the solid domain. Fixed and regular shaped influence-domains can lead to the loss of accuracy in the numerical analysis. Therefore, in order to maintain a constant connectivity along the solid domain variable influence-domains are a better solution. To illustrate this idea the number of nodes inside the influence-domain is established in $n = 14$ nodes, then performing a radial search, using the interest point as centre, the n closest nodes are found. In Fig. 3.5c are shown the variable influence-domains of two interest points. This technique permits to avoid the numerical problems identified in Fig. 3.3b and to construct shape functions with the same degree of complexity in the complete domain.

3.2.2 Influence-Cells

A new approach to establish the influence-domains in meshless methods was proposed by Belinha and co-workers in 2007 [12]. Instead of using fixed or variable blind influence-domains, the approach proposed by Belinha uses the spatial colocation of the nodes discretizing the problem domain to determine directly the influence-domains. This approach uses mathematical concepts such as the Voronoï diagrams and the Delaunay triangulation, to determine the nodal connectivity of each node belonging to the global nodal set. Since these influence-domains are determined based on the geometric and spatial relations between the Voronoï cells obtained from the Voronoï diagram of the nodal distribution, the influence-domains are called 'influence-cells'. In order to fully understand the influence-cell concept a brief presentation of the Voronoï diagram concept is required.

3.2.3 Natural Neighbours

The Voronoï diagram of a discrete nodal set is obtained using the natural neighbour mathematical concept, which was firstly introduced by Sibson for data fitting and field smoothing [13].

Consider the nodal set $\boldsymbol{N} = \{n_1, n_2, \ldots, n_N\}$ discretizing the space domain $\Omega \subset \mathbb{R}^d$ with $\boldsymbol{X} = \{\boldsymbol{x}_1, \boldsymbol{x}_2, \ldots, \boldsymbol{x}_N\} \in \Omega$. The Voronoï diagram of $\boldsymbol{N}$ is the partition of the function space discretized by $\boldsymbol{X}$ in sub-regions V_i, closed and convex. Each sub-region V_i is associated to the node n_i in a way that any point in the interior of V_i is closer to n_i than any other node $n_j \in \boldsymbol{N} \wedge j \neq i$. The set of Voronoï cells $\boldsymbol{V}$ define the Voronoï diagram, $\boldsymbol{V} = \{V_1, V_2, \ldots, V_N\}$. The Voronoï cell is defined by,

$$V_i := \left\{ \boldsymbol{x}_I \in \Omega \subset \mathbb{R}^{\mathrm{d}} : \|\boldsymbol{x}_I - \boldsymbol{x}_i\| < \left\|\boldsymbol{x}_I - \boldsymbol{x}_j\right\|, \quad \forall i \neq j \right\} \tag{3.3}$$

being $\boldsymbol{x}_I$ an interest point of the domain and $\|\cdot\|$ the Euclidian metric norm. Thus, the Voronoï cell V_i is the geometric place where all points are closer to n_i than to any other node.

The Voronoï diagrams implications are extensive, with applications from the natural sciences to engineering. In the literature it is possible to find detailed descriptions of the properties and applications of such mathematical tool [14, 15], as well as efficient algorithms to construct Voronoï tessellations [16]. Since it is easier to visualize, it is represented a two-dimensional space $\Omega \subset \mathbb{R}^2$ in order to show how the Voronoï diagram can be generically obtained. Consider the nodal set represented in Fig. 3.6a. Since the objective is to determine the Voronoï cell V_0 of node n_0, the nodes on Fig. 3.6a are chosen as potential neighbours of n_0. Then one of the nodes is selected as potential neighbour, for example node n_4, Fig. 3.6b, and the vector $\boldsymbol{u}_{40}$ is determined,

$$\boldsymbol{u}_{40} = \frac{(\boldsymbol{x}_0 - \boldsymbol{x}_4)}{\|\boldsymbol{x}_0 - \boldsymbol{x}_4\|} \tag{3.4}$$

being $\boldsymbol{u}_{40} = \{u_{40}, v_{40}, w_{40}\}$. Using the normal vector $\boldsymbol{u}_{40}$ it is possible to defined plane π_{40},

$$u_{40}x + v_{40}y + w_{40}z = (u_{40}x_4 + v_{40}y_4 + w_{40}z_4) \tag{3.5}$$

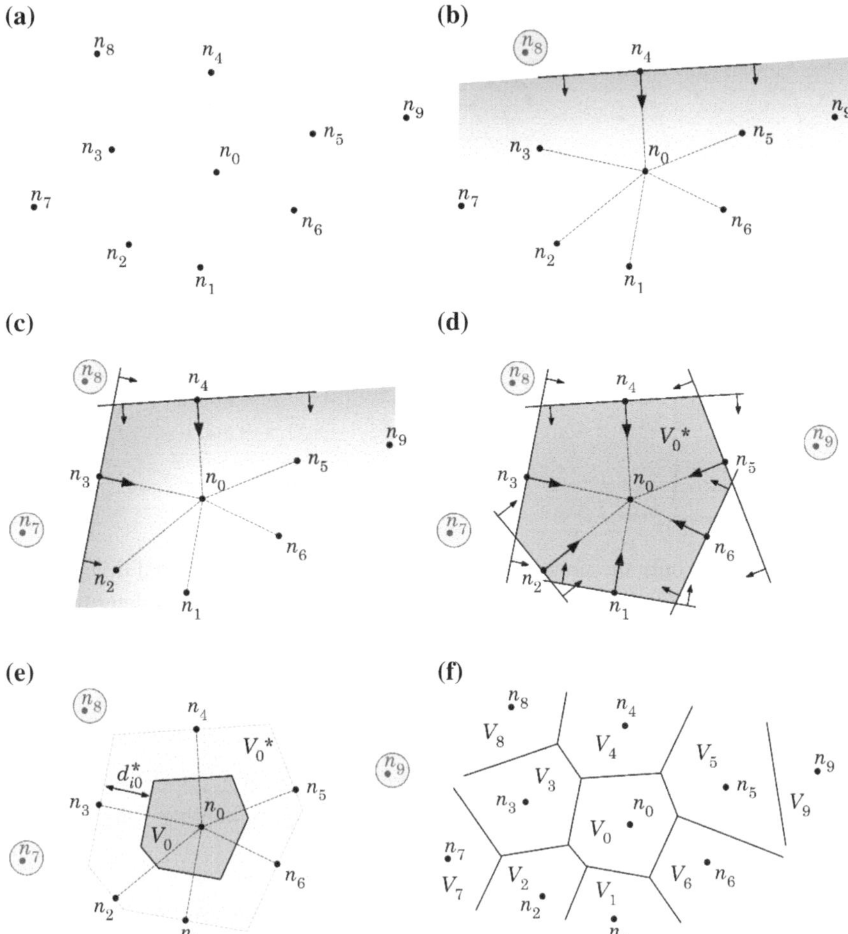

Fig. 3.6 **a** Initial nodal set of potential neighbour nodes of node n_0. **b** First trial plane. **c** Second trial plane. **d** Final trial cell containing just the natural neighbours of node n_0. **e** Node n_0 Voronoï cell V_0. **f** Voronoï diagram

After the definition of plane π_{40}, all nodes that do not respect the following condition:

$$u_{40}x + v_{40}y + w_{40}z \geq (u_{40}x_4 + v_{40}y_4 + w_{40}z_4) \tag{3.6}$$

are eliminated as natural neighbours of node n_0. In Fig. 3.6b it is possible to observe that node n_8 is not a natural neighbour of node n_0. Afterwards the process is repeated for each one of initial nodal set, Fig. 3.6c. In the end, each one of the 6 natural neighbours of node n_0, represented in Fig. 3.6d, respect simultaneously the following 6 conditions:

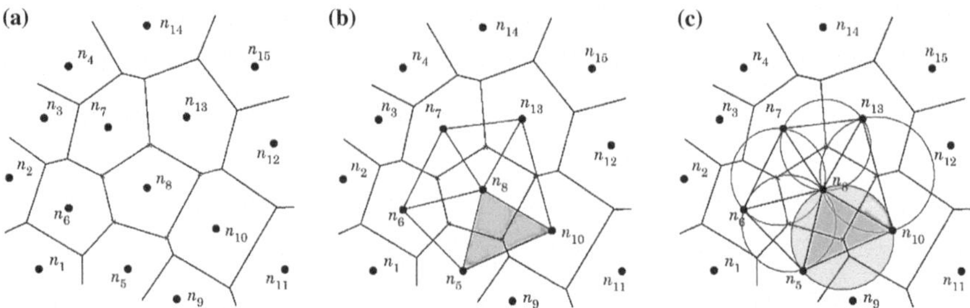

Fig. 3.7 **a** Initial Voronoï diagram. **b** Delaunay *triangulation*. **c** Natural neighbour *circumcircle*

$$\begin{cases} u_{10}x + v_{10}y + w_{10}z \geq (u_{10}x_1 + v_{10}y_1 + w_{10}z_1) \\ u_{20}x + v_{20}y + w_{20}z \geq (u_{20}x_2 + v_{20}y_2 + w_{20}z_2) \\ \vdots \\ u_{60}x + v_{60}y + w_{60}z \geq (u_{60}x_6 + v_{60}y_6 + w_{60}z_6) \end{cases} \tag{3.7}$$

By definition, only the nodes on the perimeter of the obtained final domain, V_0^*, are considered as neighbour nodes, Fig. 3.6d. The Voronoï cell V_0 is determined as Fig. 3.6e indicates. As it is possible to visualize the Voronoï cell V_0 is the homothetic form of the auxiliary domain V_0^*, being

$$d_{0i}^* = \frac{d_{0i}}{2} = \frac{\|\boldsymbol{x}_0 - \boldsymbol{x}_i\|}{2} \tag{3.8}$$

A similar procedure is applied in order to obtain the remaining Voronoï cells, Fig. 3.6f. The presented procedure can easily be extrapolated to any d-dimensional Euclidian space $\mathbb{R}^d$.

The Delaunay triangulation is the geometrical dual of the Voronoï diagram and it is constructed by connecting the nodes whose Voronoï cells have common boundaries. The duality between the Voronoï diagram and the Delaunay triangulation implies that a Delaunay edge exists between two nodes in the plane if and only if their Voronoï cells share a common edge. An important property of the Delaunay triangles is the "empty circumcircle criterion [17]. If a set of nodes $\boldsymbol{N}_t = \{n_j\, n_k\, n_l\} \in \boldsymbol{N}$ forms a Delaunay triangle then the circumcircle formed by the triangle $\boldsymbol{N}_t$ contains no other nodes of the global nodal set $\boldsymbol{N}$. In the context of the natural neighbour interpolation these circles are known as "natural neighbour circumcircles" [18]. The centre of the natural neighbour circumcircle is the vertex of the respective Voronoï cell. These features are presented in Fig. 3.7.

In the NNRPIM the Voronoï diagram is used to create the "influence-cells", which enforce the connectivity between the nodes discretizing the problem domain, $\boldsymbol{N}$. On the other hand the duality between the Voronoï cells and the Delaunay triangles is used in order to construct a nodal dependent background integration mesh.

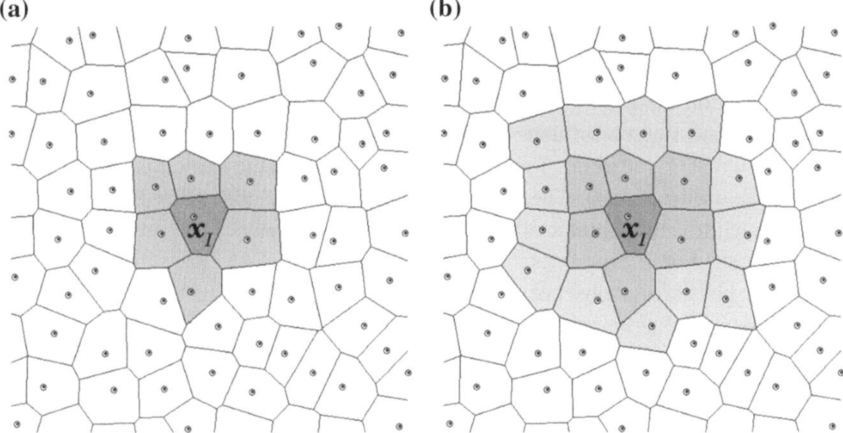

Fig. 3.8 **a** First degree influence-cell. **b** Second degree influence-cell

3.2.3.1 Defining the Influence-Cells

As already mentioned, in meshless methods the nodal connectivity is imposed by the overlap of the influence-domains of each interest point. However the approach shown in Sect. 3.2.1, using blind influence-domains, can affect the efficiency of the meshless method. The blind methodologies described in Fig. 3.5a, b are not capable to assure approximately the same number of nodes inside each influence-domain, leading to unbalanced influence-domains in highly discontinuous boundaries, such as concavities or cracks, and in irregular nodal distributions presenting sporadic locations with high nodal concentrations, Fig. 3.3b.

To respond to these difficulties the influence-cell concept was developed [12]. Presently in the literature it is only possible to find research works combining the influence-cell concept with one particular meshless method, the Natural Neighbour Radial Point Interpolation Method (NNRPIM), however this concept can be applied to all existent meshless approaches. The influence-cell approach works similarly with the influence-domain concept, since the nodal connectivity is imposed by the overlapping of the influence-cells. The influence-cell is also composed by a set of n nodes contributing to the interpolation of the interest point $\boldsymbol{x}_I$. However the set of n nodes is found using the Voronoï Diagram instead radial distances.

Since it is simpler to represent, only the determination of the 2D influence-cell is presented, nevertheless this concept is applicable to any d-dimensional Euclidian space $\mathbb{R}^d$, as shown in several research works available in the literature [19–33].

In this book two distinct types of influence-cells are used: the "first degree influence-cell", Fig. 3.8a, and the "second degree influence-cell", Fig. 3.8b.

To find the “first degree influence-cell” a point of interest, $\boldsymbol{x}_I$, searches for its neighbour nodes following the Natural Neighbour Voronoï construction presented early. Thus, the first degree influence-cell is composed by only these first natural neighbours, Fig. 3.8a.

The “second degree influence-cell” can be found following the procedure: A point of interest, $\boldsymbol{x}_I$, searches for its neighbour nodes, in the same manner as in the first degree influence-cell. Then, based on a previously constructed Voronoï diagram, the natural neighbours of the first natural neighbours of $\boldsymbol{x}_I$ are added to the influence-cell, Fig. 3.8b.

As it is possible to observe the first degree influence-cell is naturally smaller than the second degree influence-cell. Therefore, as expected, the use of second degree influence-cell generally leads to better numerical results.

3.3 Numerical Integration

The main concern of this section is to introduce the integration scheme used in the numerical examples presented in this book. In numerical methods using a variational formulation, such as the Galerkin weak formulation, the numerical integration process, required to determine the system of equations based on the integro-differential equations ruling the studied physical phenomenon, represents a significant percentage of the total computational cost of the analysis.

In the FEM the integration mesh is coincident with the element mesh. Since the FEM shape functions are known polynomial functions, the number of integration points per integration cell can be determined using accurate well-known relations [33, 34]. In meshless methods the shape function degree is generally unknown, thus it is not possible to define a priori the background integration mesh.

The scientific community recognized from the very beginning of the meshless methods development that the numerical integration within meshless methods represents a much bigger challenge than the FEM numerical integration [35]. Several approaches addressing this topic can be found in the literature [4, 36–39].

It is very important to establish for each distinct meshless approach the optimal relationship between the density of the field nodes and the density of the background integration mesh. This relationship must be studied every time the shape function construction procedure is modified and/or the influence-domains size or shape is changed. Several authors have proposed empiric expressions to determine the optimal relation between the total number of field nodes and the total number of integration nodes used to discretize the problem domain [1–3, 37].

In this book two integration schemes are addresses: the Gauss-Legendre quadrature scheme and; a flexible nodal based integration scheme.

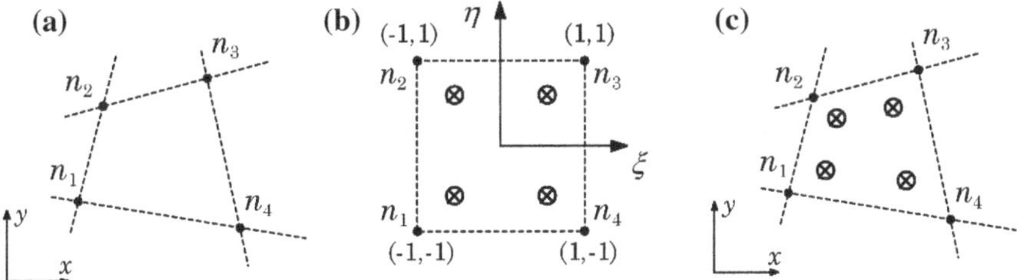

Fig. 3.9 **a** Initial quadrilateral from the grid-cell. **b** Transformation of the initial quadrilateral into an isoparametric square shape and application of the 2 × 2 quadrature point rule. **c** Return to the initial quadrilateral shape

3.3.1 Gaussian Quadrature Integration

Recently a meshless method, based on the RPI, using a stabilized nodal integration [40, 41] was successfully implemented and the obtained results proved to be better than the other meshless RPI approaches based on Gauss-Legendre integration schemes [9, 42]. However the extra time spent in stabilizing the nodal integration does not pay the increased accuracy of the final solution. Within the Gauss-Legendre integration, the solid domain is divided in a regular grid, as Fig. 3.2a or b indicates. Then each grid-cell is filled with integration points, respecting the Gauss-Legendre quadrature rule. The detailed description of the Gauss-Legendre integration procedure, which is beyond the scope of the present book, can be found in the literature [32, 34]. Nevertheless a simple example is illustrated. Assume the grid-cell present in Fig. 3.9a. The initial quadrilateral is transform in an isoparametric square, Fig. 3.9b, then Gauss-Legendre quadrature points are distributed inside the isoparametric square, in Fig. 3.9b it is used a 2 × 2 quadrature. Using isoparametric interpolation functions the Cartesian coordinates of the quadrature points are obtained, Fig. 3.9c. The integration weight of the quadrature point is obtained multiplying the isoparametric weight of the quadrature point with the inverse of the Jacobian matrix determinant of the respective grid-cell.

If the grid fits the solid domain no pos-treatment is required, Fig. 3.2a, however if the grid is larger than the solid domain all the integration points outside the solid domain have to be removed, Fig. 3.2b.

In Fig. 3.2b a blind regular quadrature integration mesh was constructed, in this case all the integration points in the grey area are removed. Although fitted integration meshes present a higher computational cost in comparison with blind regular quadrature meshes, it also produces more stable and accurate results. Generally meshless methods use regular quadrature integration meshes because it is simpler to apply.

In order to perform the numerical integration, considered the function $F(\boldsymbol{x})$ defined in the domain Ω. The global integration can be expressed as a sum,

$$\int_{\Omega} \boldsymbol{F}(\boldsymbol{x})\, d\Omega = \sum_{i=1}^{n_g} \widehat{w}_i \boldsymbol{F}(\boldsymbol{x}_i) \tag{3.9}$$

where $\widehat{w}_i$ is the weight of the integration point $\boldsymbol{x}_i$.

The Gauss-Legendre quadrature integration scheme depends on the complexity of the shape function. The moving least square approximation functions used by the EFGM require $m_c = \sqrt{N}$ integration cells, being N the total nodes discretizing the problem domain [3]. However this rule is blind and does not take in account a potential denser nodal distribution near crack tips or boundaries. Therefore the following EFGM research [37] proposed to distribute the integration cells in accordance with the support-domain of each node discretizing the problem domain. The procedure permitted to increase significantly the EFGM accuracy, however it increased the computational cost. It was also attempt to extend the nodal integration to the EFGM [36]. However the EFGM with nodal integration require a stabilization technique, which increases the computational cost, and it is not capable to achieve the same accuracy and the same convergence rate as the Gaussian quadrature integration scheme.

3.3.2 Nodal Based Integration

The nodal based integration scheme was proposed by the author and co-works to numerically integrate the integro-differential equations ruling the studied physical phenomenon. The technique was successfully applied to integrate one-, two- and three-dimensional domains using the NNRPIM formulation [19, 20, 33]. More recently the technique was used with another meshless method, the Natural Radial Element Method (NREM) [43–45], showing that the technique can be applied to any kind of meshless method.

The most important advantage of this nodal based integration scheme is that the background integration scheme is constructed using uniquely the nodal distribution spatial information. Meshless methods using this numerical integration technique are truly meshless methods, since no other information besides the spatial location of the nodes discretizing the problem domain is necessary to: establish the nodal connectivity, determine the integration points; and construct the shape functions.

Recently a meshless method, based on the RPI, using a stabilized nodal integration [40], was successfully implemented and the obtained results proved to be better than the other meshless RPI approaches based on Gauss integration schemes [9, 42]. However the extra time spent in stabilizing the nodal integration does not compensate the increased accuracy of the final solution. Thus, in this section an

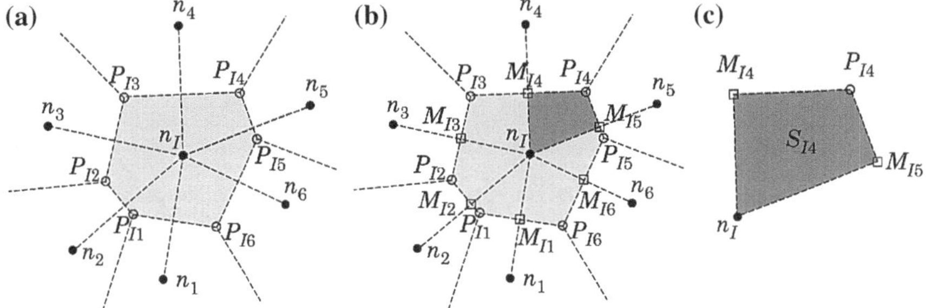

Fig. 3.10 **a** Voronoï cell and the respective P_{Ii} intersection points. **b** Middle points M_{Ii} and the respective generated quadrilaterals. **c** Quadrilateral $\overline{n_I M_{I4} P_{I4} M_{I5}}$

innovating integration scheme based on the Voronoï tessellation and the Delaunay triangulation is presented. Since it is easier to visualize, first it is demonstrated the 2D integration scheme.

3.3.2.1 2D Nodal Based Integration

In an initial phase, after the domain discretization with a regular or an irregular nodal distribution, the Voronoï cells of each node are determined. Using the obtained Voronoï cells small areas are established, which can be quadrilaterals or triangles respectively consistent with an irregular or a regular nodal discretization.

Consider a two-dimensional domain $\Omega \subset \mathbb{R}^2$ discretized by a nodal set $\boldsymbol{N} = \{n_1, n_2, \ldots, n_N\}$. The N nodes discretizing the problem domain are irregularly scattered and have the following coordinates: $\boldsymbol{X} = \{\boldsymbol{x}_1, \boldsymbol{x}_2, \ldots, \boldsymbol{x}_N\} \in \Omega$. The natural neighbours of node n_I, defined by the finite nodal set $\boldsymbol{N}_I = \{n_1\, n_2 \ldots n_6\} \in \boldsymbol{N}$, Fig. 3.10a, allow to construct the Voronoï cell V_I of node n_I and to determine the corners P_{Ii} of the polygonal shape defined by V_I, Fig. 3.10a. Afterwards the middle points, M_{Ii}, between n_I and each neighbour nodes $n_i \in \boldsymbol{N}_I$ are obtained, Fig. 3.10b. Thus, as Fig. 3.10b, c indicate, the Voronoï cells are divided in n quadrilateral sub-cells, S_{Ii}, being n the number of natural neighbour of node n_I.

If the field nodes $\boldsymbol{N} = \{n_1, n_2, \ldots, n_N\}$ discretizing the problem domain $\Omega \subset \mathbb{R}^2$ are scattered in a regular nodal distribution, the Voronoï cells are divided in triangular sub-cells, instead quadrilateral sub-cells. Consider the natural neighbours of node n_I, defined by the finite nodal set $\boldsymbol{N}_I = \{n_1\, n_2 \ldots n_8\} \in \boldsymbol{N}$, represented in Fig. 3.11a. The respective Voronoï cellV_I is a perfect square and the Voronoï polygonal only has four corners P_{Ii}, Fig. 3.11a. It is possible to obtain the middle points, M_{Ii}, between n_I and each neighbour nodes $n_i \in \boldsymbol{N}_I$ as Fig. 3.11b indicates. As it is perceptible in Fig. 3.11b, c, the Voronoï cells of regular nodal distributions are divided in n triangular sub-cells, S_{Ii}, being n the number of natural neighbour of node n_I.

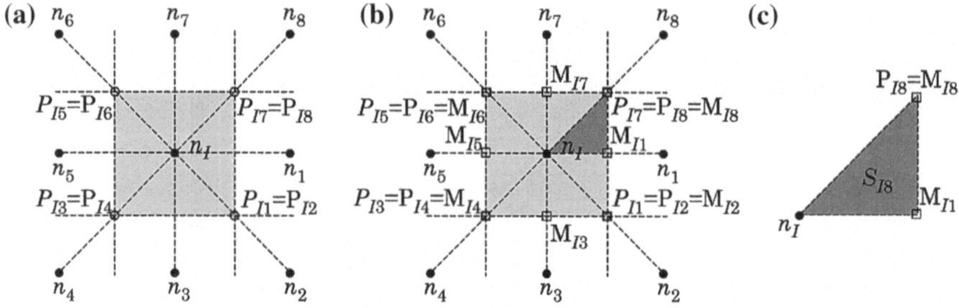

Fig. 3.11 **a** Voronoï cell and the respective P_{Ii} intersection points. **b** Middle points M_{Ii} and the respective generated triangles. **c** Triangle $\overline{n_I P_{I8} M_{I1}}$

As has been shown, it is always possible to divide the Voronoï cellV_I in n sub-cells S_{Ii}, being n the total number of natural neighbours of n_I and $V_I = \cup_{i=1}^{n} S_{Ii}$. Therefore, the size of the Voronoï cell V_I can be determined using the size of the n sub-cells S_{Ii},

$$A_{V_I} = \sum_{i=1}^{n} A_{S_{Ii}}, \ \ \forall A_{S_{Ii}} \geq 0 \tag{3.10}$$

Being A_{V_I} the size of the Voronoï cell V_I and $A_{S_{Ii}}$ the size of sub-cell S_{Ii}. For the one-dimensional domain A represents lengths, for the two-dimensional domain A stands for areas and for the three-dimensional domain A is a volume. Notice that, if the set of Voronoï cells are a partition, without gaps, of the global domain then, the set of sub-cells are also a partition, without gaps, of the global domain.

It is clear now, with Figs. 3.10 and 3.11, how the construction of the sub-cells generates two types of basic shapes - triangles or quadrilaterals. Starting with these two shapes, numerous integrations schemes can be constructed. In this book it is shown an ordered scheme, based on the Gauss-Legendre numerical integration.

Basic Integration Scheme

The simplest integration scheme that can be established, using the sub-cells triangular and quadrilateral shapes, is obtained inserting a single integration point in the barycentre of the sub-cells. Therefore, spatial location of each integration point is determined on each sub-cell, as indicated in Fig. 3.12, being the weight of each integration point the area of the respective sub-cell.

Considering Fig. 3.12, the area of the triangle shape sub-cell is defined by,

$$A_I^{\triangle} = \frac{1}{2}\left| \det \begin{bmatrix} x_2 - x_1 & y_2 - y_1 \\ x_3 - x_1 & y_3 - y_1 \end{bmatrix} \right| \tag{3.11}$$

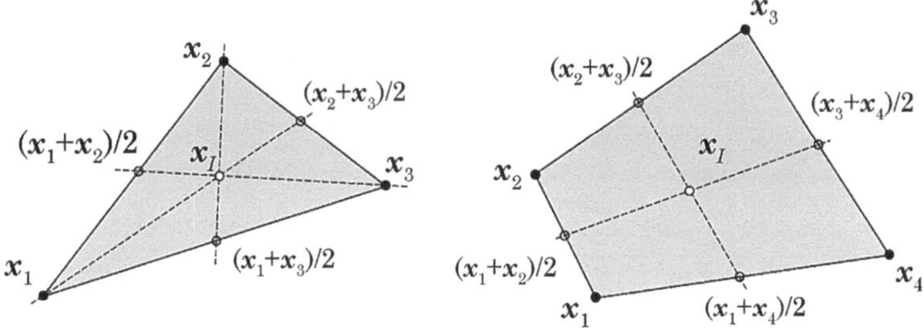

Fig. 3.12 Triangular shape and quadrilateral shape and the respective integration points $\boldsymbol{x}_I$

and for the quadrilateral shape the area is,

$$A_I^{\square} = \frac{1}{2}\left|\det\begin{bmatrix} x_2 - x_1 & y_2 - y_1 \\ x_3 - x_1 & y_3 - y_1 \end{bmatrix} + \det\begin{bmatrix} x_4 - x_1 & y_4 - y_1 \\ x_3 - x_1 & y_3 - y_1 \end{bmatrix}\right| \tag{3.12}$$

This process is equivalent with the single integration point of the Gauss-Legendre quadrature scheme for triangle and quadrilateral shapes, respectively.

Gauss-Legendre Quadrature Integration Scheme

From the previous basic integration scheme it is possible to obtain a more general integration scheme. The sub-cell basic geometric forms in Fig. 3.12 are subdivided again, however in this case only as quadrilaterals. Using the triangle and the quadrilateral sub-cells obtained in Fig. 3.12, firstly it is determined the centre of the geometric shape, $\boldsymbol{x}_C$, then middle points on the quadrilateral edges are determined, $\boldsymbol{x}_{ij} = (\boldsymbol{x}_i + \boldsymbol{x}_j)/2$, and thus new sub-quadrilaterals are defined, Fig. 3.13. It is possible to apply the Gauss-Legendre quadrature to the obtained sub-quadrilaterals in order to obtain the integration points [32, 34]. The process is briefly described in Sect. 3.3.1. This process permits to fill each sub-quadrilateral with $k \times k$ integration points. In Fig. 3.13 are shown distinct integrations schemes for the triangular and the quadrilateral sub-cell simultaneously: a 1×1 quadrature per sub-quadrilateral; and a 3×3 quadrature per sub-quadrilateral.

The integration weight of each integration point $\boldsymbol{x}_I$ is obtained using the following expression,

$$\widehat{w}_I = w_\eta \, w_\xi \left(\frac{A^{\square}}{4}\right) \tag{3.13}$$

being $A^{\square}$ the area of the respective sub-quadrilateral, which can be obtained using Eq. (3.12), and w_η and w_ξ are the Gauss-Legendre quadrature weights for an isoparametric quadrilateral cell, Fig. 3.9.

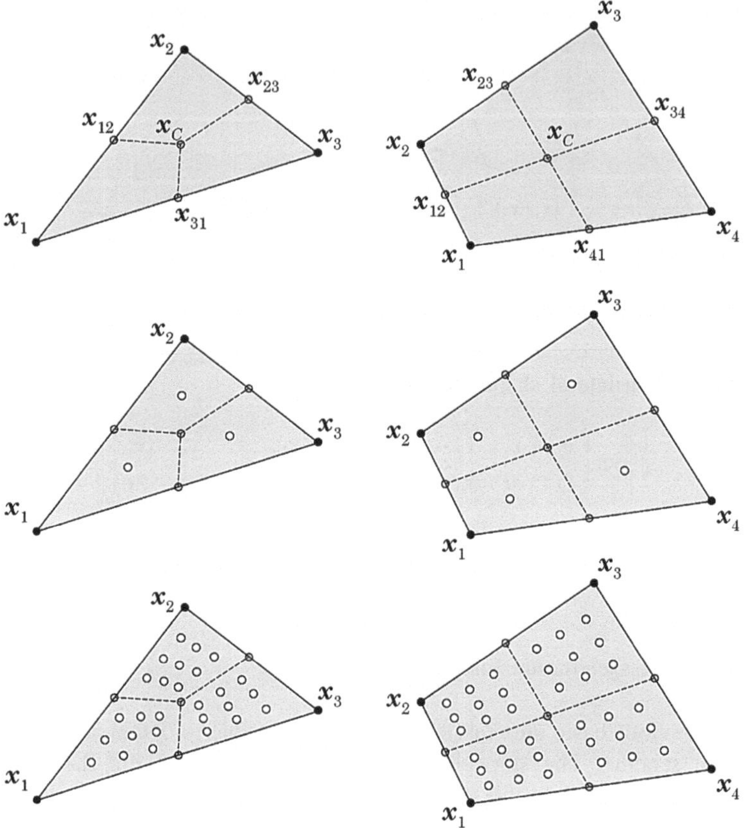

Fig. 3.13 Triangular shape and quadrilateral shape and the respective integration points $\boldsymbol{x}_I$ using the Gauss-Legendre integration scheme

Nodal Integration Scheme

The nodal integration scheme is as simple as considering the integration point, $\boldsymbol{x}_I$, coincident with barycentre of each Voronoï cell, V_I. Thus, the weight of each integration point, $\widehat{w}_I$, is the sum of each of the areas of the respective sub-cells, i.e., it is the area of the respective Voronoï cell V_I, obtained with Eq. (3.10). The spatial location of each integration point can be determined by,

$$\boldsymbol{x}_I = \frac{\sum_{i=1}^{n} \boldsymbol{x}_{S_i} A_{S_{Ii}}}{\sum_{i=1}^{n} A_{S_{Ii}}} \tag{3.14}$$

Being $\boldsymbol{x}_{S_i}$ the barycentre of the sub-cell S_{Ii} and $A_{S_{Ii}}$ is the area of sub-cell S_{Ii}. Notice that the nodal integration is not sufficient to integrate accurately the meshless shape functions, it is necessary to apply stabilization techniques [4, 36,

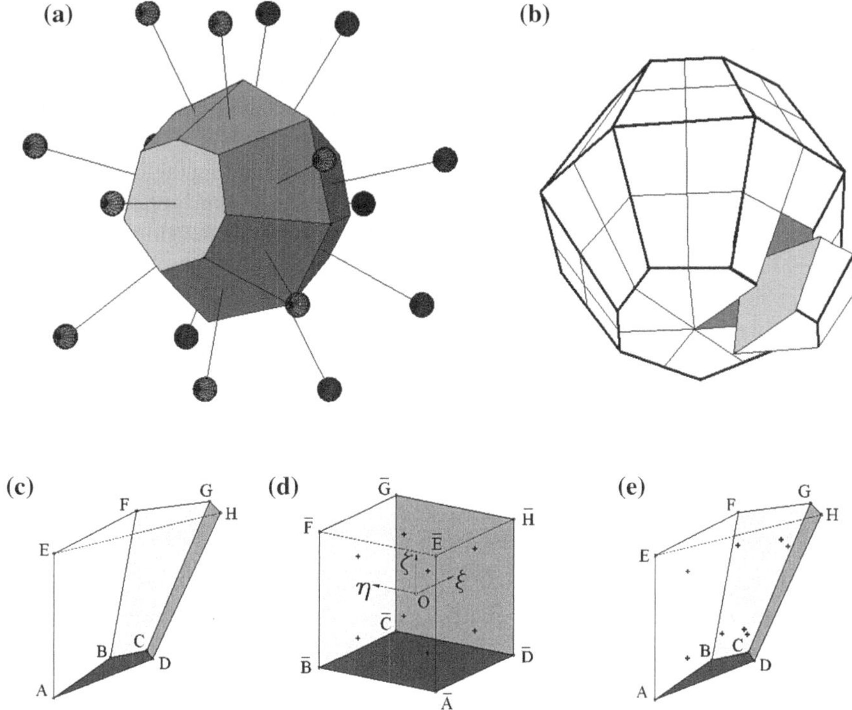

Fig. 3.14 **a** Voronoï cell and **b** Respective hexahedrons. **c** Initial hexahedron. **d** Hexahedron isoparameterization and determination of the quadrature integration points. **e** Quadrature integration points in Cartesians coordinates

39, 40], which increases the method computational cost, reducing consequently the meshless method efficiency.

3.3.2.2 3D Nodal Based Integration

The previously presented two-dimensional numerical integration scheme can be directly extended for the three-dimensional space. If irregular nodal distributions are used to discretize the problem domain, then the 3D Voronoï cell is subdivided in hexahedral sub-cells. In opposition, if the problem domain is discretized in a regular mesh, then the 3D Voronoï cell is subdivided in tetrahedral sub-cells. Afterwards, all the numerical integration schemes presented in Sect. 3.3.2.1 can be applied. In Fig. 3.14 it is illustrated the 3D procedure.

First the three-dimensional Voronoï cells are obtained for each node belonging to the set of field nodes discretizing the problem domain. In Fig. 3.14a it is presented an example of a three-dimensional Voronoï cell. The Voronoï cell is partitioned in hexahedral or tetrahedral sub-cells. In Fig. 3.14b it is shown the

Voronoï cell subdivision in hexahedral sub-cells. Then each partition, Fig. 3.14c, is isoparameterized, Fig. 3.14d, and the Gauss-Legendre quadrature scheme is applied. Afterwards the obtained integrations points spatial location and respective integration weights are transformed again to the Cartesian coordinate system, Fig. 3.14e.

The 3D analysis is computational very demanding. Regarding the NNRPIM and the NREM meshless methods, research works available in the literature [19, 43] show that, although it is possible to apply all the integration schemes referred in Sect. 3.3.2.1, there is no need to use a Gauss-Legendre quadrature scheme higher than $1 \times 1 \times 1$ per hexahedron. This integration scheme is sufficient for the 3D NNRPIM and 3D NREM meshless formulations, since higher integration schemes raise enormously the computational cost. Additionally it was noticed that the 3D nodal integration without stabilization is not sufficient to integrate the NNRPIM or the NREM interpolation functions.

3.4 Numerical Implementation

This book presents two meshless methods based on the Galerkin weak formulation, defined over the global problem domain, using the locally supported meshless shape functions that will be introduced in Chap. 4. In this section it is presented the generic numerical implementation of approximation and interpolation meshless methods based on the Galerkin weak formulation.

Consider the solid domain defined by $\Omega \subset \mathbb{R}^3$ and bounded by Γ, Fig. 3.15. The solid domain is made from a generic heterogeneous material, being the density of such material defined by the functional $\rho(\boldsymbol{x}) \in \mathbb{R}$.

Besides the body forces caused by the material density, the solid is submitted to the external forces $\bar{\boldsymbol{t}}_C(\boldsymbol{x})$ and $\bar{\boldsymbol{t}}_S(\boldsymbol{x})$ applied on the boundary Γ, respectively over the curve $\Gamma_{t(C)} \in \Gamma$ and over the surface, $\Gamma_{t(S)} \in \Gamma$. The solid displacements are constrained on the curve $\Gamma_{u(C)} \in \Gamma$ and on the surface, $\Gamma_{u(S)} \in \Gamma$.

First, the solid domain is discretized by a irregularly nodal distribution $\boldsymbol{X} = \{\boldsymbol{x}_1, \boldsymbol{x}_2, \ldots, \boldsymbol{x}_N\} \in \Omega$, being N the total number of nodes discretizing the problem domain and $\boldsymbol{x} \in \mathbb{R}^3$, Fig. 3.16a. These field nodes permit to approximate the field variable—the displacement field $\boldsymbol{u} = \{\boldsymbol{u}_1^T, \boldsymbol{u}_2^T, \ldots, \boldsymbol{u}_N^T\}^T$, being $\boldsymbol{u}_i = \{u_i, v_i, w_i\}^T \in \mathbb{R}^3 \wedge i \in \{1, 2, \ldots, N\}$.

With the nodal distribution it is possible to obtain the background integration mesh, using for instance one of the integration schemes presented in Sect. 3.3. In the end, an integration mesh discretizing the problem domain is obtained, $\boldsymbol{Q} = \{\boldsymbol{q}_1, \boldsymbol{q}_2, \ldots, \boldsymbol{q}_Q\} \in \Omega$, being Q the total number of integration points discretizing the problem domain and $\boldsymbol{q} \in \mathbb{R}^3$ the spatial coordinates. Notice that $\boldsymbol{Q} \subset \Omega \backslash \Gamma$, Fig. 3.16b, i.e., the integration points are defined only inside the solid

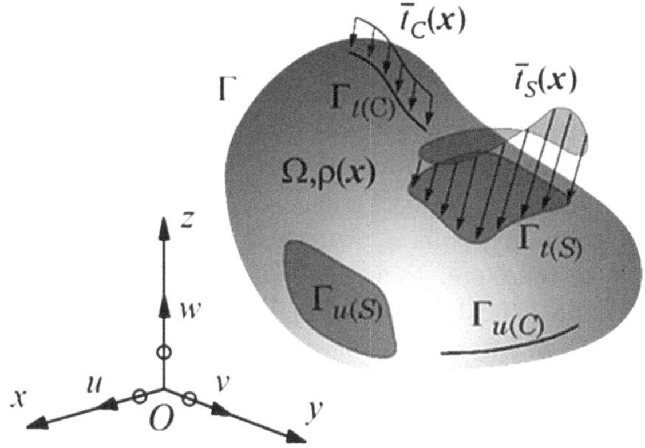

Fig. 3.15 Solid domain and applied volume and external forces and displacement constrains

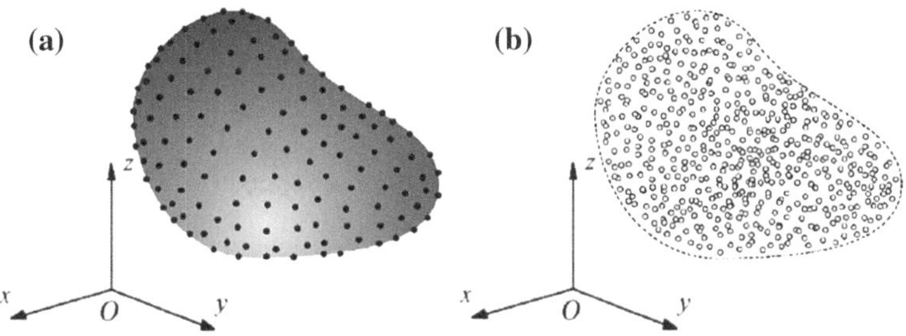

Fig. 3.16 **a** Example of a nodal distribution. **b** Example of a background integration mesh

domain Ω, the solid boundary Γ does not contains any integration point. Being $\widehat{w}_I$ the integration weight of the Ith integration point, $\boldsymbol{q}_I \in \boldsymbol{Q}$, then

$$Vol_\Omega = \int_\Omega \widehat{w}(\boldsymbol{q})\, d\Omega = \sum_{I=1}^{Q} \widehat{w}_I \tag{3.15}$$

where Vol_Ω is the volume of the solid domain. The integration weight $\widehat{w}_I$ is the infinitesimal volume of the integration point $\boldsymbol{q}_I$. The mass of the solid domain $Mass_\Omega$ is obtained by,

$$Mass_\Omega = \int_\Omega \widehat{w}(\boldsymbol{q}) \cdot \rho(\boldsymbol{q})\, d\Omega = \sum_{I=1}^{Q} \widehat{w}_I \cdot \rho(\boldsymbol{q}_I) \tag{3.16}$$

In Sect. 2.3 it was shown that the Galerkin weak formulation permits to obtain the following system of equations,

$$\boldsymbol{K}\boldsymbol{u} + \boldsymbol{M}\boldsymbol{u} = f_b + \boldsymbol{f}_t = \boldsymbol{f} \tag{3.17}$$

In the next sections it is shown how to numerically obtain the stiffness matrix$\boldsymbol{K}$, the mass matrix$\boldsymbol{M}$, the body forces vector, $\boldsymbol{f}_b$, the external forces vector, $\boldsymbol{f}_t$, and additionally how to numerically enforce the essential boundary conditions.

After the construction of the background integration mesh, $\boldsymbol{Q}$, the influence-domains of each integration point $\boldsymbol{q}_I$ is defined using one of the techniques presented in Sect. 3.3. Therefore, each integration point $\boldsymbol{q}_I \in \boldsymbol{Q}$ will possess an influence-domain with n field nodes, which will directly contribute to the construction of the shape function of the interest point $\boldsymbol{q}_I$.

$$\boldsymbol{\varphi}(\boldsymbol{q}_I)^{\mathrm{T}} = \{\, \varphi_1(\boldsymbol{q}_I) \quad \varphi_2(\boldsymbol{q}_I) \quad \cdots \quad \varphi_n(\boldsymbol{q}_I)\,\} \tag{3.18}$$

The construction procedure of the meshless shape functions is explained in detail in Chap. 4.

3.4.1 Stiffness Matrix

In Sect. 2.3.2 it was shown that the stiffness matrix depends on the contribution of the linear and nonlinear parts of the deformation matrix, $\bar{\boldsymbol{B}} = \boldsymbol{B}_0 + \boldsymbol{B}_{\mathrm{NL}}(\boldsymbol{u})$. Since the numerical examples presented in this work only assume small strains, the nonlinear deformation matrix, $\boldsymbol{B}_{\mathrm{NL}}(\boldsymbol{u})$, can be for now neglected. Therefore, the stiffness matrix defined in Sect. 2.3.2 can be simply presented as,

$$\boldsymbol{K} = \boldsymbol{K}_0 = \int_{\Omega} \boldsymbol{B}_0^{\mathrm{T}}\, \boldsymbol{c}\, \boldsymbol{B}_0 \, d\Omega \tag{3.19}$$

Following the same procedure previously presented to obtain the solid volume and the solid mass, the integral can be substituted by a discrete sum.

$$\underbrace{\boldsymbol{K}}_{[3N\times 3N]} = \int_{\Omega} \boldsymbol{B}^{\mathrm{T}}\, \boldsymbol{c}\, \boldsymbol{B}\, d\Omega = \sum_{I=1}^{Q} \widehat{w}_I \underbrace{\boldsymbol{B}_I^{\mathrm{T}}}_{[3N\times 6]} \underbrace{\boldsymbol{c}(\boldsymbol{q}_I)}_{[6\times 6]} \underbrace{\boldsymbol{B}_I}_{[6\times 3N]} \tag{3.20}$$

Being the solid made of a generic heterogeneous material, the mechanical properties will also depend on the spatial position, therefore the constitutive matrix is a functional depending on the spatial coordinates of the interest point, $\boldsymbol{c}(\boldsymbol{q}_I)$. If the solid material is homogeneous the constrictive matrix is constant over the entire domain, $\boldsymbol{c}$. The global deformation matrix $\boldsymbol{B}_I$ of the interest point $\boldsymbol{q}_I$ is firstly defined locally with the shape function constructed for the interest point $\boldsymbol{q}_I$ using only the n field nodes inside the influence-domain of $\boldsymbol{q}_I$. The local deformation matrix is defined as,

$$\underbrace{\boldsymbol{B}_I^{local}}_{[6\times 3n]} = \left[\begin{array}{ccc|ccc|c|ccc} \frac{\partial\varphi_1}{\partial x} & 0 & 0 & \frac{\partial\varphi_2}{\partial x} & 0 & 0 & \cdots & \frac{\partial\varphi_n}{\partial x} & 0 & 0 \\ 0 & \frac{\partial\varphi_1}{\partial y} & 0 & 0 & \frac{\partial\varphi_2}{\partial y} & 0 & \cdots & 0 & \frac{\partial\varphi_n}{\partial y} & 0 \\ 0 & 0 & \frac{\partial\varphi_1}{\partial z} & 0 & 0 & \frac{\partial\varphi_2}{\partial z} & \cdots & 0 & 0 & \frac{\partial\varphi_n}{\partial z} \\ \frac{\partial\varphi_1}{\partial y} & \frac{\partial\varphi_1}{\partial x} & 0 & \frac{\partial\varphi_2}{\partial y} & \frac{\partial\varphi_2}{\partial x} & 0 & \cdots & \frac{\partial\varphi_n}{\partial y} & \frac{\partial\varphi_n}{\partial x} & 0 \\ 0 & \frac{\partial\varphi_1}{\partial z} & \frac{\partial\varphi_1}{\partial y} & 0 & \frac{\partial\varphi_2}{\partial z} & \frac{\partial\varphi_2}{\partial y} & \cdots & 0 & \frac{\partial\varphi_n}{\partial z} & \frac{\partial\varphi_n}{\partial y} \\ \frac{\partial\varphi_1}{\partial z} & 0 & \frac{\partial\varphi_1}{\partial x} & \frac{\partial\varphi_2}{\partial z} & 0 & \frac{\partial\varphi_2}{\partial x} & \cdots & \frac{\partial\varphi_n}{\partial z} & 0 & \frac{\partial\varphi_n}{\partial x} \end{array}\right] \tag{3.21}$$

Then the local deformation matrix, $\boldsymbol{B}_I^{local}$, is assembled to the global deformation matrix $\boldsymbol{B}_I$. The process is exemplified in Fig. 3.17. Consider the integration point $\boldsymbol{q}_I \subset \Omega$ with an influence-domain with 5 nodes $\{n_4, n_{11}, n_{16}, n_{17}, n_{23}\}$ represented in Fig. 3.17. The meshless shape function, and the respective partial derivatives, are constructed for the integration point $\boldsymbol{q}_I$ using the 5 nodes belonging to the influence-domain. Then, it is possible to construct the local deformation matrix, $\boldsymbol{B}_I^{local}$, and afterwards the global deformation matrix, $\boldsymbol{B}_I = \boldsymbol{B}_I^{global}$.

Naturally, this assemblage process can be optimized to haste the computational procedure [46, 47], however the purpose of this book is just to introduce the meshless method concept and not to introduce efficient computational techniques for meshless methods.

In this book only two meshless techniques are presented: the radial point interpolator (RPI) meshless method or moving least square (MLS) meshless method. Since these meshless methods use the "support-domain" concept to construct the meshless shape function, both the RPI and the MLS lead to a sparse stiffness matrix. The nodes outside the influence-domain of the interest point $\boldsymbol{q}_I$ do not contribute to the local stiffness matrix of the interest point $\boldsymbol{q}_I$. Notice that if the field nodes are properly numbered then it is possible to obtain a banded stiffness matrix. Additionally, the stiffness matrix obtained with the RPI or the MLS meshless methods is symmetric. From Eq. (3.20) it is possible to understand that being $\boldsymbol{c}(\boldsymbol{q}_I)$ a symmetric matrix then $\boldsymbol{B}_I^{\mathrm{T}}\boldsymbol{c}(\boldsymbol{q}_I)\boldsymbol{B}_I$ is also symmetric. The combination of these three properties: sparse; banded; and symmetry, permit to solve more efficiently the meshless discrete system of equations.

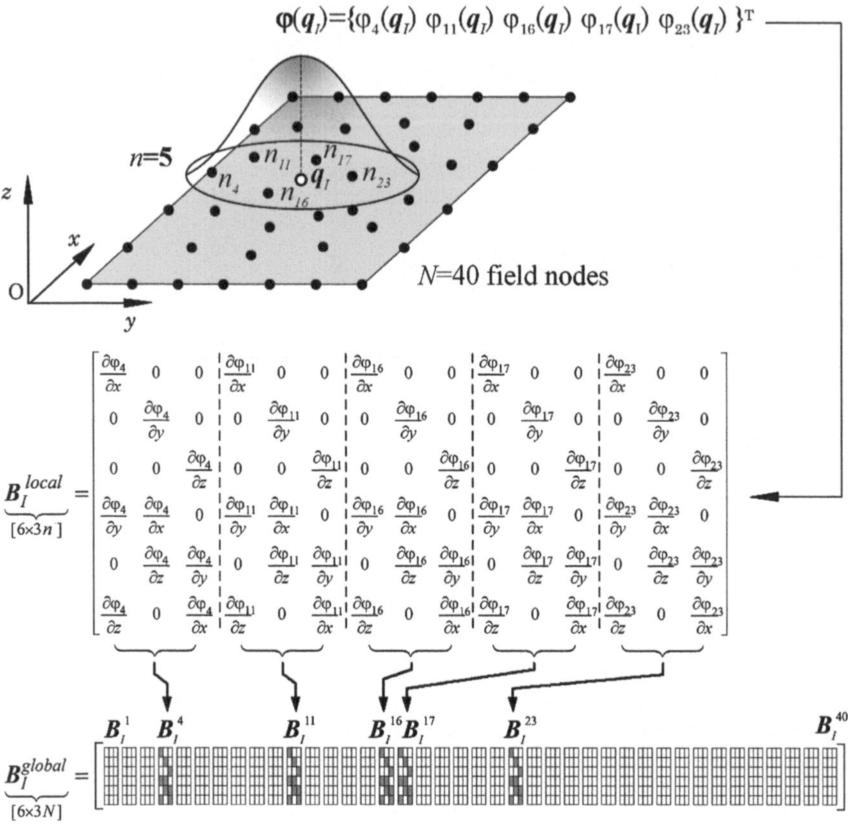

Fig. 3.17 Assemblage of the deformation matrix

3.4.2 Mass Matrix

The procedure to obtain the mass matrix, using the Galerkin weak formulation, is described in Sect. 2.3.3. Since the continuous solid domain is discretized by a background integration mesh, the continuous integral over the solid domain can be substituted by the following discrete sum,

$$\underbrace{\mathbf{M}}_{[3N\times 3N]} = \int_{\Omega} \mathbf{H}^{\mathrm{T}} \rho \mathbf{H}\, d\Omega = \sum_{I=1}^{Q} \hat{w}_I \underbrace{\mathbf{H}_I^{\mathrm{T}}}_{[3N\times 3]} \underbrace{\rho(\mathbf{q}_I)\mathbf{I}}_{[3\times 3]} \underbrace{\mathbf{H}_I}_{[3\times 3N]} \tag{3.22}$$

Being $\mathbf{I}$ an identity matrix with dimension $[3 \times 3]$ for the present three-dimensional case. For the generic case of a heterogeneous material, the material density is not constant along the solid domain, therefore it can be represented as a functional depending on the spatial coordinates of the interest point, $\rho(\mathbf{q}_I) \in \mathbb{R}$.

For a simpler homogeneous material case, the solid density is obviously constant, $\rho(\boldsymbol{q}_I) = \rho, \forall \boldsymbol{q}_I \in \boldsymbol{Q}$.

The diagonal shape function matrix $\boldsymbol{H}_I$ of the interest point $\boldsymbol{q}_I$ is determined using the meshless shape function obtained for the interest point $\boldsymbol{q}_I$ considering only the n field nodes belonging to the influence-domain of $\boldsymbol{q}_I$. Following the same procedure described previously for the deformation matrix$\boldsymbol{B}_I$, firstly the local diagonal shape function matrix is determined, $\boldsymbol{H}_I^{local}$, and then it is assembled to the global diagonal shape function matrix $\boldsymbol{H}_I$. The process is similar with the example shown in Fig. 3.17. The local diagonal shape function matrix is defined as,

$$\underbrace{\boldsymbol{H}_I^{local}}_{[3\times 3n]} = \left[\begin{array}{ccc|ccc|c|ccc} \varphi_1 & 0 & 0 & \varphi_2 & 0 & 0 & \cdots & \varphi_n & 0 & 0 \\ 0 & \varphi_1 & 0 & 0 & \varphi_2 & 0 & \cdots & 0 & \varphi_n & 0 \\ 0 & 0 & \varphi_1 & 0 & 0 & \varphi_2 & \cdots & 0 & 0 & \varphi_n \end{array}\right] \tag{3.23}$$

The mass matrix obtained with the RPI or the MLS meshless methods is sparse, banded and symmetric.

3.4.3 Body Force Vector

In Sect. 2.3.4 it was shown how the body force vector could be obtained for the continuous solid domain. Using the background integration points discretizing the continuous solid domain, it is possible to obtain the body force vector with,

$$\underbrace{\boldsymbol{f}_b}_{[3N\times 1]} = \int_{\Omega} \boldsymbol{H}^T \boldsymbol{b}\, d\Omega = \sum_{I=1}^{Q} \hat{w}_I \underbrace{\boldsymbol{H}_I^T}_{[3N\times 3]} \underbrace{\rho(\boldsymbol{q}_I) g}_{\text{scalar}} \underbrace{\boldsymbol{g}}_{[3\times 1]} \tag{3.24}$$

being g the magnitude of the acceleration due the gravity and $\boldsymbol{g} = \{g_x, g_y, g_z\}$ a non-dimensional unit vector, $||\boldsymbol{g}|| = 1$, defining the gravity direction. The construction of the global diagonal shape function matrix $\boldsymbol{H}_I$ is described in the previous Sect. 3.4.2.

3.4.4 External Force Vector

In addition to the body forces, other forces can be applied in the solid boundary. Consider the external forces $\bar{\boldsymbol{t}}_C(\boldsymbol{x})$ applied along the boundary curve $\Gamma_{t(C)} \in \Gamma$, Fig. 3.15. First, the field nodes on the boundary curve $\Gamma_{t(C)} \in \Gamma$ are identified:

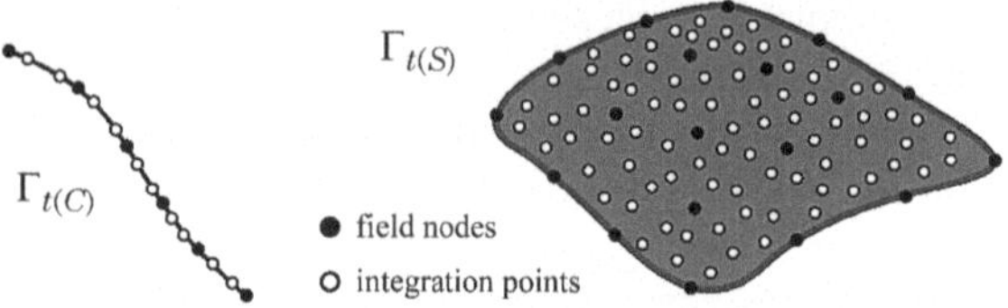

Fig. 3.18 Representation of the natural boundaries

$\boldsymbol{X}_{\Gamma} = \{\boldsymbol{x}_1, \boldsymbol{x}_2, \ldots, \boldsymbol{x}_{N\Gamma}\} \in \Gamma_{t(C)}$, being $\boldsymbol{X}_{\Gamma} \subset \boldsymbol{X}$ and $N\Gamma$ the number of nodes along the boundary curve $\Gamma_{t(C)}$. Then new integration points have to be determined on the boundary curve $\Gamma_{t(C)}$, since the integration points discretizing the entire solid domain, $\boldsymbol{Q}$, are not valid to numerically integrate a functional along the boundary curve $\Gamma_{t(C)}$. From Sect. 3.3 it is possible to understand that $\boldsymbol{Q} \subset \Omega \backslash \Gamma$, as illustrated in Fig. 3.16b. Notice that the set $\boldsymbol{Q}$ discretizes a volume and now a curve discretization is required.

Therefore, based on the nodal discretization $\boldsymbol{X}_{\Gamma}$, a new set of integration points $\boldsymbol{Q}_{\Gamma} = \{\boldsymbol{q}_1^{\Gamma}, \boldsymbol{q}_2^{\Gamma}, \ldots, \boldsymbol{q}_{Q\Gamma}^{\Gamma}\} \in \Gamma_{t(C)}$ is defined, in which $Q\Gamma$ represents the number of integration points along the boundary curve $\Gamma_{t(C)}$. In opposition to the set $\boldsymbol{Q}$, in which the integration points represent infinitesimal volumes, the integration points on set $\boldsymbol{Q}_{\Gamma}$ represent infinitesimal curves. Thus, the integration weight$\widehat{w}_I^{\Gamma}$ of each integration point $\boldsymbol{q}_I^{\Gamma}$ represent a length. In Fig. 3.18 are represented the field nodes and the integration points along the boundary curve $\Gamma_{t(C)}$.

Since new integration points were determined, $\boldsymbol{Q}_{\Gamma}$, new influence-domains have to be established for each one $\boldsymbol{q}_I^{\Gamma} \in \boldsymbol{Q}_{\Gamma}$. However the nodes of the new influence-domains have to belong to the field nodes on the boundary curve $\Gamma_{t(C)}$. It is not permitted the inclusion of any other field node. Hence, the meshless shape functions determined for each one integration point $\boldsymbol{q}_I^{\Gamma}$ are constructed using only the nodes on the boundary curve $\Gamma_{t(C)}$.

The global force vector for the boundary curve $\Gamma_{t(C)}$ is obtained with the following expression,

$$\underbrace{\boldsymbol{f}_C}_{[3N\times 1]} = \int_{\Gamma_{t(C)}} \boldsymbol{H}^{\mathrm{T}} \bar{\boldsymbol{t}}_C(\boldsymbol{x})\, d\Gamma_{t(C)} = \sum_{I=1}^{Q} \widehat{w}_I^{\Gamma} \underbrace{\boldsymbol{H}_I^{\mathrm{T}}}_{[3N\times 3]} \underbrace{\bar{\boldsymbol{t}}_C(\boldsymbol{q}_I)}_{[3\times 1]} \tag{3.25}$$

The generic vector $\bar{\boldsymbol{t}}_C(\boldsymbol{q}_I)$ depends on the integration point spatial position and it is defined by $\bar{\boldsymbol{t}}_C(\boldsymbol{q}_I) = \{\bar{t}_C(\boldsymbol{q}_I)_x, \bar{t}_C(\boldsymbol{q}_I)_y, \bar{t}_C(\boldsymbol{q}_I)_z\}$. The global diagonal shape function matrix $\boldsymbol{H}_I$ is obtained following the process described in Sect. 3.4.2.

If the external forces $\bar{\boldsymbol{t}}_S(\boldsymbol{x})$ are applied along a boundary surface $\Gamma_{t(S)} \in \Gamma$, Fig. 3.15, first the field nodes belonging to $\Gamma_{t(S)}$ have to be identified: $\boldsymbol{X}_{\Gamma} = \{\boldsymbol{x}_1, \boldsymbol{x}_2, \ldots, \boldsymbol{x}_{N\Gamma}\} \in \Gamma_{t(S)}$, being $\boldsymbol{X}_{\Gamma} \subset \boldsymbol{X}$. Next, new integration points have to be determined on the surface domain $\Gamma_{t(S)}$ using only the spatial information of the field nodes belonging to $\Gamma_{t(S)}$. Thus, an additional set of integration points $\boldsymbol{Q}_\Gamma = \{\boldsymbol{q}_1^\Gamma, \boldsymbol{q}_2^\Gamma, \ldots, \boldsymbol{q}_{Q\Gamma}^\Gamma\} \in \Gamma_{t(S)}$ is obtained just to integrate numerically the external forces functional $\bar{\boldsymbol{t}}_S(\boldsymbol{x})$ on the boundary surface $\Gamma_{t(S)}$. Notice that now the integration points on set $\boldsymbol{Q}_\Gamma$ represent infinitesimal surfaces, which means that the integration weight $\widehat{w}_I^\Gamma$ of each integration point $\boldsymbol{q}_I^\Gamma \in \boldsymbol{Q}_\Gamma$ represent a surface area. A schematic example of the field nodes and integration points along the boundary surface $\Gamma_{t(S)}$ is presented in Fig. 3.18.

Once again, using only the field nodes $\boldsymbol{X}_\Gamma$ on the boundary surface $\Gamma_{t(S)}$, new influence-domains have to be determined for each $\boldsymbol{q}_I^\Gamma \in \boldsymbol{Q}_\Gamma$. Then, it is possible to construct the meshless shape functions of each integration point $\boldsymbol{q}_I^\Gamma$ and to obtain the global force vector on the boundary surface $\Gamma_{t(S)}$,

$$\underbrace{\boldsymbol{f}_S}_{[3N\times 1]} = \int\limits_{\Gamma_{t(S)}} \boldsymbol{H}^{\mathrm{T}} \bar{\boldsymbol{t}}_S(\boldsymbol{x})\, d\Gamma_{t(S)} = \sum_{I=1}^{Q} \widehat{w}_I^\Gamma \underbrace{\boldsymbol{H}_I^{\mathrm{T}}}_{[3N\times 3]} \underbrace{\bar{\boldsymbol{t}}_S(\boldsymbol{q}_I)}_{[3\times 1]} \tag{3.26}$$

being the generic vector $\bar{\boldsymbol{t}}_S(\boldsymbol{q}_I)$ defined by $\bar{\boldsymbol{t}}_S(\boldsymbol{q}_I) = \{\bar{t}_S(\boldsymbol{q}_I)_x, \bar{t}_S(\boldsymbol{q}_I)_y, \bar{t}_S(\boldsymbol{q}_I)_z\}$ and the global diagonal shape function matrix $\boldsymbol{H}_I$ defined as in Sect. 3.4.2.

After the definition of all force vectors it is possible to combine the three load vectors in a global force vector,

$$\underbrace{\boldsymbol{f}}_{[3N\times 1]} = \underbrace{\boldsymbol{f}_b}_{[3N\times 1]} + \underbrace{\boldsymbol{f}_C}_{[3N\times 1]} + \underbrace{\boldsymbol{f}_S}_{[3N\times 1]} \tag{3.27}$$

3.4.5 Essential Boundary Conditions

In meshless methods based on the Galerkin weak formulation, the body forces and the external forces are naturally formulated into the discretized system of equations. Thus, the body forces and the external forces are generally named as natural boundary conditions. On the other hand, the boundary conditions regarding displacement constrains are not included in the weak form. Thus, it is essential to impose explicitly the displacement constrains on the discretized system of equations. This is the reason why the displacement constrains are called essential boundary conditions.

In the case of meshless methods approximations, the essential conditions can be exactly satisfied but the natural conditions can only be satisfied up to the order of the shape function of the meshless method.

If the meshless shape function possess the Kronecker delta property the essential boundary conditions can be imposed using the same simple techniques used in the finite element method. However, if the meshless shape functions is an approximation functions, i.e., it does not possess the Kronecker delta property, the imposition of the essential boundary conditions requires complex methodologies.

To simplify the exposition, only the imposition of essential boundary conditions on elastostatic problems is addressed,

$$\boldsymbol{Ku} = \boldsymbol{f} \tag{3.28}$$

In the literature [34] it is possible to find detailed descriptions on the imposition of essential boundary conditions for elastodynamic problems.

3.4.5.1 Interpolating Meshless Methods

If the meshless shape functions used in the construction of the discretized system of equations are interpolating shape functions, such as the RPI shape functions, then the "direct imposition method" or the "penalty method" can be used to enforce the essential boundary conditions.

Considering the three-dimensional problem presented in Fig. 3.15, first it is necessary to identify the field nodes with displacement constrains. Figure 3.15 indicates that the solid displacements are constrained on the curve $\Gamma_{u(C)} \in \Gamma$ and on the surface $\Gamma_{u(S)} \in \Gamma$. Therefore, all field nodes belonging to $\Gamma_{u(C)}$ or to $\Gamma_{u(S)}$ have to be selected: $\boldsymbol{X}_{\Gamma} = \{\boldsymbol{x}_1, \boldsymbol{x}_2, \ldots, \boldsymbol{x}_{N\Gamma}\} \in \left(\Gamma_{u(C)} \cup \Gamma_{u(S)}\right)$, being $\boldsymbol{X}_{\Gamma} \subset \boldsymbol{X}$ and $N\Gamma$ the total number of field nodes on the essential boundaries. Afterwards, the displacement constrains in each field node belonging to $\boldsymbol{X}_{\Gamma}$ must be determined. If each field node discretizing the solid domain possess m degrees of freedom then a field node $\boldsymbol{x}_i \in \boldsymbol{X}_{\Gamma}$ can have from just one displacement constrain up to m displacement constrains.

Direct Imposition Method

It is consider that the field node $\boldsymbol{x}_i \in \boldsymbol{X}_{\Gamma} \wedge \boldsymbol{x}_i \in \mathbb{R}^3$ presents the following displacement constrains: along the direction of the ox axis can move freely; along the direction of the oy axis the displacement component is prescribed by $\bar{u}_y$; and along the direction of the oz axis the displacement component is prescribed by $\bar{u}_z$.

These essential boundary conditions can be enforced directly on the discrete system of equations, $\boldsymbol{Ku} = \boldsymbol{f}$, by modifying directly the stiffness matrix and the global force vector. Being a three-dimensional problem, in which each field node possess three degrees of freedom, the initial stiffness matrix presents the following disposition,

$$
\boldsymbol{K}=\left[\begin{array}{ccc|c|ccc|c|ccc}
K_{11} & K_{12} & K_{13} & \cdots & K_{1(3i-2)} & K_{1(3i-1)} & K_{1(3i-0)} & \cdots & K_{1(3N-2)} & K_{1(3N-1)} & K_{1(3N-0)} \\
K_{21} & K_{22} & K_{23} & \cdots & K_{2(3i-2)} & K_{2(3i-1)} & K_{2(3i-0)} & \cdots & K_{2(3N-2)} & K_{2(3N-1)} & K_{2(3N-0)} \\
K_{31} & K_{32} & K_{33} & \cdots & K_{3(3i-2)} & K_{3(3i-1)} & K_{3(3i-0)} & \cdots & K_{3(3N-2)} & K_{3(3N-1)} & K_{3(3N-0)} \\
\hline
\vdots & \vdots & \vdots & & \vdots & \vdots & \vdots & & \vdots & \vdots & \vdots \\
\hline
K_{(3i-2)1} & K_{(3i-2)2} & K_{(3i-2)3} & \cdots & K_{(3i-2)(3i-2)} & K_{(3i-2)(3i-1)} & K_{(3i-2)(3i-0)} & \cdots & K_{(3i-2)(3N-2)} & K_{(3i-2)(3N-1)} & K_{(3i-2)(3N-0)} \\
K_{(3i-1)1} & K_{(3i-1)2} & K_{(3i-1)3} & \cdots & K_{(3i-1)(3i-2)} & K_{(3i-1)(3i-1)} & K_{(3i-1)(3i-0)} & \cdots & K_{(3i-1)(3N-2)} & K_{(3i-1)(3N-1)} & K_{(3i-1)(3N-0)} \\
K_{(3i-0)1} & K_{(3i-0)2} & K_{(3i-0)3} & \cdots & K_{(3i-0)(3i-2)} & K_{(3i-0)(3i-1)} & K_{(3i-0)(3i-0)} & \cdots & K_{(3i-0)(3N-2)} & K_{(3i-0)(3N-1)} & K_{(3i-0)(3N-0)} \\
\hline
\vdots & \vdots & \vdots & & \vdots & \vdots & \vdots & & \vdots & \vdots & \vdots \\
\hline
K_{(3N-2)1} & K_{(3N-2)2} & K_{(3N-2)3} & \cdots & K_{(3N-2)(3i-2)} & K_{(3N-2)(3i-1)} & K_{(3N-2)(3i-0)} & \cdots & K_{(3N-2)(3N-2)} & K_{(3N-2)(3N-1)} & K_{(3N-2)(3N-0)} \\
K_{(3N-1)1} & K_{(3N-1)2} & K_{(3N-1)3} & \cdots & K_{(3N-1)(3i-2)} & K_{(3N-1)(3i-1)} & K_{(3N-1)(3i-0)} & \cdots & K_{(3N-1)(3N-2)} & K_{(3N-1)(3N-1)} & K_{(3N-1)(3N-0)} \\
K_{(3N-0)1} & K_{(3N-0)2} & K_{(3N-0)3} & \cdots & K_{(3N-0)(3i-2)} & K_{(3N-0)(3i-1)} & K_{(3N-0)(3i-0)} & \cdots & K_{(3N-0)(3N-2)} & K_{(3N-0)(3N-1)} & K_{(3N-0)(3N-0)}
\end{array}\right] \tag{3.29}
$$

In order to impose the mentioned displacements constrains the following two matrix lines $(3i-1)$ and $(3i-0)$ are modified,

$$
\boldsymbol{K}=\left[\begin{array}{ccc|c|ccc|c|ccc}
K_{11} & K_{12} & K_{13} & \cdots & K_{1(3i-2)} & K_{1(3i-1)} & K_{1(3i-0)} & \cdots & K_{1(3N-2)} & K_{1(3N-1)} & K_{1(3N-0)} \\
K_{21} & K_{22} & K_{23} & \cdots & K_{2(3i-2)} & K_{2(3i-1)} & K_{2(3i-0)} & \cdots & K_{2(3N-2)} & K_{2(3N-1)} & K_{2(3N-0)} \\
K_{31} & K_{32} & K_{33} & \cdots & K_{3(3i-2)} & K_{3(3i-1)} & K_{3(3i-0)} & \cdots & K_{3(3N-2)} & K_{3(3N-1)} & K_{3(3N-0)} \\
\hline
\vdots & \vdots & \vdots & & \vdots & \vdots & \vdots & & \vdots & \vdots & \vdots \\
\hline
K_{(3i-2)1} & K_{(3i-2)2} & K_{(3i-2)3} & \cdots & K_{(3i-2)(3i-2)} & K_{(3i-2)(3i-1)} & K_{(3i-2)(3i-0)} & \cdots & K_{(3i-2)(3N-2)} & K_{(3i-2)(3N-1)} & K_{(3i-2)(3N-0)} \\
0 & 0 & 0 & \cdots & 0 & 1 & 0 & \cdots & 0 & 0 & 0 \\
0 & 0 & 0 & \cdots & 0 & 0 & 1 & \cdots & 0 & 0 & 0 \\
\hline
\vdots & \vdots & \vdots & & \vdots & \vdots & \vdots & & \vdots & \vdots & \vdots \\
\hline
K_{(3N-2)1} & K_{(3N-2)2} & K_{(3N-2)3} & \cdots & K_{(3N-2)(3i-2)} & K_{(3N-2)(3i-1)} & K_{(3N-2)(3i-0)} & \cdots & K_{(3N-2)(3N-2)} & K_{(3N-2)(3N-1)} & K_{(3N-2)(3N-0)} \\
K_{(3N-1)1} & K_{(3N-1)2} & K_{(3N-1)3} & \cdots & K_{(3N-1)(3i-2)} & K_{(3N-1)(3i-1)} & K_{(3N-1)(3i-0)} & \cdots & K_{(3N-1)(3N-2)} & K_{(3N-1)(3N-1)} & K_{(3N-1)(3N-0)} \\
K_{(3N-0)1} & K_{(3N-0)2} & K_{(3N-0)3} & \cdots & K_{(3N-0)(3i-2)} & K_{(3N-0)(3i-1)} & K_{(3N-0)(3i-0)} & \cdots & K_{(3N-0)(3N-2)} & K_{(3N-0)(3N-1)} & K_{(3N-0)(3N-0)}
\end{array}\right] \tag{3.30}
$$

The initial global force vector presents the following components,

$$
\boldsymbol{f}=\left\{ f_1 \quad f_2 \quad f_3 \;\middle|\; \;\middle|\; f_{(3i-2)} \quad f_{(3i-1)} \quad f_{(3i-0)} \;\middle|\; \;\middle|\; f_{(3N-2)} \quad f_{(3N-1)} \quad f_{(3N-0)} \right\}^{\mathrm{T}} \tag{3.31}
$$

The imposition of the essential boundary conditions require that the components $(3i-1)$ and $(3i-0)$ of the global force vector are substituted respectively by $\bar{u}_y$ and $\bar{u}_z$,

$$
\boldsymbol{f}=\left\{ f_1 \quad f_2 \quad f_3 \;\middle|\; \;\middle|\; f_{(3i-2)} \quad \bar{u}_y \quad \bar{u}_z \;\middle|\; \;\middle|\; f_{(3N-2)} \quad f_{(3N-1)} \quad f_{(3N-0)} \right\}^{\mathrm{T}} \tag{3.32}
$$

Using the modified stiffness matrix and global force vector the discrete system of equation $\boldsymbol{Ku}=\boldsymbol{f}$ is solved and the displacement constrains $\bar{u}_y$ and $\bar{u}_z$ are exactly satisfied.

This simple procedure can be applied to impose the essential boundary conditions in any other solid mechanics formulations, assuming other dimension orders and degrees of freedom. Consider now the general case in which each field node $\boldsymbol{x} \in \mathbb{R}^d$ possesses m degrees of freedom. The field node $\boldsymbol{x}_I \in \boldsymbol{X}_\Gamma$ presents a displacement constrain $\bar{u}$ on the Jth degree of freedom. Thus, the essential boundary condition can be imposed by,

$$K_{(m \cdot i-(m-n))(m \cdot j-(m-k))} = \begin{cases} 0 & \text{if } i = I \wedge n = J \\ 1 & \text{if } i = j = I \wedge n = k = J \\ K_{(m \cdot i-(m-n))(m \cdot j-(m-k))} & \text{if } i \neq I \vee (i = I \wedge n \neq J) \end{cases} \tag{3.33}$$

and

$$f_{(m \cdot i-(m-n))} = \begin{cases} \bar{u} & \text{if } i = I \wedge n = J \\ f_{(m \cdot i-(m-n))} & \text{if } i \neq I \vee (i = I \wedge n \neq J) \end{cases} \tag{3.34}$$

being $\{i,j\} = \{1, 2, \ldots, N\}$ and $\{n, k\} = \{1, 2, \ldots, m\}$. If the meshless shape function possesses the Kronecker delta property, then the direct imposition method is able to enforce exactly the essential boundary conditions.

Since the RPIM and the NNRPIM are interpolating meshless methods, the numerical examples solved in Chaps. 5, 7 use the direct imposition method to enforce the essential boundary conditions.

Penalty Method

The penalty method is another simple and efficient technique to enforce the essential boundary conditions. In the penalty method the diagonal stiffness components corresponding to the constrained degree of freedom are multiplied by a penalty coefficient α, which is a scalar value much larger than the biggest component of the stiffness matrix, $\alpha \gg \max(K_{ij})$. In the FEM the penalty coefficient usually varies between 10^4 and 10^8 [34, 48], nevertheless in meshless methods the same magnitude range is acceptable [47]. A detail description of the penalty method can be found in the classic FEM literature [34, 48].

Consider again the field node $\boldsymbol{x}_i \in \boldsymbol{X}_\Gamma \wedge \boldsymbol{x}_i \in \mathbb{R}^3$ presenting the same displacement constrains referred in "Direct imposition method": free along ox; constrained by $\bar{u}_y$ along oy; and constrained by $\bar{u}_z$ along oz.

Using the penalty technique, the referred displacements constrains are enforced by multiplying the respective diagonal components of the stiffness matrix, $K_{(3i-1)(3i-1)}$ and $K_{(3i-0)(3i-0)}$, with the penalty number α,

$$\boldsymbol{K}=\begin{bmatrix} K_{11} & \cdots & K_{1(3i-2)} & K_{1(3i-1)} & K_{1(3i-0)} & \cdots & K_{1(3N-0)} \\ \vdots & & \vdots & \vdots & \vdots & & \vdots \\ K_{(3i-2)1} & \cdots & K_{(3i-2)(3i-2)} & K_{(3i-2)(3i-1)} & K_{(3i-2)(3i-0)} & \cdots & K_{(3i-2)(3N-0)} \\ K_{(3i-1)1} & \cdots & K_{(3i-1)(3i-2)} & \alpha \cdot K_{(3i-1)(3i-1)} & K_{(3i-1)(3i-0)} & \cdots & K_{(3i-1)(3N-0)} \\ K_{(3i-0)1} & \cdots & K_{(3i-0)(3i-2)} & K_{(3i-0)(3i-1)} & \alpha \cdot K_{(3i-0)(3i-0)} & \cdots & K_{(3i-0)(3N-0)} \\ \vdots & & \vdots & \vdots & \vdots & & \vdots \\ K_{(3N-0)1} & \cdots & K_{(3N-0)(3i-2)} & K_{(3N-0)(3i-1)} & K_{(3N-0)(3i-0)} & \cdots & K_{(3N-0)(3N-0)} \end{bmatrix} \quad (3.35)$$

Notice that in the penalty method, in opposition to the direct imposition method, only the diagonal components are modified. Additionally, it is necessary to substitute the components $f_{(3i-1)}$ and $f_{(3i-0)}$ of the global force vector by $\alpha \cdot K_{(3i-1)(3i-1)} \cdot \bar{u}_y$ and $\alpha \cdot K_{(3i-0)(3i-0)} \cdot \bar{u}_z$,

$$\boldsymbol{f}=\left\{ f_1 \quad \cdots \quad f_{(3i-2)} \quad \alpha \cdot K_{(3i-1)(3i-1)} \cdot \bar{u}_y \quad \alpha \cdot K_{(3i-0)(3i-0)} \cdot \bar{u}_z \quad \cdots \quad f_{(3N-0)} \right\}^{\mathrm{T}} \quad (3.36)$$

The penalty method does not permit to satisfy exactly the essential boundary conditions. The accuracy of the solution depends on the relative magnitude of the penalty coefficient α [34, 48]. Therefore, the displacement constrains $\bar{u}_y$ and $\bar{u}_z$ are approximately satisfied solving the discrete system of equation $\boldsymbol{Ku}=\boldsymbol{f}$ considering the modified stiffness matrix and global force vector.

Due to the simplicity of the penalty method, several authors prefer this methodology to numerically enforce the essential boundary conditions, regardless the studied solid mechanics formulation of the problem, the solid dimension or the number of degrees of freedom per node.

As in previous section, consider now that each field node $\boldsymbol{x} \in \mathbb{R}^d$ possesses m degrees of freedom and that a generic field node $\boldsymbol{x}_I \in \boldsymbol{X}_\Gamma$ has a displacement constrain $\bar{u}$ on the Jth degree of freedom. Using the penalty method the essential boundary condition can be imposed by substituting only the stiffness matrix diagonal component $K_{(m \cdot I-(m-J))(m \cdot I-(m-J))}$ by $\alpha \cdot K_{(m \cdot I-(m-J))(m \cdot I-(m-J))}$ and by replacing only the global force vector component $f_{(m \cdot I-(m-J))}$ by $\alpha \cdot K_{(m \cdot I-(m-J))(m \cdot I-(m-J))} \cdot \bar{u}$. All the other stiffness matrix and global force vector components maintain the original value.

3.4.5.2 Approximating Meshless Methods

Meshless methods using approximation shape functions lacking the Kronecker delta function property, such as the MLS shape functions, cannot impose the essential boundary conditions using the same methodologies as the ones used in interpolating meshless methods. Additionally, the shape functions obtained with the MLS approximation are not polynomials. Thus, the enforcement of the essential boundary conditions in the EFG methods is not straightforward as it is in interpolating meshless methods.

Several approaches have been studied for enforcing the essential boundary conditions in the EFG method [49]. In the first EFGM papers [3, 50] the essential boundary conditions were imposed using the Lagrange multipliers method. Afterwards, other approaches were investigated, such as using tractions as Lagrange multipliers [51, 52] or using the weak form of the essential boundary conditions [53]. Another popular approach is to combine the FEM with the EFG [54, 55]. Other authors were able to enforce the essential boundary conditions using the penalty method [56].

In this book two distinct methodologies are presented: the "Lagrange multipliers" and the "penalty method".

Lagrange Multipliers

The Lagrange multiplier method is a very popular technique to impose the essential boundary conditions in meshless methods using MLS shape functions. The essential boundary condition$\boldsymbol{u} - \bar{\boldsymbol{u}} = 0$ is represented by a functional, which can be written as an integral using the Lagrange multiplier λ,

$$\int_{\Gamma u} \lambda^{\mathrm{T}} (\boldsymbol{u} - \bar{\boldsymbol{u}}) d\Gamma_u \tag{3.37}$$

Being $\Gamma_u \in \Gamma$ a boundary on the solid domain with displacements constrains. The integral form indicates that somehow it will be require to numerically integrate the subdomain Γ_u. Therefore, it is necessary to discretize Γ_u with integration points. The procedure is quite similar with the numeric integration of the traction forces, Sect. 3.4.4.

The three-dimensional example presented in Fig. 3.15 shows two types of essential boundaries: an essential boundary curve $\Gamma_{u(C)} \in \Gamma$ and an essential boundary surface $\Gamma_{u(S)} \in \Gamma$. In a first step the field nodes on the essential boundaries $\Gamma_{u(C)} \in \Gamma$ and $\Gamma_{u(S)} \in \Gamma$ must be identified separately: $\boldsymbol{X}_{\Gamma u(C)} = \{\boldsymbol{x}_1, \boldsymbol{x}_2, \ldots, \boldsymbol{x}_{NC}\} \in \Gamma_{u(C)}$ and $\boldsymbol{X}_{\Gamma u(S)} = \{\boldsymbol{x}_1, \boldsymbol{x}_2, \ldots, \boldsymbol{x}_{NS}\} \in \Gamma_{u(S)}$, being NC and NS the number of field nodes along the boundary curve $\Gamma_{u(C)}$ and the boundary surface $\Gamma_{u(S)}$ respectively. Additionally, both sets $\boldsymbol{X}_{\Gamma u(C)}$ and $\boldsymbol{X}_{\Gamma u(S)}$ respect the following conditions: $\boldsymbol{X}_{\Gamma u(C)} \subset \boldsymbol{X}$ and $\boldsymbol{X}_{\Gamma u(S)} \subset \boldsymbol{X}$, being $\boldsymbol{X}$ the complete nodal distribution discretizing the problem domain $\Omega \in \mathbb{R}^3$. Then, new integration points have to be determined on the boundary curve $\Gamma_{u(C)}$ and on the boundary surface $\Gamma_{u(S)}$.

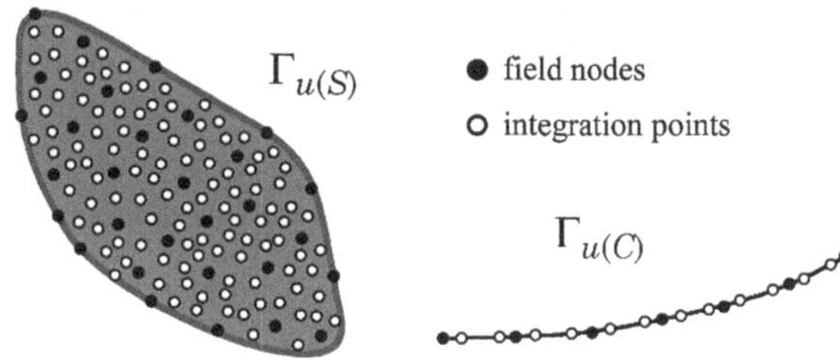

Fig. 3.19 Representation of the essential boundaries

Using the nodal sets $\boldsymbol{X}_{\Gamma u(C)}$ and $\boldsymbol{X}_{\Gamma u(S)}$, two new sets of integration points $\boldsymbol{Q}_{\Gamma u(C)} = \left\{\boldsymbol{q}_1^{\Gamma}, \boldsymbol{q}_2^{\Gamma}, \ldots, \boldsymbol{q}_{QC}^{\Gamma}\right\} \in \Gamma_{u(C)}$ and $\boldsymbol{Q}_{\Gamma u(S)} = \{\boldsymbol{q}_1^{\Gamma}, \boldsymbol{q}_2^{\Gamma}, \ldots, \boldsymbol{q}_{QS}^{\Gamma}\} \in \Gamma_{u(S)}$ are respectively defined, in which QC represents the number of integration points along the boundary curve $\Gamma_{u(C)}$ and QS stands for the number of integration point inside of the area of the essential boundary surface $\Gamma_{u(S)}$. As mentioned in Sect. 3.4.4, each integration point on set $\boldsymbol{Q}_{\Gamma u(C)}$ represents an infinitesimal curve and the integration points on set $\boldsymbol{Q}_{\Gamma u(S)}$ represent infinitesimal surface areas. Therefore, the integration weight $\hat{w}_I^{\Gamma}$ of each integration point represents a length if $\boldsymbol{q}_I^{\Gamma} \in \boldsymbol{Q}_{\Gamma u(C)}$, or a surface area if $\boldsymbol{q}_I^{\Gamma} \in \boldsymbol{Q}_{\Gamma u(S)}$. In Fig. 3.19 are presented the field nodes and the integration points along the essential boundary curve $\Gamma_{u(C)}$ and the essential boundary surface $\Gamma_{u(S)}$.

Additionally, it is required to establish new influence-domains for each one interest point $\boldsymbol{q}_I^{\Gamma}$ discretizing the essential boundary. However, the nodes making the new influence-domains have to belong to the essential boundary curve $\Gamma_{u(C)}$ if $\boldsymbol{q}_I^{\Gamma} \in \boldsymbol{Q}_{\Gamma u(C)}$, or to the essential boundary surface $\Gamma_{u(S)}$ if $\boldsymbol{q}_I^{\Gamma} \in \boldsymbol{Q}_{\Gamma u(S)}$. Notice that the meshless shape functions constructed to approximate the field variable on the boundary for each one integration point $\boldsymbol{q}_I^{\Gamma}$ are constructed using only the field nodes on the essential boundary.

Neglecting the dynamic term of the Galerkin weak form presented in Eq. (2.65) and adding the Lagrange multipliers, the Galerkin weak form expression can be written as,

$$\int_{\Omega} \delta(\boldsymbol{L}\boldsymbol{u})^{\mathrm{T}}\boldsymbol{c}(\boldsymbol{L}\boldsymbol{u})\mathrm{d}\Omega - \int_{\Omega} \delta\boldsymbol{u}^{\mathrm{T}}\boldsymbol{b}\,\mathrm{d}\Omega - \int_{\Gamma_t} \delta\boldsymbol{u}^{\mathrm{T}}\bar{\boldsymbol{t}}\,\mathrm{d}\Gamma_t - \delta\left[\int_{\Gamma_u} \lambda^{\mathrm{T}}(\boldsymbol{u} - \bar{\boldsymbol{u}})\mathrm{d}\Gamma_u\right] = 0 \tag{3.38}$$

The last variational term results from the Lagrange multipliers method and it is included in the expression to ensure: $\boldsymbol{u} - \bar{\boldsymbol{u}} = 0$. The last term can be developed,

$$\begin{aligned}\delta\left[\int_{\Gamma u} \lambda^{\mathrm{T}}(\boldsymbol{u}-\bar{\boldsymbol{u}})\mathrm{d}\Gamma_u\right] &= \int_{\Gamma u} \delta\lambda^{\mathrm{T}}(\boldsymbol{u}-\bar{\boldsymbol{u}})\mathrm{d}\Gamma_u + \int_{\Gamma u} \lambda^{\mathrm{T}}\delta(\boldsymbol{u}-\bar{\boldsymbol{u}})\mathrm{d}\Gamma_u \\ &= \int_{\Gamma u} \delta\lambda^{\mathrm{T}}(\boldsymbol{u}-\bar{\boldsymbol{u}})\mathrm{d}\Gamma_u + \int_{\Gamma u} \lambda^{\mathrm{T}}(\delta\boldsymbol{u} - \underbrace{\delta\bar{\boldsymbol{u}}}_{0})\mathrm{d}\Gamma_u \\ &= \int_{\Gamma u} \delta\lambda^{\mathrm{T}}(\boldsymbol{u}-\bar{\boldsymbol{u}})\mathrm{d}\Gamma_u + \int_{\Gamma u} \delta\boldsymbol{u}^{\mathrm{T}}\lambda\mathrm{d}\Gamma_u\end{aligned} \tag{3.39}$$

Permitting to re-write the Eq. (2.65) as,

$$\begin{aligned}&\int_{\Omega} \delta(\boldsymbol{L}\boldsymbol{u})^{\mathrm{T}}\boldsymbol{c}(\boldsymbol{L}\boldsymbol{u})\mathrm{d}\Omega - \int_{\Omega} \delta\boldsymbol{u}^{\mathrm{T}}\boldsymbol{b}\,\mathrm{d}\Omega - \int_{\Gamma t} \delta\boldsymbol{u}^{\mathrm{T}}\bar{\boldsymbol{t}}\,\mathrm{d}\Gamma_t - \int_{\Gamma u} \delta\lambda^{\mathrm{T}}(\boldsymbol{u}-\bar{\boldsymbol{u}})d\Gamma_u \\ &- \int_{\Gamma u} \delta\boldsymbol{u}^{\mathrm{T}}\lambda\mathrm{d}\Gamma_u = 0\end{aligned} \tag{3.40}$$

From Eq. (3.1) it is possible to determine the virtual displacement approximation of the interest point $\boldsymbol{q}_I \in \boldsymbol{Q}$,

$$\delta\boldsymbol{u}(\boldsymbol{q}_I) = \sum_{i=1}^{n} \varphi_i(\boldsymbol{q}_I)\,\delta\boldsymbol{u}(\boldsymbol{x}_i) = \underbrace{\boldsymbol{H}_I^{local}}_{[3\times 3n]}\,\underbrace{\delta\boldsymbol{u}(\boldsymbol{x})}_{[3n\times 1]} \tag{3.41}$$

being n the number of nodes inside the influence-domain of the interest point $\boldsymbol{q}_I$. The virtual displacement on node $\boldsymbol{x}_i \in \boldsymbol{X}$ is defined by $\delta\boldsymbol{u}(\boldsymbol{x}_i)$ and the local diagonal shape function matrix $\boldsymbol{H}_I^{local}$ defined by Eq. (3.23). Using the procedure described in Fig. 3.17 it is possible to present Eq. (3.41) using the global arrays,

$$\delta\boldsymbol{u}(\boldsymbol{q}_I) = \underbrace{\boldsymbol{H}_I}_{[3\times 3N]}\,\underbrace{\delta\boldsymbol{u}(\boldsymbol{x})}_{[3N\times 1]} \tag{3.42}$$

In the classic three-dimensional deformation theory assuming small strains, the linear differential operator $\boldsymbol{L}$ is defined as,

$$\boldsymbol{L}^{\mathrm{T}} = \begin{bmatrix} \frac{\partial}{\partial x} & 0 & 0 & \frac{\partial}{\partial y} & 0 & \frac{\partial}{\partial z} \\ 0 & \frac{\partial}{\partial y} & 0 & \frac{\partial}{\partial x} & \frac{\partial}{\partial z} & 0 \\ 0 & 0 & \frac{\partial}{\partial z} & 0 & \frac{\partial}{\partial y} & \frac{\partial}{\partial x} \end{bmatrix} \tag{3.43}$$

Thus, using Eqs. (3.41) and (3.43) it is possible to present the first term of Eq. (3.40) as,

$$\begin{aligned}
\int_{\Omega} \delta\left(\underbrace{L}_{[6\times3]}\underbrace{u(q_I)}_{[3\times1]}\right)^{\mathrm{T}}\underbrace{c}_{[6\times6]}\left(\underbrace{L}_{[6\times3]}\underbrace{u(q_I)}_{[3\times1]}\right)\mathrm{d}\Omega &= \int_{\Omega}\delta\left(\underbrace{L}_{[6\times3]}\underbrace{H_I}_{[3\times3N]}\underbrace{u(x)}_{[3N\times1]}\right)^{\mathrm{T}}\underbrace{c}_{[6\times6]}\left(\underbrace{L}_{[6\times3]}\underbrace{H_I}_{[3\times3N]}\underbrace{u(x)}_{[3N\times1]}\right)\mathrm{d}\Omega\\
&= \int_{\Omega}\delta\left(\underbrace{u(x)^{\mathrm{T}}}_{[1\times3N]}\underbrace{H_I^{\mathrm{T}}}_{[3N\times3]}\underbrace{L^{\mathrm{T}}}_{[3\times6]}\right)\underbrace{c}_{[6\times6]}\left(\underbrace{L}_{[6\times3]}\underbrace{H_I}_{[3\times3N]}\underbrace{u(x)}_{[3N\times1]}\right)\mathrm{d}\Omega\\
&= \underbrace{\delta u(x)^{\mathrm{T}}}_{[1\times3N]}\left[\int_{\Omega}\left(\underbrace{H_I^{\mathrm{T}}}_{[3N\times3]}\underbrace{L^{\mathrm{T}}}_{[3\times6]}\right)\underbrace{c}_{[6\times6]}\left(\underbrace{L}_{[6\times3]}\underbrace{H_I}_{[3\times3N]}\right)\mathrm{d}\Omega\right]\underbrace{u(x)}_{[3N\times1]}\\
&= \underbrace{\delta u(x)^{\mathrm{T}}}_{[1\times3N]}\left[\int_{\Omega}\underbrace{B_I^{\mathrm{T}}}_{[3N\times6]}\underbrace{c}_{[6\times6]}\underbrace{B_I}_{[6\times3N]}\mathrm{d}\Omega\right]\underbrace{u(x)}_{[3N\times1]}\\
&= \underbrace{\delta u(x)^{\mathrm{T}}}_{[1\times3N]}\underbrace{K}_{[3N\times3N]}\underbrace{u(x)}_{[3N\times1]}
\end{aligned} \tag{3.44}$$

being the local deformation matrix defined by Eq. (3.21) and the global deformation matrix B_I obtained using the assemblage shown in Fig. 3.17. The second term of Eq. (3.40) can be explicitly written as,

$$\begin{aligned}
\int_{\Omega}\underbrace{\delta u(q_I)^{\mathrm{T}}}_{[3\times1]}\underbrace{b(q_I)}_{[3\times1]}\mathrm{d}\Omega &= \int_{\Omega}\left(\underbrace{H_I}_{[3\times3N]}\underbrace{\delta u(x)}_{[3N\times1]}\right)^{\mathrm{T}}\underbrace{b(q_I)}_{[3\times1]}\mathrm{d}\Omega\\
&= \int_{\Omega}\underbrace{\delta u(x)^{\mathrm{T}}}_{[1\times3N]}\underbrace{H_I^{\mathrm{T}}}_{[3N\times3]}\underbrace{b(q_I)}_{[3\times1]}\mathrm{d}\Omega\\
&= \underbrace{\delta u(x)^{\mathrm{T}}}_{[1\times3N]}\left[\int_{\Omega}\underbrace{H_I^{\mathrm{T}}}_{[3N\times3]}\underbrace{b(q_I)}_{[3\times1]}\mathrm{d}\Omega\right] = \underbrace{\delta u(x)^{\mathrm{T}}}_{[1\times3N]}\underbrace{f_b}_{[3N\times1]}
\end{aligned} \tag{3.45}$$

The third term of Eq. (3.40) follows the same development,

$$\begin{aligned}
\int_{\Gamma_t}\underbrace{\delta u(q_I^{\Gamma})^{\mathrm{T}}}_{[3\times1]}\underbrace{\bar{t}(q_I^{\Gamma})}_{[3\times1]}\mathrm{d}\Gamma_t &= \int_{\Gamma_t}\left(\underbrace{H_I}_{[3\times3N]}\underbrace{\delta u(x)}_{[3N\times1]}\right)^{\mathrm{T}}\underbrace{\bar{t}(q_I^{\Gamma})}_{[3\times1]}\mathrm{d}\Gamma_t\\
&= \int_{\Gamma_t}\underbrace{\delta u(x)^{\mathrm{T}}}_{[1\times3N]}\underbrace{H_I^{\mathrm{T}}}_{[3N\times3]}\underbrace{\bar{t}(q_I^{\Gamma})}_{[3\times1]}\mathrm{d}\Gamma_t\\
&= \underbrace{\delta u(x)^{\mathrm{T}}}_{[1\times3N]}\left[\int_{\Gamma_t}\underbrace{H_I^{\mathrm{T}}}_{[3N\times3]}\underbrace{\bar{t}(q_I^{\Gamma})}_{[3\times1]}\mathrm{d}\Gamma_t\right] = \underbrace{\delta u(x)^{\mathrm{T}}}_{[1\times3N]}\underbrace{f_e}_{[3N\times1]}
\end{aligned} \tag{3.46}$$

Being $\boldsymbol{q}_I^{\Gamma}$ an interest point discretizing the natural boundary curve, $\boldsymbol{Q}_{\Gamma t(C)}$, or the natural boundary surface, $\boldsymbol{Q}_{\Gamma t(S)}$, as represented in Fig. 3.18.

A closer look on Eq. (3.40) permit find some resemblances between the force integrals and the Lagrange multipliers integrals. In fact the Lagrange multipliers can be seen as "smart forces" forcing $\boldsymbol{u} - \bar{\boldsymbol{u}} = 0$. In order to obtain the discretized formulation of the last two terms in Eq. (3.40), the Lagrange multipliers have to be approximated using the respective nodal values and the shape functions for the field nodes along the essential boundaries, sets $\boldsymbol{X}_{\Gamma u(C)}$ and $\boldsymbol{X}_{\Gamma u(S)}$. Thus, since the Lagrange multipliers are unknown functions of the coordinates variables, the Lagrange multipliers approximation on an interest point $\boldsymbol{q}_I^{\Gamma}$, belonging to the essential boundary ($\boldsymbol{q}_I^{\Gamma} \in \boldsymbol{Q}_{\Gamma u(C)}$ or $\boldsymbol{q}_I^{\Gamma} \in \boldsymbol{Q}_{\Gamma u(S)}$), can be determined with,

$$\delta\lambda(\boldsymbol{q}_I^{\Gamma}) = \sum_{i=1}^{n_\lambda} \Phi_i(\boldsymbol{q}_I^{\Gamma})\, \delta\lambda(\boldsymbol{x}_i) = \underbrace{\mathbf{N}_I^{local}}_{[3\times 3n_\lambda]} \cdot \underbrace{\delta\lambda(\boldsymbol{x})}_{[3n_\lambda \times 1]}, \quad \boldsymbol{x} \in \Gamma_u \tag{3.47}$$

being n_λ the number of field nodes used for this interpolation. The vector of the virtual Lagrange multipliers nodal values, considering only the nodes on the essential boundary Γ_u, can be represented for the classical three-dimensional deformation theory as,

$$\delta\lambda(\boldsymbol{x}) = \left\{ \delta\lambda(\boldsymbol{x}_1)_x \quad \delta\lambda(\boldsymbol{x}_1)_y \quad \delta\lambda(\boldsymbol{x}_1)_z \quad \cdots \quad \delta\lambda(\boldsymbol{x}_{n_\lambda})_x \quad \delta\lambda(\boldsymbol{x}_{n_\lambda})_y \quad \delta\lambda(\boldsymbol{x}_{n_\lambda})_z \right\} \tag{3.48}$$

The local diagonal interpolation matrix $\mathbf{N}_I^{local}$ is defined by,

$$\underbrace{\boldsymbol{N}_I^{local}}_{[3\times 3n_\lambda]} = \left[\begin{array}{ccc|ccc|c|ccc} \Phi_1 & 0 & 0 & \Phi_2 & 0 & 0 & \cdots & \Phi_{n_\lambda} & 0 & 0 \\ 0 & \Phi_1 & 0 & 0 & \Phi_2 & 0 & \cdots & 0 & \Phi_{n_\lambda} & 0 \\ 0 & 0 & \Phi_1 & 0 & 0 & \Phi_2 & \cdots & 0 & 0 & \Phi_{n_\lambda} \end{array}\right] \tag{3.49}$$

The interpolation function value $\Phi_i(\boldsymbol{q}_I^{\Gamma})$ is not obtained using the MLS approximation, alternatively $\Phi_i(\boldsymbol{q}_I^{\Gamma})$ it is obtained using, for example, the Lagrange interpolants.

This is the point where the Lagrange multipliers methodology becomes more delicate. Each integration point $\boldsymbol{q}_I^{\Gamma} \in \boldsymbol{Q}_{\Gamma u}$ possesses two influence-domains: one with n field nodes belonging to Γ_u used to construct the MLS approximation shape functions $\varphi(\boldsymbol{q}_I^{\Gamma})$, Eq. (3.18), and another with n_λ field nodes, also belonging to Γ_u, used to construct the interpolation function$\boldsymbol{\Phi}(\boldsymbol{q}_I^{\Gamma})$,

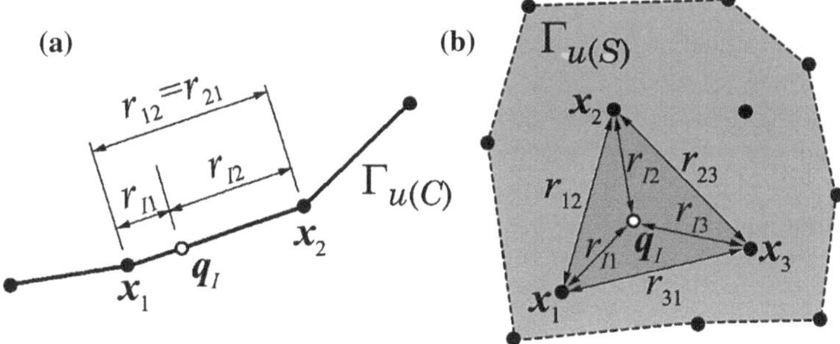

Fig. 3.20 **a** Essential boundary curve. **b** Essential boundary surface

$$\boldsymbol{\Phi}(\boldsymbol{q}_I^\Gamma)^{\mathrm{T}} = \left\{ \Phi_1(\boldsymbol{q}_I^\Gamma) \quad \Phi_2(\boldsymbol{q}_I^\Gamma) \quad \ldots \quad \Phi_{n_\lambda}(\boldsymbol{q}_I^\Gamma) \right\} \tag{3.50}$$

The local diagonal interpolation matrix $\mathbf{N}_I^{local}$ can be assembled in a global diagonal interpolation matrix $\mathbf{N}_I$ using the process in Fig. 3.17.

$$\delta\lambda(\boldsymbol{q}_I^\Gamma) = \underbrace{\boldsymbol{N}_I}_{[3\times 3N_\lambda]} \underbrace{\delta\lambda(\boldsymbol{x})}_{[3N_\lambda\times 1]}, \quad \boldsymbol{x} \in \Gamma_u \tag{3.51}$$

Notice that the dimension of the global diagonal interpolation matrix $\mathbf{N}_I$ is not $[3 \times 3N]$ as the global diagonal shape function matrix $\boldsymbol{H}_I$, but instead it is $[3 \times 3N_\lambda]$, being N_λ the total number of nodes belonging to the essential boundary domain Γ_u.

As already mentioned, the interpolation function$\boldsymbol{\Phi}(\boldsymbol{q}_I^\Gamma)$ can be obtained using the Lagrange interpolants used in the conventional FEM [3, 46, 47, 50]. However the author prefer to use a simpler radial point interpolation to obtain $\boldsymbol{\Phi}(\boldsymbol{q}_I^\Gamma)$ [43–45].

The works available in the literature refer that the first order Lagrange interpolation function is sufficient [3, 46, 47, 50]. That is to say that: the essential boundary curve is discretized using line segments and each integration point possesses only two nodes on the influence-domain ($n_\lambda = 2$); and the essential boundary surface is discretized using triangular patches and each integration point possesses three nodes on the influence-domain ($n_\lambda = 3$).

In order to obtain the interpolation function on the essential boundary curve $\Gamma_{u(C)}$ using the radial point interpolation, first it is necessary to identify the two closest field nodes from the integration point $\boldsymbol{q}_I^\Gamma$. Notice that the two field nodes $\{\boldsymbol{x}_1, \boldsymbol{x}_2\} \in \Gamma_{u(C)}$ form a segment line, containing the integration point $\boldsymbol{q}_I^\Gamma$, Fig. 3.20a. The radial distance $r_{ij} \in \mathbb{R}$ between to spatial points $\{\boldsymbol{x}_i, \boldsymbol{x}_j\} \in \mathbb{R}^3$ is determined by the Euclidean norm,

$$r_{ij} = \|\boldsymbol{x}_j - \boldsymbol{x}_i\| = \sqrt{(x_j - x_i)^2 + (y_j - y_i)^2 + (z_j - z_i)^2} \tag{3.52}$$

Being a radial distance: $r_{ij} = r_{ji}$. The radial distances between the interest point $\boldsymbol{q}_I^{\Gamma}$ and the two field nodes $\boldsymbol{x}_1$ and $\boldsymbol{x}_2$ are determined and arranged in the vector $\boldsymbol{r}_I$,

$$\boldsymbol{r}_I = \{r_{I1}, r_{I2}\}^{\mathrm{T}} \tag{3.53}$$

Then, the radial distances between the two nodes $\boldsymbol{x}_1$ and $\boldsymbol{x}_2$ are inserted in a momentum matrix $\boldsymbol{M}$,

$$\boldsymbol{M} = \begin{bmatrix} r_{11} & r_{12} \\ r_{21} & r_{22} \end{bmatrix} = \begin{bmatrix} 0 & r_{12} \\ r_{12} & 0 \end{bmatrix} \tag{3.54}$$

Afterwards the interpolation function $\boldsymbol{\Phi}(\boldsymbol{q}_I^{\Gamma})$ can be obtained with,

$$\boldsymbol{\Phi}(\boldsymbol{q}_I^{\Gamma}) = \boldsymbol{M}^{-1}\boldsymbol{r}_I = \begin{bmatrix} 0 & r_{12} \\ r_{12} & 0 \end{bmatrix}^{-1} \begin{Bmatrix} r_{I1} \\ r_{I2} \end{Bmatrix} = \begin{bmatrix} 0 & \frac{1}{r_{12}} \\ \frac{1}{r_{12}} & 0 \end{bmatrix} \begin{Bmatrix} r_{I1} \\ r_{I2} \end{Bmatrix} = \begin{Bmatrix} \frac{r_{I2}}{r_{12}} \\ \frac{r_{I1}}{r_{12}} \end{Bmatrix} \tag{3.55}$$

In this book the radial point interpolation technique is shown with detail in Chap. 4.

It is also possible to obtain the interpolation function on the essential boundary surface $\Gamma_{u(S)}$ with the radial point interpolation methodology. To increase the computational efficiency, the interpolation function $\boldsymbol{\Phi}(\boldsymbol{q}_I^{\Gamma})$ of each integration point $\boldsymbol{q}_I^{\Gamma}$ is constructed using small influence-domains, containing each one only the three closest field nodes with relation to $\boldsymbol{q}_I^{\Gamma}$. These three field nodes $\{\boldsymbol{x}_1, \boldsymbol{x}_2, \boldsymbol{x}_3\} \in \Gamma_{u(S)}$ cannot be collinear and form a triangular surface, containing the integration point $\boldsymbol{q}_I^{\Gamma}$, Fig. 3.20b.

Following the same process used to obtain the interpolation function $\boldsymbol{\Phi}(\boldsymbol{q}_I^{\Gamma})$ on the curve essential boundary $\Gamma_{u(C)}$, it is possible to obtain the following expression to define the interpolation function,

$$\begin{aligned} \boldsymbol{\Phi}(\boldsymbol{q}_I^{\Gamma}) = \boldsymbol{M}^{-1}\boldsymbol{r}_I &= \begin{bmatrix} 0 & r_{12} & r_{13} \\ r_{12} & 0 & r_{23} \\ r_{13} & r_{23} & 0 \end{bmatrix}^{-1} \begin{Bmatrix} r_{I1} \\ r_{I2} \\ r_{I3} \end{Bmatrix} \\ &= \begin{bmatrix} -\frac{r_{23}}{2r_{12}r_{13}} & \frac{1}{2r_{12}} & \frac{1}{2r_{13}} \\ \frac{1}{2r_{12}} & -\frac{r_{13}}{2r_{12}r_{23}} & \frac{1}{2r_{23}} \\ \frac{1}{2r_{13}} & \frac{1}{2r_{23}} & -\frac{r_{12}}{2r_{13}r_{23}} \end{bmatrix} \begin{Bmatrix} r_{I1} \\ r_{I2} \\ r_{I3} \end{Bmatrix} = \begin{Bmatrix} -\frac{r_{23}r_{I1}}{2r_{12}r_{13}} + \frac{r_{I2}}{2r_{12}} + \frac{r_{I3}}{2r_{13}} \\ \frac{r_{I1}}{2r_{12}} - \frac{r_{13}r_{I2}}{2r_{12}r_{23}} + \frac{r_{I3}}{2r_{23}} \\ \frac{r_{I1}}{2r_{13}} + \frac{r_{I2}}{2r_{23}} - \frac{r_{12}r_{I3}}{2r_{13}r_{23}} \end{Bmatrix} \end{aligned} \tag{3.56}$$

To increase the accuracy and the stability of the interpolation extra field nodes could be added to the influence-domain. However, the experience indicates that for essential boundary curves $n_{\lambda} = 2$ is sufficient and for essential boundary surfaces influence-domains with $n_{\lambda} = 3$ field nodes are enough.

After the appropriate discretization of the essential boundaries and the determination of the interpolation functions $\mathbf{\Phi}(\boldsymbol{q}_I^{\Gamma})$ of each integration point, it is possible to write explicitly the last two terms of Eq. (3.40). Thus, using the approximations presented in Eqs. (3.42) and (3.51) it is possible to develop the fourth term of Eq. (3.40),

$$\begin{aligned}\int_{\Gamma u} \delta\lambda^{\mathrm{T}}(\boldsymbol{u}-\bar{\boldsymbol{u}})\mathrm{d}\Gamma_u &= \int_{\Gamma u} \delta\lambda^{\mathrm{T}}\boldsymbol{u}\,\mathrm{d}\Gamma_u - \int_{\Gamma u} \delta\lambda^{\mathrm{T}}\bar{\boldsymbol{u}}\,\mathrm{d}\Gamma_u \\ &= \int_{\Gamma u} \left[\underbrace{\boldsymbol{N}_I}_{[3\times 3N_\lambda]} \underbrace{\delta\lambda(\boldsymbol{x})}_{[3N_\lambda\times 1]}\right]^{\mathrm{T}} \left[\underbrace{\boldsymbol{H}_I}_{[3\times 3N]} \underbrace{\boldsymbol{u}(\boldsymbol{x})}_{[3N\times 1]}\right] \mathrm{d}\Gamma_u - \int_{\Gamma u} \left[\underbrace{\boldsymbol{N}_I}_{[3\times 3N_\lambda]} \underbrace{\delta\lambda(\boldsymbol{x})}_{[3N_\lambda\times 1]}\right]^{\mathrm{T}} \underbrace{\bar{\boldsymbol{u}}(\boldsymbol{q}_I)}_{[3\times 1]} \mathrm{d}\Gamma_u \\ &= \int_{\Gamma u} \underbrace{\delta\lambda(\boldsymbol{x})^{\mathrm{T}}}_{[1\times 3N_\lambda]} \underbrace{\boldsymbol{N}_I^{\mathrm{T}}}_{[3N_\lambda\times 3]} \underbrace{\boldsymbol{H}_I}_{[3\times 3N]} \underbrace{\boldsymbol{u}(\boldsymbol{x})}_{[3N\times 1]} \mathrm{d}\Gamma_u - \int_{\Gamma u} \underbrace{\delta\lambda(\boldsymbol{x})^{\mathrm{T}}}_{[1\times 3N_\lambda]} \underbrace{\boldsymbol{N}_I^{\mathrm{T}}}_{[3N_\lambda\times 3]} \underbrace{\bar{\boldsymbol{u}}(\boldsymbol{q}_I)}_{[3\times 1]} \mathrm{d}\Gamma_u \\ &= \underbrace{\delta\lambda(\boldsymbol{x})^{\mathrm{T}}}_{[1\times 3N_\lambda]} \underbrace{\left[\int_{\Gamma u} \underbrace{\boldsymbol{N}_I^{\mathrm{T}}}_{[3N_\lambda\times 3]} \underbrace{\boldsymbol{H}_I}_{[3\times 3N]} \mathrm{d}\Gamma_u\right]}_{-\boldsymbol{G}_\lambda^{\mathrm{T}}} \underbrace{\boldsymbol{u}(\boldsymbol{x})}_{[3N\times 1]} - \underbrace{\delta\lambda(\boldsymbol{x})^{\mathrm{T}}}_{[1\times 3N_\lambda]} \underbrace{\left[\int_{\Gamma u} \underbrace{\boldsymbol{N}_I^{\mathrm{T}}}_{[3N_\lambda\times 3]} \underbrace{\bar{\boldsymbol{u}}(\boldsymbol{q}_I)}_{[3\times 1]} \mathrm{d}\Gamma_u\right]}_{-\boldsymbol{g}_\lambda}\end{aligned} \tag{3.57}$$

The displacement constrain approximation $\bar{\boldsymbol{u}}(\boldsymbol{q}_I)$ is obtained using only the nodes belonging to the essential boundary Γ_u. Being the essential boundary discretized by a set of integration points, $\boldsymbol{Q}_{\Gamma u} = \left\{\boldsymbol{q}_1^{\Gamma}, \boldsymbol{q}_2^{\Gamma}, \ldots, \boldsymbol{q}_{Q\Gamma}^{\Gamma}\right\} \in \Gamma_u$, it is possible to obtain the $\boldsymbol{G}_\lambda$ matrix by substituting the continuous integral over the essential boundary domain Γ_u with the following discrete sum,

$$\underbrace{\boldsymbol{G}_\lambda^{\mathrm{T}}}_{[3N_\lambda\times 3N]} = -\int_{\Gamma u} \underbrace{\boldsymbol{N}_I^{\mathrm{T}}}_{[3N_\lambda\times 3]} \underbrace{\boldsymbol{H}_I}_{[3\times 3N]} \mathrm{d}\Gamma_u = -\sum_{I=1}^{Q\Gamma} \hat{w}_I^{\Gamma} \underbrace{\boldsymbol{N}_I^{\mathrm{T}}}_{[3N_\lambda\times 3]} \underbrace{\boldsymbol{H}_I}_{[3\times 3N]} \tag{3.58}$$

Being $\hat{w}_I^{\Gamma}$ the integration weight of $\boldsymbol{q}_I^{\Gamma} \in \boldsymbol{Q}_{\Gamma u}$. The vector $\boldsymbol{g}_\lambda$ is obtained using a similar expression,

$$\underbrace{\boldsymbol{g}_\lambda}_{[3N_\lambda\times 1]} = -\int_{\Gamma u} \underbrace{\boldsymbol{N}_I^{\mathrm{T}}}_{[3N_\lambda\times 3]} \underbrace{\bar{\boldsymbol{u}}(\boldsymbol{q}_I)}_{[3\times 1]} \mathrm{d}\Gamma_u = -\sum_{I=1}^{Q\Gamma} \hat{w}_I^{\Gamma} \underbrace{\boldsymbol{N}_I^{\mathrm{T}}}_{[3N_\lambda\times 3]} \underbrace{\bar{\boldsymbol{u}}(\boldsymbol{q}_I)}_{[3\times 1]} \tag{3.59}$$

Thus, the fourth term Eq. (3.40) can be presented as,

$$\int_{\Gamma u} \delta\lambda^{\mathrm{T}}(\boldsymbol{u}-\bar{\boldsymbol{u}})\mathrm{d}\Gamma_u = -\underbrace{\delta\lambda^{\mathrm{T}}}_{[1\times 3N_\lambda]} \underbrace{\boldsymbol{G}_\lambda^{\mathrm{T}}}_{[3N_\lambda\times 3N]} \underbrace{\boldsymbol{u}}_{[3N\times 1]} + \underbrace{\delta\lambda^{\mathrm{T}}}_{[1\times 3N_\lambda]} \underbrace{\boldsymbol{g}_\lambda}_{[3N_\lambda\times 1]} \tag{3.60}$$

The fifth term of Eq. (3.40) can be developed using the same procedure,

$$\int_{\Gamma u} \delta \boldsymbol{u}^{\mathrm{T}} \lambda \mathrm{d}\Gamma_u = \int_{\Gamma u} \left[\underbrace{\boldsymbol{H}_I}_{[3\times 3N]} \underbrace{\delta \boldsymbol{u}(\boldsymbol{x})}_{[3N\times 1]} \right]^{\mathrm{T}} \left[\underbrace{\boldsymbol{N}_I}_{[3\times 3N_\lambda]} \underbrace{\lambda(\boldsymbol{x})}_{[3N_\lambda\times 1]} \right] \mathrm{d}\Gamma_u$$

$$= \int_{\Gamma u} \underbrace{\delta \boldsymbol{u}(\boldsymbol{x})^{\mathrm{T}}}_{[1\times 3N]} \underbrace{\boldsymbol{H}_I^{\mathrm{T}}}_{[3N\times 3]} \underbrace{\boldsymbol{N}_I}_{[3\times 3N_\lambda]} \underbrace{\lambda(\boldsymbol{x})}_{[3N_\lambda\times 1]} \mathrm{d}\Gamma_u = \underbrace{\delta \boldsymbol{u}(\boldsymbol{x})^{\mathrm{T}}}_{[1\times 3N]} \underbrace{\left[\int_{\Gamma u} \underbrace{\boldsymbol{H}_I^{\mathrm{T}}}_{[3N\times 3]} \underbrace{\boldsymbol{N}_I}_{[3\times 3N_\lambda]} \mathrm{d}\Gamma_u \right]}_{-\boldsymbol{G}_\lambda} \underbrace{\lambda(\boldsymbol{x})}_{[3N_\lambda\times 1]} \tag{3.61}$$

Substituting the results from Eqs. (3.44), (3.45), (3.46), (3.60) and (3.61) in the Galerkin weak form expression, Eq. (3.40), the following expression is obtained,

$$\underbrace{\delta \boldsymbol{u}^{\mathrm{T}}}_{[1\times 3N]} \underbrace{\boldsymbol{K}}_{[3N\times 3N]} \underbrace{\boldsymbol{u}}_{[3N\times 1]} - \underbrace{\delta \boldsymbol{u}^{\mathrm{T}}}_{[1\times 3N]} \underbrace{\boldsymbol{f}_b}_{[3N\times 1]} - \underbrace{\delta \boldsymbol{u}^{\mathrm{T}}}_{[1\times 3N]} \underbrace{\boldsymbol{f}_e}_{[3N\times 1]} + \underbrace{\delta \lambda^{\mathrm{T}}}_{[1\times 3N_\lambda]} \underbrace{\boldsymbol{G}_\lambda^{\mathrm{T}}}_{[3N_\lambda\times 3N]} \underbrace{\boldsymbol{u}}_{[3N\times 1]} - \underbrace{\delta \lambda^{\mathrm{T}}}_{[1\times 3N_\lambda]} \underbrace{\boldsymbol{g}_\lambda}_{[3N_\lambda\times 1]}$$
$$+ \underbrace{\delta \boldsymbol{u}^{\mathrm{T}}}_{[1\times 3N]} \underbrace{\boldsymbol{G}_\lambda}_{[3N\times 3N_\lambda]} \underbrace{\lambda}_{[3N_\lambda\times 1]} = 0 \tag{3.62}$$

Which can be re-written as,

$$\underbrace{\delta \boldsymbol{u}^{\mathrm{T}}}_{[1\times 3N]} \left[\underbrace{\boldsymbol{K}}_{[3N\times 3N]} \underbrace{\boldsymbol{u}}_{[3N\times 1]} + \underbrace{\boldsymbol{G}_\lambda}_{[3N\times 3N_\lambda]} \underbrace{\lambda}_{[3N_\lambda\times 1]} - \underbrace{\boldsymbol{f}}_{[3N\times 1]} \right] + \underbrace{\delta \lambda^{\mathrm{T}}}_{[1\times 3N_\lambda]} \left[\underbrace{\boldsymbol{G}_\lambda^{\mathrm{T}}}_{[3N_\lambda\times 3N]} \underbrace{\boldsymbol{u}}_{[3N\times 1]} - \underbrace{\boldsymbol{g}_\lambda}_{[3N_\lambda\times 1]} \right]$$
$$= 0 \tag{3.63}$$

Being $\boldsymbol{f} = \boldsymbol{f}_b + \boldsymbol{f}_e$. Since $\delta \boldsymbol{u}$ and $\delta\lambda$ are arbitrary Eq. (3.63) can only be satisfied with,

$$\begin{cases} \underbrace{\boldsymbol{K}}_{[3N\times 3N]} \underbrace{\boldsymbol{u}}_{[3N\times 1]} + \underbrace{\boldsymbol{G}_\lambda}_{[3N\times 3N_\lambda]} \underbrace{\lambda}_{[3N_\lambda\times 1]} - \underbrace{\boldsymbol{f}}_{[3N\times 1]} = 0 \\ \underbrace{\boldsymbol{G}_\lambda^{\mathrm{T}}}_{[3N_\lambda\times 3N]} \underbrace{\boldsymbol{u}}_{[3N\times 1]} - \underbrace{\boldsymbol{g}_\lambda}_{[3N_\lambda\times 1]} = 0 \end{cases} \tag{3.64}$$

The discrete system of equations represented in Eq. (3.64) can be organized in the following matrix form,

$$\underbrace{\begin{bmatrix} \underbrace{\boldsymbol{K}}_{[3N\times 3N]} & \underbrace{\boldsymbol{G}_\lambda}_{[3N\times 3N_\lambda]} \\ \underbrace{\boldsymbol{G}_\lambda^{\mathrm{T}}}_{[3N_\lambda\times 3N]} & \underbrace{\boldsymbol{0}}_{[3N_\lambda\times 3N_\lambda]} \end{bmatrix}}_{[(3N+3N_\lambda)\times(3N+3N_\lambda)]} \underbrace{\begin{Bmatrix} \underbrace{\boldsymbol{u}}_{[3N\times 1]} \\ \underbrace{\boldsymbol{\lambda}}_{[3N_\lambda\times 1]} \end{Bmatrix}}_{[(3N+3N_\lambda)\times 1]} = \underbrace{\begin{Bmatrix} \underbrace{\boldsymbol{f}}_{[3N\times 1]} \\ \underbrace{\boldsymbol{g}_\lambda}_{[3N_\lambda\times 1]} \end{Bmatrix}}_{[(3N+3N_\lambda)\times 1]} \tag{3.65}$$

which is the obtained final discrete system of equations when the Lagrange multipliers method is used to enforce the essential boundary conditions. The displacement nodal parameters vector $\boldsymbol{u}$ is determined with Eq. (3.65) and then, the displacement of any interest point $\boldsymbol{x}_I \in \Omega$ can be obtained with Eq. (3.1).

In structural analysis it is very common to use the Lagrange multipliers to impose the essential boundary conditions when approximation meshless methods are used, such as the EFGM. This technique is accurate, permitting to impose exactly the essential boundary conditions. However, as Eq. (3.65) shows, the Lagrange multipliers method increases the number of unknowns of the initial system of equations from $3N$ to $3N + 3N_\lambda$. If the essential boundary contains a high percentage of field nodes, then the efficiency of the meshless analysis is certainly compromised. Additionally, the linear system of equations is no longer positive definite [49] and it loses the banded property, increasing the computational cost of the analysis. Nevertheless, the final discrete system of equations preserves the symmetry.

Penalty Method

The penalty method is an alternative efficient numerical technique capable to impose the essential boundary conditions in approximation meshless methods [49]. Since approximation shape functions, such as the MLS shape functions, do not possess the Kronecker delta property, the Galerkin weak form expression can be presented as,

$$\begin{aligned} &\int_{\Omega} \delta(\boldsymbol{L}\boldsymbol{u})^{\mathrm{T}}\boldsymbol{c}(\boldsymbol{L}\boldsymbol{u})\mathrm{d}\Omega - \int_{\Omega} \delta\boldsymbol{u}^{\mathrm{T}}\boldsymbol{b}\,\mathrm{d}\Omega - \int_{\Gamma_t} \delta\boldsymbol{u}^{\mathrm{T}}\bar{\boldsymbol{t}}\,\mathrm{d}\Gamma_t \\ &\quad + \delta\left[\int_{\Gamma_u} \frac{1}{2}(\boldsymbol{u}-\bar{\boldsymbol{u}})^{\mathrm{T}}\boldsymbol{\alpha}(\boldsymbol{u}-\bar{\boldsymbol{u}})\mathrm{d}\Gamma_u\right] \\ &= 0 \end{aligned} \tag{3.66}$$

Notice that the dynamic term was neglected and it was included an additional term to enforce the displacement constrain $\boldsymbol{u} - \bar{\boldsymbol{u}} = 0$ by the penalty method. Matrix $\boldsymbol{\alpha}$ is defined as,

$$\boldsymbol{\alpha} = \begin{bmatrix} \alpha(\boldsymbol{x})_1 & 0 & \cdots & 0 \\ 0 & \alpha(\boldsymbol{x})_2 & \cdots & 0 \\ \vdots & \vdots & \ddots & \vdots \\ 0 & 0 & \cdots & \alpha(\boldsymbol{x})_m \end{bmatrix} \tag{3.67}$$

where $\alpha(\boldsymbol{x})_i$are penalty factors and m is the number of degrees of freedom in each field node $\boldsymbol{x} \in \mathbb{R}^d$required by the studied formulation. The magnitudes of the penalty factors $\alpha(\boldsymbol{x})_i$ may be assumed as a functional of the spatial location of the interest point. Additionally, for the same interest point, distinct $\alpha(\boldsymbol{x})_i$ can be considered for distinct degrees of freedom. Nevertheless, generally the penalty factors $\alpha(\boldsymbol{x})_i$ are large positive values, constant along the solid discretized domain and equal for all degrees of freedom: $\alpha(\boldsymbol{x})_i = \alpha$. The magnitude of the penalty factor is addressed in Section "Penalty Method".

The last term of Eq. (3.66) represents an integral over the essential boundary $\Gamma_u \in \Gamma$, thus once again it will be necessary to numerically integrate the subdomain Γ_u. Therefore the essential boundary subdomain has to be discretized with integration points. The procedure is fully described in "Lagrange multipliers". First the $N\Gamma$ field nodes on the essential boundary are identified: $\boldsymbol{X}_{\Gamma u} = \{\boldsymbol{x}_1, \boldsymbol{x}_2, \ldots, \boldsymbol{x}_{N\Gamma}\} \in \Gamma_u$, being $\boldsymbol{X}_{\Gamma u} \in \boldsymbol{X}$ and $\boldsymbol{X}$ the complete nodal set, Fig. 3.16a, discretizing the problem domain $\Omega \in \mathbb{R}^3$, Fig. 3.15. Afterwards, new integration points are determined on the boundary subdomain Γ_u: $\boldsymbol{Q}_{\Gamma u} = \left\{\boldsymbol{q}_1^\Gamma, \boldsymbol{q}_2^\Gamma, \ldots, \boldsymbol{q}_{Q\Gamma}^\Gamma\right\} \in \Gamma_u$, being $Q\Gamma$ the number of integration points discretizing the essential boundary Γ_u. To each integration point $\boldsymbol{q}_I^\Gamma \in \boldsymbol{Q}_{\Gamma u}$ it is associated an integration weight$\widehat{w}_I^\Gamma$, representing the dimensional size of $\boldsymbol{q}_I^\Gamma \in \boldsymbol{Q}_{\Gamma u}$. In addition, for each integration point $\boldsymbol{q}_I^\Gamma$ a new influence-domain is determined, considering only field nodes $\boldsymbol{x}_i$ belonging to $\boldsymbol{X}_{\Gamma u} \in \boldsymbol{X}$. Meaning that the meshless shape functions of the integration points $\boldsymbol{q}_I^\Gamma \in \boldsymbol{Q}_{\Gamma u}$ are constructed just using field nodes on the essential boundary: $\boldsymbol{X}_{\Gamma u} \in \boldsymbol{X}$.

The last term of Eq. (3.66), included in the expression to ensure: $\boldsymbol{u} - \bar{\boldsymbol{u}} = 0$, can be developed,

$$
\begin{aligned}
\delta\left[\int_{\Gamma u} \frac{1}{2}(\boldsymbol{u}-\bar{\boldsymbol{u}})^{\mathrm{T}}\boldsymbol{\alpha}(\boldsymbol{u}-\bar{\boldsymbol{u}})\mathrm{d}\Gamma_u\right] &= \int_{\Gamma u} \frac{1}{2}\delta(\boldsymbol{u}-\bar{\boldsymbol{u}})^{\mathrm{T}}\boldsymbol{\alpha}(\boldsymbol{u}-\bar{\boldsymbol{u}})\mathrm{d}\Gamma_u \\
&= \int_{\Gamma u} \frac{1}{2}(\delta\boldsymbol{u}-\underbrace{\delta\bar{\boldsymbol{u}}}_{0})^{\mathrm{T}}\boldsymbol{\alpha}(\boldsymbol{u}-\bar{\boldsymbol{u}})\mathrm{d}\Gamma_u \\
&= \int_{\Gamma u} \frac{1}{2}\delta\boldsymbol{u}^{\mathrm{T}}\boldsymbol{\alpha}\,\boldsymbol{u}\,\mathrm{d}\Gamma_u - \int_{\Gamma u} \frac{1}{2}\delta\boldsymbol{u}^{\mathrm{T}}\boldsymbol{\alpha}\,\bar{\boldsymbol{u}}\,\mathrm{d}\Gamma_u
\end{aligned}
\tag{3.68}
$$

Recall that the virtual displacement approximation $\delta\boldsymbol{u}$ of the integration point $\boldsymbol{q}_I \in \boldsymbol{Q}$ is obtained with Eq. (3.41) and the approximated displacement components $\boldsymbol{u}_I = (u_x\, u_y\, u_z)$of the integration point $\boldsymbol{q}_I \in \boldsymbol{Q}$ are obtained with Eq. (3.1). Both expressions Eqs. (3.41) and (3.1) permit to obtain local approximations which can be extended to global approximations using Eq. (3.42) and following the process described in Fig. 3.17. Thus, using the global approximation presented in Eq. (3.42) it is possible to develop Eq. (3.68),

$$
\begin{aligned}
\int_{\Gamma u} \frac{1}{2}\delta\boldsymbol{u}^{\mathrm{T}}\boldsymbol{\alpha}\,\boldsymbol{u}\,\mathrm{d}\Gamma_u - \int_{\Gamma u} \frac{1}{2}\delta\boldsymbol{u}^{\mathrm{T}}\boldsymbol{\alpha}\,\bar{\boldsymbol{u}}\,\mathrm{d}\Gamma_u &= \int_{\Gamma u} \frac{1}{2}\left[\underbrace{\boldsymbol{H}_I}_{[3\times 3N]}\underbrace{\delta\boldsymbol{u}(\boldsymbol{x})}_{[3N\times 1]}\right]^{\mathrm{T}}\underbrace{\boldsymbol{\alpha}}_{[3\times 3]}\left[\underbrace{\boldsymbol{H}_I}_{[3\times 3N]}\underbrace{\boldsymbol{u}(\boldsymbol{x})}_{[3N\times 1]}\right]\mathrm{d}\Gamma_u \\
&\quad - \int_{\Gamma u} \frac{1}{2}\left[\underbrace{\boldsymbol{H}_I}_{[3\times 3N]}\underbrace{\delta\boldsymbol{u}(\boldsymbol{x})}_{[3N\times 1]}\right]^{\mathrm{T}}\underbrace{\boldsymbol{\alpha}}_{[3\times 3]}\underbrace{\bar{\boldsymbol{u}}(\boldsymbol{q}_I)}_{[3\times 1]}\mathrm{d}\Gamma_u \\
&= \int_{\Gamma u} \frac{1}{2}\underbrace{\delta\boldsymbol{u}(\boldsymbol{x})^{\mathrm{T}}}_{[1\times 3N]}\underbrace{\boldsymbol{H}_I^{\mathrm{T}}}_{[3N\times 3]}\underbrace{\boldsymbol{\alpha}}_{[3\times 3]}\underbrace{\boldsymbol{H}_I}_{[3\times 3N]}\underbrace{\boldsymbol{u}(\boldsymbol{x})}_{[3N\times 1]}\mathrm{d}\Gamma_u \\
&\quad - \int_{\Gamma u} \frac{1}{2}\underbrace{\delta\boldsymbol{u}(\boldsymbol{x})^{\mathrm{T}}}_{[1\times 3N]}\underbrace{\boldsymbol{H}_I^{\mathrm{T}}}_{[3N\times 3]}\underbrace{\boldsymbol{\alpha}}_{[3\times 3]}\underbrace{\bar{\boldsymbol{u}}(\boldsymbol{q}_I)}_{[3\times 1]}\mathrm{d}\Gamma_u \\
&= \underbrace{\delta\boldsymbol{u}(\boldsymbol{x})^{\mathrm{T}}}_{[1\times 3N]}\underbrace{\left[\int_{\Gamma u}\underbrace{\boldsymbol{H}_I^{\mathrm{T}}}_{[3N\times 3]}\underbrace{\frac{1}{2}\boldsymbol{\alpha}}_{[3\times 3]}\underbrace{\boldsymbol{H}_I}_{[3\times 3N]}\mathrm{d}\Gamma_u\right]}_{\boldsymbol{K}^{\alpha}}\underbrace{\boldsymbol{u}(\boldsymbol{x})}_{[3N\times 1]} \\
&\quad - \underbrace{\delta\boldsymbol{u}(\boldsymbol{x})^{\mathrm{T}}}_{[1\times 3N]}\underbrace{\left[\int_{\Gamma u}\underbrace{\boldsymbol{H}_I^{\mathrm{T}}}_{[3N\times 3]}\underbrace{\frac{1}{2}\boldsymbol{\alpha}}_{[3\times 3]}\underbrace{\bar{\boldsymbol{u}}(\boldsymbol{q}_I)}_{[3\times 1]}\mathrm{d}\Gamma_u\right]}_{\boldsymbol{f}^{\alpha}}
\end{aligned}
\tag{3.69}
$$

being $\bar{\boldsymbol{u}}(\boldsymbol{q}_I)$the approximation of the displacement constrain obtained for the integration point $\boldsymbol{q}_I^{\Gamma} \in \boldsymbol{Q}_{\Gamma u}$only considering field nodes from $\boldsymbol{X}_{\Gamma u} \in \boldsymbol{X}$. Using the integration points discretizing the essential boundary domain Γ_u: $\boldsymbol{Q}_{\Gamma u} = \left\{\boldsymbol{q}_1^{\Gamma}, \boldsymbol{q}_2^{\Gamma}, \ldots, \boldsymbol{q}_{Q\Gamma}^{\Gamma}\right\} \in \Gamma_u$, it is possible to substitute the continuous integral over Γ_u by a discrete sum and determine the global penalty stiffness matrix $\boldsymbol{K}^{\alpha}$,

$$\underbrace{\boldsymbol{K}^{\alpha}}_{[3N\times 3N]} = \int_{\Gamma u} \underbrace{\boldsymbol{H}_I^{\mathrm{T}}}_{[3N\times 3]} \underbrace{\frac{1}{2}\boldsymbol{\alpha}}_{[3\times 3]} \underbrace{\boldsymbol{H}_I}_{[3\times 3N]} \mathrm{d}\Gamma_u = \sum_{I=1}^{Q\Gamma} \widehat{w}_I^{\Gamma} \underbrace{\boldsymbol{H}_I^{\mathrm{T}}}_{[3N\times 3]} \underbrace{\frac{1}{2}\boldsymbol{\alpha}}_{[3\times 3]} \underbrace{\boldsymbol{H}_I}_{[3\times 3N]} \tag{3.70}$$

Being $\widehat{w}_I^{\Gamma}$ the integration weight of $\boldsymbol{q}_I^{\Gamma} \in \boldsymbol{Q}_{\Gamma u}$. Additionally, the penalty force vector $\boldsymbol{f}^{\alpha}$, caused by the essential boundary conditions, is obtained with,

$$\underbrace{\boldsymbol{f}^{\alpha}}_{[3N\times 1]} = \int_{\Gamma u} \underbrace{\boldsymbol{H}_I^{\mathrm{T}}}_{[3N\times 3]} \underbrace{\frac{1}{2}\boldsymbol{\alpha}}_{[3\times 3]} \underbrace{\bar{\boldsymbol{u}}(\boldsymbol{q}_I)}_{[3\times 1]} \mathrm{d}\Gamma_u = \sum_{I=1}^{Q\Gamma} \widehat{w}_I^{\Gamma} \underbrace{\boldsymbol{H}_I^{T}}_{[3N\times 3]} \underbrace{\frac{1}{2}\boldsymbol{\alpha}}_{[3\times 3]} \underbrace{\bar{\boldsymbol{u}}(\boldsymbol{q}_I)}_{[3\times 1]} \tag{3.71}$$

Thus, the fourth term Eq. (3.66) can be presented as,

$$\delta\left[\int_{\Gamma u} \frac{1}{2}(\boldsymbol{u}-\bar{\boldsymbol{u}})^{\mathrm{T}}\boldsymbol{\alpha}(\boldsymbol{u}-\bar{\boldsymbol{u}})\mathrm{d}\Gamma_u\right] = \underbrace{\delta\boldsymbol{u}^{\mathrm{T}}}_{[1\times 3N]} \underbrace{\boldsymbol{K}^{\alpha}}_{[3N\times 3N]} \underbrace{\boldsymbol{u}}_{[3N\times 1]} - \underbrace{\delta\boldsymbol{u}^{\mathrm{T}}}_{[1\times 3N]} \underbrace{\boldsymbol{f}^{\alpha}}_{[3N\times 1]} \tag{3.72}$$

and Eq. (3.66) can be re-written as,

$$\underbrace{\delta\boldsymbol{u}^{\mathrm{T}}}_{[1\times 3N]} \underbrace{\boldsymbol{K}}_{[3N\times 3N]} \underbrace{\boldsymbol{u}}_{[3N\times 1]} - \underbrace{\delta\boldsymbol{u}^{\mathrm{T}}}_{[1\times 3N]} \underbrace{\boldsymbol{f}_b}_{[3N\times 1]} - \underbrace{\delta\boldsymbol{u}^{\mathrm{T}}}_{[1\times 3N]} \underbrace{\boldsymbol{f}_e}_{[3N\times 1]} + \underbrace{\delta\boldsymbol{u}^{\mathrm{T}}}_{[1\times 3N]} \underbrace{\boldsymbol{K}^{\alpha}}_{[3N\times 3N]} \underbrace{\boldsymbol{u}}_{[3N\times 1]} - \underbrace{\delta\boldsymbol{u}^{\mathrm{T}}}_{[1\times 3N]} \underbrace{\boldsymbol{f}^{\alpha}}_{[3N\times 1]} = 0 \tag{3.73}$$

Yielding the following global discretized system of equation,

$$\underbrace{\left[\underbrace{\boldsymbol{K}}_{[3N\times 3N]} + \underbrace{\boldsymbol{K}^{\alpha}}_{[3N\times 3N]}\right]}_{[3N\times 3N]} \underbrace{\left\{\underbrace{\boldsymbol{u}}_{[3N\times 1]}\right\}}_{[3N\times 1]} = \underbrace{\left\{\underbrace{\boldsymbol{f}_b}_{[3N\times 1]} + \underbrace{\boldsymbol{f}_e}_{[3N\times 1]} + \underbrace{\boldsymbol{f}^{\alpha}}_{[3N\times 1]}\right\}}_{[3N\times 1]} \tag{3.74}$$

which is the obtained final discrete system of equations when the penalty method is used to enforce the essential boundary conditions. This methodology, in opposition with the Lagrange multipliers method, permits to maintain the initial size of the discrete system of equations: $[mN \times mN]$. With the penalty method, if the used penalty factors are positive, then the positive definite property of discrete system of equations is preserved. Additionally, as it is perceptible, the Galerkin procedure produces symmetric and banded stiffness matrices $\boldsymbol{K}$ and $\boldsymbol{K}^{\alpha}$.

Nevertheless, the penalty method presents some disadvantages. It is not possible to enforce exactly the essential boundary conditions with the penalty method. The accuracy of the solution depends on the magnitude of the penalty factors and the obtained results are less accurate than the results obtained with the Lagrange multipliers method. Experience shows that higher penalty factors lead to more accurate results, however larger penalty factors could lead to ill-conditioned stiffness matrix. In addition, the appropriate magnitude of the penalty factors vary with the analysed problem, therefore it is not possible to define a universally acceptable penalty factor.

References

1. Liu GR (2002) A point assembly method for stress analysis for two-dimensional solids. Int J Solid Struct 39:261–276
2. Liu GR (2002) Mesh free methods-moving beyond the finite element method. CRC Press, Boca Raton
3. Belytschko T, Lu YY, Gu L (1994) Element-free galerkin method. Int J Numer Meth Eng 37:229–256
4. Chen JS, Wu CT, Yoon S, You Y (2001) A stabilized conforming nodal integration for Galerkin mesh-free methods. Int J Numer Methods Eng 50(2):435–466
5. Sze KY, Chen JS, Sheng N, Liu XH (2004) Stabilized conforming nodal integration: exactness and variational. Finite Elem Anal Des 41(2):147–171
6. Elmer W, Chen JS, Puso M, Taciroglu E (2012) A stable, meshfree, nodal integration method for nearly incompressible solids. Finite Elem Anal Des 51:81–85
7. Liu WK, Jun S, Zhang YF (1995) Reproducing kernel particle methods. Int J Numer Meth Fluids 20(6):1081–1106
8. Atluri SN, Zhu T (1998) A new meshless local Petrov-Galerkin (MLPG) approach in computational mechanics. Comput Mech 22(2):117–127
9. Wang JG, Liu GR (2002) A point interpolation meshless method based on radial basis functions. Int J Numer Meth Eng 54:1623–1648
10. Nguyen VP, Rabczuk T, Bordas S, Duflot M (2008) Meshless methods: a review and computer implementation aspects. Math Comput Simul 79(3):763–813
11. Liu GR, Gu YT (2001) A point interpolation method for two-dimensional solids. Int J Numer Meth Eng 50:937–951
12. Dinis LMJS, Jorge RMN, Belinha J (2007) Analysis of 3D solids using the natural neighbour radial point interpolation method. Comput Methods Appl Mech Eng 196(13–16):2009–2028
13. Sibson R (1981) A brief description of natural neighbor interpolation. In: Barnett V (ed) Interpreting multivariate data. Wiley, Chichester, pp 21–36
14. Boots BN (1986) Voronoï (Thiessen) polygons. Geo Books, Norwich
15. Preparata FP, Shamos MI (1985) Computational geometry—an introduction. Springer, New York
16. Okabe A, Boots BN, Sugihara K, Chiu SN (2000) Spatial tessellations: concepts and applications of Voronoï diagrams, 2nd edn. Wiley, Chichester
17. Lawson CL (1977) Software for C1 surface interpolation. In: Rice JR (ed) Mathematical software III, 3rd edn. Academic Press, New York

18. Watson DF (1992) Contouring: a guide to the analysis and display of spatial data. Pergamon Press, Oxford
19. Dinis LMJS, Jorge RMN, Belinha J (2007) Analysis of 3D solids using the natural neighbour radial point interpolation method. Comput Methods Appl Mech Eng 196(13–16):2009–2028
20. Dinis LMJS, Jorge RMN, Belinha J (2008) Analysis of plates and laminates using the natural neighbour radial point interpolation method. Eng Anal Bound Elem 32(3):267–279
21. Dinis LMJS, Jorge RMN, Belinha J (2008) The radial natural neighbour interpolators extended to elastoplasticity. In: Ferreira AJM, Kansa EJ, Fasshauer GE, Leitao VMA (eds) Progress on meshless methods. Springer, Netherlands, pp 175–198
22. Dinis LMJS, Jorge RMN, Belinha J (2009) The natural neighbour radial point interpolation method: dynamic applications. Eng Comput 26(8):911–949
23. Dinis LMJS, Jorge RMN, Belinha J (2009) Large deformation applications with the radial natural neighbours interpolators. Comput Modell Eng Sci 44(1):1–34
24. Dinis LMJS, Jorge RMN, Belinha J (2010) An unconstrained third-order plate theory applied to functionally graded plates using a meshless method. Mech Adv Mater Struct 17:1–26
25. Dinis LMJS, Jorge RMN, Belinha J (2010) Composite laminated plates: a 3D natural neighbour radial point interpolation method approach. J Sandwich Struct Mater 12(2):119–138
26. Dinis LMJS, Jorge RMN, Belinha J (2010) A 3D shell-like approach using a natural neighbour meshless method: isotropic and orthotropic thin structures. Compos Struct 92(5):1132–1142
27. Dinis LMJS, Jorge RMN, Belinha J (2011) The dynamic analysis of thin structures using a radial interpolator meshless method. In: Vasques CMA, Dias Rodrigues J (eds) Vibration and strucutural acoustics analysis. Springer, Netherlands, pp 1–20
28. Dinis LMJS, Jorge RMN, Belinha J (2011) Static and dynamic analysis of laminated plates based on an unconstrained third order theory and using a radial point interpolator meshless method. Comput Struct 89(19–20):1771–1784
29. Dinis LMJS, Jorge RMN, Belinha J (2011) A natural neighbour meshless method with a 3D shell-like approach in the dynamic analysis of thin 3D structures. Thin-Walled Struct 49(1):185–196
30. Belinha J, Jorge RMN, Dinis LMJS (2013) A meshless microscale bone tissue trabecular remodelling analysis considering a new anisotropic bone tissue material law. Comput Methods Biomech Biomed Eng 16(11):1170–1184
31. Belinha J, Jorge RMN, Dinis LMJS (2012) Bone tissue remodelling analysis considering a radial point interpolator meshless method. Eng Anal Boundary Elem 36(11):1660–1670
32. Zienkiewicz OC, Taylor RL (1994) The finite element method, 4th edn. McGraw-Hill, London
33. Moreira S, Belinha J, Dinis LMJS, Jorge RMN (2014) Analysis of laminated beams using the natural neighbour radial point interpolation method. Revista Internacional de Métodos Numéricos para Cálculo y Diseño en Ingeniería 2014. http://dx.doi.org/10.1016/j.rimni.2013.02.002
34. Bathe KJ (1996) Finite element procedures. Prentice-Hall, Englewood Cliffs
35. Babuška I, Banerjee U, Osborn JE, Zhang Q (2009) Effect of numerical integration on meshless methods. Comput Methods Appl Mech Eng 198(37–40):2886–2897
36. Beissel S, Belytschko T (1996) Nodal integration of the element-free Galerkin method. Comput Methods Appl Mech Eng 139(1–4):49–74
37. Dolbow J, Belytschko T (1999) Numerical integration of the Galerkin weak form in meshfree methods. Comput Mech 23:219–230
38. De S, Bathe KJ (2001) The method of finite spheres with improved numerical integration. Comput Struct 79(22–25):2183–2196
39. Chen JS, Yoon S, Wu CT (2002) Non-linear version of stabilized conforming nodal integration Galerkin mesh-free methods. Int J Numer Meth Eng 53(12):2587–2615
40. Dai KY, Liu GR, Han X, Li Y (2006) Inelastic analysis of 2D solids using a weak-form RPIM based on deformation theory. Comput Methods Appl Mech Eng 195:4179–4193

41. Liu GR, Zhang GY, Wang YY, Zhong ZH, Li GY, Han X (2007) A nodal integration technique for meshfree radial point interpolation method (NI-RPIM). Int J Solids Struct 44(11–12):3840–3860
42. Wang JG, Liu GR (2002) On the optimal shape parameters of radial basis functions used for 2-D meshless methods. Comput Methods Appl Mech Eng 191:2611–2630
43. Belinha J, Jorge RMN, Dinis LMJS (2013) The natural radial element method. Int J Numer Meth Eng 93(12):1286–1313
44. Belinha J, Jorge RMN, Dinis LMJS (2013) Composite laminated plate analysis using the natural radial element method. Compos Struct 103(1):50–67
45. Belinha J, Jorge RMN, Dinis LMJS (2013) Analysis of thick plates by the natural radial element method. Int J Mech Sci 76(1):33–48
46. Dolbow J, Belytschko T (1998) An introduction to programming the meshless element free Galerkin method. Arch Comput Mech 5(3):207–241
47. Liu GR, Gu YT (2005) An introduction to meshfree methods and their programming. Springer, Netherlands
48. Zienkiewicz OC, Taylor RL (1994) The finite element method, 4th edn. McGraw-Hill, London
49. Zhu T, Atluri SN (1998) A modified collocation method and a penalty formulation for enforcing the essential boundary conditions in the element free Galerkin method. Comput Mech 21:211–222
50. Belytschko T, Gu L, Lu YY (1994) Fracture and crack growth by element free Galerkin methods. Modell Simul Mater Sci Eng 2(3A):519–534
51. Lu YY, Belytschko T, Gu L (1994) A new implementation of the element free Galerkin method. Comput Methods Appl Mech Eng 113(3–4):397–414
52. Mukherjee YX, Mukherjee S (1997) On boundary conditions in the element-free Galerkin method. Comput Mech 19(4):264–270
53. Lu YY, Belytschko T, Tabbara M (1995) Element-free Galerkin method for wave propagation and dynamic fracture. Comput Methods Appl Mech Eng 126(1–2):131–153
54. Krongauz Y, Belytschko T (1996) Enforcement of essential boundary conditions in meshless approximations using finite elements. Comput Methods Appl Mech Eng 131(1–2):133–145
55. Hegen D (1996) Element-free Galerkin methods in combination with finite element approaches. Comput Methods Appl Mech Eng 135:143–166
56. Gavete L, Benito JJ, Falcón S, Ruiz A (2000) Penalty functions in constrained variational principles for element free Galerkin method. Eur J Mech A Solids 19(4):699–720

Chapter 4
Shape Functions

Abstract This chapter explicitly shows how to construct shape functions for meshless methods. The chapter starts with the introduction of the "support-domain" concept which permits to identify the field nodes contributing to the construction of the shape function. Afterwards, the most popular approximation function is presented: the moving least square (MLS) approximation function. The construction procedure is presented in detail as well as the most important numerical properties of the MLS approximation function. Additionally the importance of the weight function used in the construction of the MLS shape function is shown. Then, the radial point interpolation (RPI) functions are presented. Again, an exhaustive description of the RPI shape function construction is presented supported by examples and explicative algorithms. The most important numerical properties of the RPI shape function are demonstrated. In addition, it is shown the relevance of the radial basis function (RBF) used to construct the RPI shape function, as well as the influence of the RBF shape parameters on the final solution.

4.1 Introduction

In order to obtain a numerical solution of a physical phenomenon ruled by partial differential equations, first it is require to approximate the unknown field functions using trial functions.

In the Finite Element Method (FEM) shape functions are obtained using the stationary element based interpolation. Fixed sets of nodes, respecting the same quantity and relative nodal spatial configuration, are combined to form elements. Then, the FEM shape functions are created using interpolation techniques based on polynomial series [1, 2].

In meshless methods the problem domain is not discretized in elements as in the FEM. Generally, the problem domain within meshless methods is discretized in a

J. Belinha, *Meshless Methods in Biomechanics*, Lecture Notes in Computational Vision and Biomechanics 16, DOI: 10.1007/978-3-319-06400-0_4,

nodal mesh, which can follow a regular or irregular distribution. In meshless methods the absence of elements, which pre-establish the nodal connectivity in the FEM, requires the application of an interpolation or approximation technique, based on a moving local nodal domain, to permit the construction of the meshless shape function for the approximation of the field variable.

In this chapter two distinct meshless shape function construction techniques are studied in detail: an approximation shape function approach and an interpolation shape function approach. Both presented shape functions techniques are designated as locally supported, since the shape functions are constructed for an arbitrary interest point $\boldsymbol{x}_I$ using only a small set of field nodes spatially localized in the vicinity of $\boldsymbol{x}_I$. This small set of field nodes in the vicinity of $\boldsymbol{x}_I$ is called the support-domain of the shape function. Generally, the support-domain of the shape function constructed for an arbitrary interest point $\boldsymbol{x}_I$ is coincident with the influence-domain of $\boldsymbol{x}_I$. The constructed shape function assumes non-zero values inside the support-domain and is null outside the support-domain.

The construction and development of shape functions assume great importance in meshless methods [3], since the shape functions construction methodology should be able to use only the nodes discretizing the domain without the need of any pre-established mesh providing the nodal connectivity. The shape functions construction methodology should also be computationally efficient, with the purpose of being a proficient FEM alternative, and capable to deal effortlessly with random nodal distributions, in order to solve practical engineering problems. Additionally the methodology should be: numerically stable; present a certain order of consistency; compactly supported; compatible; and, if possible, satisfy the Kronecker delta property [4, 5].

In the following sections the moving least squares approximation (MLS) methodology and the radial point interpolation (RPI) technique for the construction of meshless shape functions are described in detail. Presently, the MLS and the RPI are among the most popular techniques used to construct meshless shape functions. Nonetheless, other appropriate methodologies to construct meshless shape functions are available in the literature and extensively described in [4, 5].

4.2 Support-Domain

Most of the procedures used to construct the meshless shape function apply the support-domain concept. The shape function support-domain can be defined as the set of field nodes that directly contribute to the construction of the shape function. Usually the shape function support-domain of an interest point is coincident with the interest point influence-domain.

The size and shape of the support-domain should be carefully selected, since the accuracy of the approximation and the computational efficiency of the shape function construction strongly depends on this parameter. To illustrate this last remark, consider the two-dimensional domain $\Omega \subset \mathbb{R}^2$ discretized with a nodal set

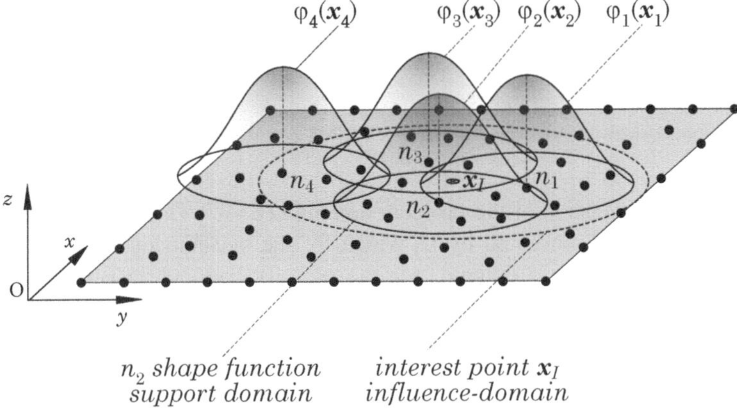

Fig. 4.1 Schematic representation of a generic influence-domain

$X = \{x_1, x_2, \ldots, x_N\} \in \Omega \wedge x_i \in \mathbb{R}^2$ represented in Fig. 4.1. As indicated in Fig. 4.1 the influence-domain of a generic interest point x_I contains several field nodes. It is possible to observe that compared with the influence-domain of x_I, the support-domain of the shape functions is much smaller. This size discrepancy reduces the computational efficiency of the shape function construction procedure. Notice that the contribution of node x_4 for the shape functions of interest point x_I is zero, $\varphi_4(x_I) = 0$. In opposition the other nodes in the vicinity of x_I, such are nodes x_1, x_2 and x_3, contribute with non-zero values, $\varphi_1(x_I) \neq 0$, $\varphi_2(x_I) \neq 0$ and $\varphi_3(x_I) \neq 0$. If the size of the influence-domain was coincident with the size of the support-domain, all nodes belonging to the influence-domain of x_I would actively contribute for the construction of the shape function.

Similar to the influence-domain, usually the support-domain is centred in an interest point, which can be a node or an integration point, and it can assume several distinct geometric shapes and sizes. The most usual shapes are the circular (as in Fig. 4.1) and the rectangular shape.

It is possible to defined the support-domain size of an interest point x_I with the following expression,

$$d_s = \beta_s d_a \tag{4.1}$$

Being β_s a dimensionless parameter governing the size of the support-domain and d_a is the average nodal spacing inside the support-domain. The parameter β_s has to be pre-established by the program user, before the analysis. The value adopted for β_s depends on the used meshless shape function and it should be determined by preliminary numerical studies. Commonly benchmark examples, for which the exact solutions are known, are used to determine the optimal value of β_s. Previous works on the MLS shape functions and RPI shape functions indicate that β_s parameters ranging between $\beta_s = [2.0, 3.0]$ lead to stable and accurate solutions.

The average nodal spacing d_a of the n nodes inside the support-domain of $\boldsymbol{x}_I$ can be determined using a simple expression [5],

$$d_a = \frac{D^{\left(\frac{1}{d}\right)}}{n^{\left(\frac{1}{d}\right)} - 1} \tag{4.2}$$

where d is the dimension of the problem domain $\Omega \subset \mathbb{R}^d$. The physical size of the support-domain is represented by D, which for the one-dimensional case defines the length of the support-domain, for the two-dimensional case represents the support-domain area and for the three-dimensional case symbolizes the volume of the support-domain.

The NNRPIM does not require the determination of the support-domain. The nodal set that constitutes the influence-cell of interest point $\boldsymbol{x}_I$ is the same nodal set used to construct the RPI shape functions for the interest point $\boldsymbol{x}_I$.

4.3 Moving Least Squares

The moving least squares (MLS) approximation was developed by Lancaster and Salkauskas [6] to smoothly approximate scattered data. Due to the MLS simplicity and low computational effort, several meshless methods use the MLS approximation to construct the shape functions [7–10]. Additionally, the use of MLS shape functions permit to approximate smoothly and continuously the field variables along the entire discretized domain.

The MLS approximation is constructed using three components: a weight function with compact support associated to each interest point; a basis, which usually consists of polynomial functions, and; a set of coefficients dependent on the interest point position.

4.3.1 MLS Shape Functions

Consider the function space T on $\Omega \subset \mathbb{R}^d$. The finite dimensional function space $T_H \subset T$ which discretize the domain Ω is defined by,

$$T_H := p_k(\boldsymbol{x}) \tag{4.3}$$

where $p_k : \mathbb{R}^d \mapsto \mathbb{R}$ is defined in the space of polynomials of degree less than k. The set of N nodes discretizing the space domain is defined by $\boldsymbol{X} = \{\boldsymbol{x}_1, \boldsymbol{x}_2, \ldots, \boldsymbol{x}_N\} \in \Omega \wedge \boldsymbol{x}_i \in \mathbb{R}^d$. The mesh density of $\boldsymbol{X}$ is identified by,

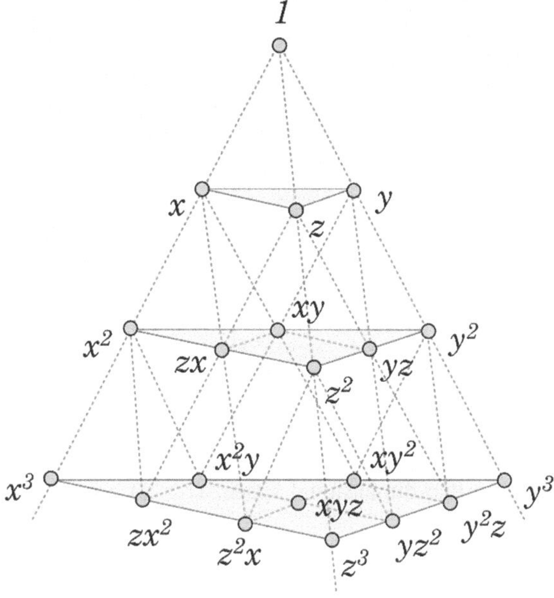

Fig. 4.2 Triangle of Pascal for polynomial monomials

$$h = \min\left\|\boldsymbol{x}_j - \boldsymbol{x}_i\right\|, \forall\{i,j\} \in \mathbb{N} : \{i,j\} \leq N \wedge i \neq j \tag{4.4}$$

where $\|\cdot\|$ is the L^2-*norm*, i.e., the Euclidean norm,

$$\|\boldsymbol{x}\| = \left(\sum_{i=1}^{d} |x_i|^2\right)^{1/2} \tag{4.5}$$

being d the space dimension. Consider now a continuous scalar function $u(\boldsymbol{x})$, being $u \in T$. It is possible to define for an interest point $\boldsymbol{x}_I \in \mathbb{R}^d$, not necessarily coincident with $\boldsymbol{X}$, the MLS approximation of $u(\boldsymbol{x}_I)$ as,

$$u^h(\boldsymbol{x}_I) = \sum_{i=1}^{m} p_i(\boldsymbol{x}_I) b_i(\boldsymbol{x}_I) = \boldsymbol{p}(\boldsymbol{x}_I)^{\mathrm{T}} \boldsymbol{b}(\boldsymbol{x}_I) \tag{4.6}$$

being $b_i(\boldsymbol{x}_I)$ the non-constant coefficients of $p_i(\boldsymbol{x}_I)$,

$$\boldsymbol{b}(\boldsymbol{x}_I)^{\mathrm{T}} = \{b_1(\boldsymbol{x}_I) \quad b_2(\boldsymbol{x}_I) \quad \ldots \quad b_m(\boldsymbol{x}_I)\} \tag{4.7}$$

The monomials of the polynomial basis are defined by $p_i(\boldsymbol{x}_I)$ and m is the basis monomial number. The polynomial basis $\boldsymbol{p}(\boldsymbol{x}_I)$ can be constructed using monomials from the triangle of Pascal, Fig. 4.2.

Therefore, for a one-dimensional space, considering an interest point $\boldsymbol{x}_I = \{x_I\}$, a quadratic polynomial basis is defined as,

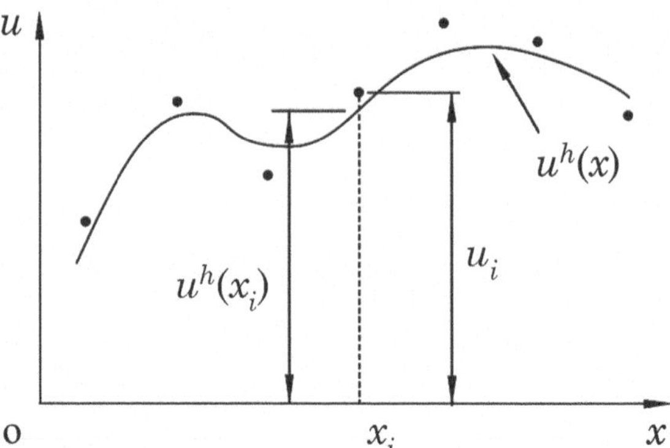

Fig. 4.3 The discrete nodal parameters $u(\boldsymbol{x}_i)$ and the MLS approximation function $u^h(\boldsymbol{x})$

$$\boldsymbol{p}(\boldsymbol{x}_I)^{\mathrm{T}} = \{\, 1 \quad x_I \quad x_I^2 \,\}, \quad m = 3 \tag{4.8}$$

For a two-dimensional space, $\boldsymbol{x}_I = \{x_i\,,\, y_I\}$, the quadratic polynomial basis is obtained,

$$\boldsymbol{p}(\boldsymbol{x}_I)^{\mathrm{T}} = \{1 \quad x_I \quad y_I \quad x_I^2 \quad x_I y_I \quad y_I^2\}, \quad m = 6 \tag{4.9}$$

and for a three-dimensional space, $\boldsymbol{x}_I = \{x_I\,,\, y_I\,,\, z_I\}$,

$$\boldsymbol{p}(\boldsymbol{x}_I)^{\mathrm{T}} = \{1 \quad x_I \quad y_I \quad z_I \quad x_I^2 \quad x_I y_I \quad y_I^2 \quad y_I z_I \quad z_I^2 \quad z_I x_I\}, \quad m = 10 \tag{4.10}$$

In order to improve the MLS approximation function performance, Eq. (4.6) can be enriched with additional functions [11]. This enrichment technique permit to capture with a higher accuracy the stress fields in the vicinity of crack tips and in the interface of distinct materials. In this book only pure polynomial basis are considered.

Ideally the MLS approximation function $u^h(\boldsymbol{x}_i)$ should match the continuous scalar function $u(\boldsymbol{x})$, however generally $u^h(\boldsymbol{x}) \neq u(\boldsymbol{x})$. This feature is represented in Fig. 4.3. Usually the MLS approximation function it is not capable to achieve the discrete values of $u(\boldsymbol{x})$, since the number of nodes n used in the MLS approximation is generally much larger than m, the number of unknowns coefficients of $\boldsymbol{b}(\boldsymbol{x}_I)$.

In order to adjust the approximation function $u^h(\boldsymbol{x}_i)$ to the n discrete nodal values $u(\boldsymbol{x}_i) = u_i$ inside the influence-domain of interest point $\boldsymbol{x}_I$, the following weighted residual functional is established,

$$J = \sum_{i=1}^{n} W(\boldsymbol{x}_i - \boldsymbol{x}_I)\left[u^h(\boldsymbol{x}_i) - u(\boldsymbol{x}_i)\right]^2 \tag{4.11}$$

being $W(\boldsymbol{x}_i - \boldsymbol{x}_I)$ the weight function, which is presented in detail in Sect. 4.3.2. The inclusion of the weight function permits to attribute distinct weights to the several nodes inside the support-domain: higher weights to nodes near $\boldsymbol{x}_I$ and; lower weights to nodes far from $\boldsymbol{x}_I$. The weight function also allows nodes to smoothly enter or leave the support-domain as a result of the $\boldsymbol{x}_I$ spatial variation. This feature permits to the MLS shape functions to satisfy the compatibility condition. The discrete quadratic norm represented in Eq. (4.11) can be developed using Eq. (4.6),

$$J = \sum_{i=1}^{n} W(\boldsymbol{x}_i - \boldsymbol{x}_I)\left[\boldsymbol{p}(\boldsymbol{x}_i)^{\mathrm{T}}\boldsymbol{b}(\boldsymbol{x}_I) - u_i\right]^2 \tag{4.12}$$

The non-constant coefficients $\boldsymbol{b}(\boldsymbol{x}_I)$ can be obtained minimizing the L_2*-norm* from Eq. (4.12) in respect to $\boldsymbol{b}(\boldsymbol{x}_I)$,

$$\frac{\partial J}{\partial \boldsymbol{b}(\boldsymbol{x}_I)} = \frac{\partial\left(\sum_{i=1}^{n} W(\boldsymbol{x}_i - \boldsymbol{x}_I)\left[\boldsymbol{p}(\boldsymbol{x}_i)^{\mathrm{T}}\boldsymbol{b}(\boldsymbol{x}_I) - u_i\right]^2\right)}{\partial \boldsymbol{b}(\boldsymbol{x}_I)} = 0 \tag{4.13}$$

developing,

$$\frac{\partial\left(\sum_{i=1}^{n} W(\boldsymbol{x}_i - \boldsymbol{x}_I)\left[\left(\boldsymbol{p}(\boldsymbol{x}_i)^{\mathrm{T}}\boldsymbol{b}(\boldsymbol{x}_I)\right)\left(\boldsymbol{p}(\boldsymbol{x}_i)^{\mathrm{T}}\boldsymbol{b}(\boldsymbol{x}_I)\right) - 2\boldsymbol{p}(\boldsymbol{x}_i)^{\mathrm{T}}\boldsymbol{b}(\boldsymbol{x}_I)u_i + u_i^2\right]\right)}{\partial \boldsymbol{b}(\boldsymbol{x}_I)} = 0 \tag{4.14}$$

being $\boldsymbol{p}(\boldsymbol{x}_i)^{\mathrm{T}}\boldsymbol{b}(\boldsymbol{x}_I)$ a scalar the vector relative position in Eq. (4.14) can be changed to $\boldsymbol{b}(\boldsymbol{x}_I)^{\mathrm{T}}\boldsymbol{p}(\boldsymbol{x}_i)$,

$$\frac{\partial\left(\sum_{i=1}^{n} W(\boldsymbol{x}_i - \boldsymbol{x}_I)\left[\left(\boldsymbol{b}(\boldsymbol{x}_I)^{\mathrm{T}}\boldsymbol{p}(\boldsymbol{x}_i)\right)\left(\boldsymbol{p}(\boldsymbol{x}_i)^{\mathrm{T}}\boldsymbol{b}(\boldsymbol{x}_I)\right) - 2\,\boldsymbol{b}(\boldsymbol{x}_I)^{\mathrm{T}}\boldsymbol{p}(\boldsymbol{x}_i)\,u_i + u_i^2\right]\right)}{\partial \boldsymbol{b}(\boldsymbol{x}_I)} = 0 \tag{4.15}$$

which leads to,

$$\sum_{i=1}^{n} W(\boldsymbol{x}_i - \boldsymbol{x}_I)\left[2\boldsymbol{p}(\boldsymbol{x}_i)\boldsymbol{p}(\boldsymbol{x}_i)^{\mathrm{T}}\boldsymbol{b}(\boldsymbol{x}_I) - 2\boldsymbol{p}(\boldsymbol{x}_i)\,u_i\right] = 0 \tag{4.16}$$

Notice that the scalar numerator 2 can vanish and the weight functions can be distributed, permitting the write the following expression,

$$\sum_{i=1}^{n} W(\boldsymbol{x}_i - \boldsymbol{x}_I)\boldsymbol{p}(\boldsymbol{x}_i)\boldsymbol{p}(\boldsymbol{x}_i)^{\mathrm{T}}\boldsymbol{b}(\boldsymbol{x}_I) - \sum_{i=1}^{n} W(\boldsymbol{x}_i - \boldsymbol{x}_I)\boldsymbol{p}(\boldsymbol{x}_i)^{\mathrm{T}}u_i = 0 \tag{4.17}$$

since $\boldsymbol{b}(\boldsymbol{x}_I)$ does not depend on x_i,

$$\left(\sum_{i=1}^{n} W(\boldsymbol{x}_i - \boldsymbol{x}_I)\boldsymbol{p}(\boldsymbol{x}_i)\boldsymbol{p}(\boldsymbol{x}_i)^{\mathrm{T}}\right)\boldsymbol{b}(\boldsymbol{x}_I) = \sum_{i=1}^{n} W(\boldsymbol{x}_i - \boldsymbol{x}_I)\boldsymbol{p}(\boldsymbol{x}_i)^{\mathrm{T}} u_i \tag{4.18}$$

Leading to the following linear system of equations,

$$\boldsymbol{A}(\boldsymbol{x}_I)\boldsymbol{b}(\boldsymbol{x}_I) = \boldsymbol{B}(\boldsymbol{x}_I)\boldsymbol{u}_s \tag{4.19}$$

being $\boldsymbol{u}_s$ the vector with the field function nodal parameters for each node inside the support-domain of the MLS shape function,

$$\boldsymbol{u}_s^{\mathrm{T}} = \{ u_1 \quad u_2 \quad \ldots \quad u_n \} \tag{4.20}$$

The weighted moment matrix $\boldsymbol{A}(\boldsymbol{x}_I)$ can be defined as,

$$\boldsymbol{A}(\boldsymbol{x}_I) = \sum_{i=1}^{n} W(\boldsymbol{x}_i - \boldsymbol{x}_I)\boldsymbol{p}(\boldsymbol{x}_i)\boldsymbol{p}(\boldsymbol{x}_i)^{\mathrm{T}} \tag{4.21}$$

Considering the three-dimensional space, $\boldsymbol{x} = \{x, y, z\}$, and using a linear polynomial basis, $\boldsymbol{p}(\boldsymbol{x})^{\mathrm{T}} = \{ 1 \quad x \quad y \quad z \}$ with $m = 4$, the moment matrix $\boldsymbol{A}(\boldsymbol{x}_I)$ can be explicitly defined as,

$$\begin{aligned}\boldsymbol{A}(\boldsymbol{x}_I) &= \sum_{i=1}^{n} W(\boldsymbol{x}_i - \boldsymbol{x}_I)\begin{bmatrix} 1 \\ x_i \\ y_i \\ z_i \end{bmatrix}[1 \quad x_i \quad y_i \quad z_i] \\ &= W(\boldsymbol{x}_1 - \boldsymbol{x}_I)\begin{bmatrix} 1 & x_1 & y_1 & z_1 \\ x_1 & x_1^2 & x_1 y_1 & x_1 z_1 \\ y_1 & y_1 x_1 & y_1^2 & y_1 z_1 \\ z_1 & z_1 x_1 & z_1 y_1 & z_1^2 \end{bmatrix} + \cdots + W(\boldsymbol{x}_n - \boldsymbol{x}_1)\begin{bmatrix} 1 & x_n & y_n & z_n \\ x_n & x_n^2 & x_n y_n & x_n z_n \\ y_n & y_n x_n & y_n^2 & y_n z_n \\ z_n & z_n x_n & z_n y_n & z_n^2 \end{bmatrix}\end{aligned} \tag{4.22}$$

The weighted polynomial matrix, $\boldsymbol{B}(\boldsymbol{x}_I)$, can be defined as,

$$\boldsymbol{B}(\boldsymbol{x}_I) = [\, W(\boldsymbol{x}_1 - \boldsymbol{x}_I)\boldsymbol{p}(\boldsymbol{x}_1) \quad W(\boldsymbol{x}_2 - \boldsymbol{x}_I)\boldsymbol{p}(\boldsymbol{x}_2) \quad \cdots \quad W(\boldsymbol{x}_n - \boldsymbol{x}_I)\boldsymbol{p}(\boldsymbol{x}_n)\,] \tag{4.23}$$

Therefore, considering the same previous spatial conditions and polynomial basis, it is possible to defined $\boldsymbol{B}(\boldsymbol{x}_I)$ explicitly as,

$$\boldsymbol{B}(\boldsymbol{x}_I) = \left[W(\boldsymbol{x}_1 - \boldsymbol{x}_I)\begin{bmatrix} 1 \\ x_1 \\ y_1 \\ z_1 \end{bmatrix} \quad W(\boldsymbol{x}_2 - \boldsymbol{x}_I)\begin{bmatrix} 1 \\ x_2 \\ y_2 \\ z_2 \end{bmatrix} \quad \ldots \quad W(\boldsymbol{x}_n - \boldsymbol{x}_I)\begin{bmatrix} 1 \\ x_n \\ y_n \\ z_n \end{bmatrix} \right] \tag{4.24}$$

The non-constant coefficients $\boldsymbol{b}(\boldsymbol{x}_I)$ can be obtained with Eq. (4.19),

$$\boldsymbol{b}(\boldsymbol{x}_I) = \boldsymbol{A}(\boldsymbol{x}_I)^{-1}\boldsymbol{B}(\boldsymbol{x}_I)\boldsymbol{u}_s \tag{4.25}$$

By back substitution in Eq. (4.6) it is possible to write,

$$u^h(\boldsymbol{x}_I) = \boldsymbol{p}(\boldsymbol{x}_I)\left(\boldsymbol{A}(\boldsymbol{x}_I)^{-1}\boldsymbol{B}(\boldsymbol{x}_I)\boldsymbol{u}_s\right) \tag{4.26}$$

Recovering the summation which originates the $\boldsymbol{B}(\boldsymbol{x}_I)\boldsymbol{u}_s$ operation, Eq. (4.26) can be represented as,

$$u^h(\boldsymbol{x}_I) = \boldsymbol{p}(\boldsymbol{x}_I)^{\mathrm{T}}\boldsymbol{A}(\boldsymbol{x}_I)^{-1}\sum_{i=1}^{n} W(\boldsymbol{x}_i - \boldsymbol{x}_I)\boldsymbol{p}(\boldsymbol{x}_i)^{\mathrm{T}}u_i \tag{4.27}$$

Notice that interest point polynomial vector $\boldsymbol{p}(\boldsymbol{x}_I)$ and the moment matrix $\boldsymbol{A}(\boldsymbol{x}_I)$ can be moved inside the summation,

$$u^h(\boldsymbol{x}_I) = \sum_{i=1}^{n} \boldsymbol{p}(\boldsymbol{x}_I)^{\mathrm{T}}\boldsymbol{A}(\boldsymbol{x}_I)^{-1}W(\boldsymbol{x}_i - \boldsymbol{x}_I)\boldsymbol{p}(\boldsymbol{x}_i)^{\mathrm{T}}u_i \tag{4.28}$$

Since the field variable for an interest point $\boldsymbol{x}_I$ is approximated using shape function values obtained at the nodes inside the support-domain of the interest point $\boldsymbol{x}_I$,

$$u^h(\boldsymbol{x}_I) = \sum_{i=1}^{n} \varphi_i(\boldsymbol{x}_I)u_i = \boldsymbol{\varphi}(\boldsymbol{x}_I)^{\mathrm{T}}\boldsymbol{u}_s \tag{4.29}$$

It is possible to recognize the MLS shape function $\varphi_i(\boldsymbol{x}_I)$,

$$\varphi_i(\boldsymbol{x}_I) = \boldsymbol{p}(\boldsymbol{x}_I)^{\mathrm{T}}\boldsymbol{A}(\boldsymbol{x}_I)^{-1}W(\boldsymbol{x}_i - \boldsymbol{x}_I)\boldsymbol{p}(\boldsymbol{x}_i)^{\mathrm{T}} \tag{4.30}$$

being $\varphi_i(\boldsymbol{x}_I)$ the shape function value of interest point $\boldsymbol{x}_I$ on the ith node. $\varphi_i(\boldsymbol{x}_I)$ is obtained considering the nodes inside the support-domain of interest point $\boldsymbol{x}_I$. The MLS shape function vector for the n nodes inside the support-domain of $\boldsymbol{x}_I$ is defined as,

$$\boldsymbol{\varphi}(\boldsymbol{x}_I)^{\mathrm{T}} = \{\, \varphi_1(\boldsymbol{x}_I) \quad \varphi_2(\boldsymbol{x}_I) \quad \dots \quad \varphi_n(\boldsymbol{x}_I) \,\} = \boldsymbol{p}(\boldsymbol{x}_I)^{\mathrm{T}}\boldsymbol{A}(\boldsymbol{x}_I)^{-1}\boldsymbol{B}(\boldsymbol{x}_I) \tag{4.31}$$

Notice that the approximation function $u^h(\boldsymbol{x})$ is defined for a specific interest point $\boldsymbol{x}_I$ possessing a particular support-domain with n nodes. Consider two distinct interest points $\{\boldsymbol{x}_I, \boldsymbol{x}_J\} \in \Omega \wedge \boldsymbol{x}_i \in \mathbb{R}^d$, being $\boldsymbol{x}_J \neq \boldsymbol{x}_I$. Both interest points shape functions possess the same support-domain nodal set $\boldsymbol{N}_I = \boldsymbol{N}_J = \{n_1, n_2, \dots, n_n\} \subset \boldsymbol{N}$, being $\boldsymbol{N}$ the complete nodal set discretizing the problem domain. Although

$\boldsymbol{N}_I = \boldsymbol{N}_J$, since the non-constant coefficient depend on the interest point spatial position, Eq. (4.25) will lead to $\boldsymbol{b}(\boldsymbol{x}_J) \neq \boldsymbol{b}(\boldsymbol{x}_I)$. Therefore, the non-constant coefficients $\boldsymbol{b}(\boldsymbol{x})$ have to be computed for each interest point on the problem domain.

Being the non-constant coefficients $\boldsymbol{b}(\boldsymbol{x})$ dependent of the interest point $\boldsymbol{x}_I$ spatial location and the weight function bell-shape, the MLS approximation shape function is able to move continuously along the discretized domain. This global continuity is an asset when the equations system of the analysed problem is obtained using a global weak-formulation.

In order to guarantee the non-singularity of matrix $\boldsymbol{A}$, defined in Eq. (4.21), the number of nodes within the support-domain of the interest point $\boldsymbol{x}_I$, n, should be much larger than the monomial number of the polynomial basis, m. The $n \gg m$ requirement assures that matrix $\boldsymbol{A}$ is well-conditioned and invertible.

To obtain the approximation field function derivatives from the field variables, Eq. (4.29), it is necessary to determine the shape functions derivatives. The approximation field function partial derivatives with respect to ξ (which is a generic variable representing x, y or z) are obtained with,

$$\frac{\partial u^h(\boldsymbol{x}_I)}{\partial \xi} = \sum_{i=1}^{n} \frac{\partial \varphi_i(\boldsymbol{x}_I)}{\partial \xi} u_i = \boldsymbol{\varphi}(\boldsymbol{x}_I)_{,\xi}^{\mathrm{T}} \boldsymbol{u}_s \tag{4.32}$$

and the spatial partial derivatives with respect to ξ of the MLS shape function are obtained with,

$$\begin{aligned}\boldsymbol{\varphi}(\boldsymbol{x}_I)_{,\xi} &= \left(\boldsymbol{p}(\boldsymbol{x}_I)^{\mathrm{T}}\boldsymbol{A}(\boldsymbol{x}_I)^{-1}\boldsymbol{B}(\boldsymbol{x}_I)\right)_{,\xi} \\ &= \left(\boldsymbol{p}(\boldsymbol{x}_I)_{,\xi}^{\mathrm{T}}\boldsymbol{A}(\boldsymbol{x}_I)^{-1}\boldsymbol{B}(\boldsymbol{x}_I) + \boldsymbol{p}(\boldsymbol{x}_I)^{\mathrm{T}}\boldsymbol{A}(\boldsymbol{x}_I)_{,\xi}^{-1}\boldsymbol{B}(\boldsymbol{x}_I) + \boldsymbol{p}(\boldsymbol{x}_I)^{\mathrm{T}}\boldsymbol{A}(\boldsymbol{x}_I)^{-1}\boldsymbol{B}(\boldsymbol{x}_I)_{,\xi}\right)\end{aligned} \tag{4.33}$$

Considering again the three-dimensional space, $\boldsymbol{x} = \{x, y, z\}$, and using a linear polynomial basis, $\boldsymbol{p}(\boldsymbol{x})^{\mathrm{T}} = \{1 \quad x \quad y \quad z\}$ with $m = 4$. The polynomial basis partial derivatives with respect to x, y and z are defined as,

$$\begin{aligned}\boldsymbol{p}(\boldsymbol{x}_I)_{,x} &= \{0 \quad 1 \quad 0 \quad 0\}^{\mathrm{T}} \\ \boldsymbol{p}(\boldsymbol{x}_I)_{,y} &= \{0 \quad 0 \quad 1 \quad 0\}^{\mathrm{T}} \\ \boldsymbol{p}(\boldsymbol{x}_I)_{,z} &= \{0 \quad 0 \quad 0 \quad 1\}^{\mathrm{T}}\end{aligned} \tag{4.34}$$

The partial derivative of matrix $\boldsymbol{B}(\boldsymbol{x}_I)$ with respect to ξ is obtained with,

$$\boldsymbol{B}(\boldsymbol{x}_I)_{,\xi} = \left[\tfrac{\partial W(\boldsymbol{x}_1 - \boldsymbol{x}_I)}{\partial \xi}\boldsymbol{p}(\boldsymbol{x}_1) \quad \tfrac{\partial W(\boldsymbol{x}_2 - \boldsymbol{x}_I)}{\partial \xi}\boldsymbol{p}(\boldsymbol{x}_2) \quad \cdots \quad \tfrac{\partial W(\boldsymbol{x}_n - \boldsymbol{x}_I)}{\partial \xi}\boldsymbol{p}(\boldsymbol{x}_n)\right] \tag{4.35}$$

which can be explicitly defined as,

$$\boldsymbol{B}(\boldsymbol{x}_I)_{,\xi} = \begin{bmatrix} \frac{\partial W(\boldsymbol{x}_1-\boldsymbol{x}_I)}{\partial \xi}\begin{bmatrix}1\\x_1\\y_1\\z_1\end{bmatrix} & \frac{\partial W(\boldsymbol{x}_2-\boldsymbol{x}_I)}{\partial \xi}\begin{bmatrix}1\\x_2\\y_2\\z_2\end{bmatrix} & \cdots & \frac{\partial W(\boldsymbol{x}_n-\boldsymbol{x}_I)}{\partial \xi}\begin{bmatrix}1\\x_n\\y_n\\z_n\end{bmatrix}\end{bmatrix} \tag{4.36}$$

and $\boldsymbol{A}(\boldsymbol{x}_I)_{,\xi}^{-1}$ is computed as,

$$\boldsymbol{A}(\boldsymbol{x}_I)_{,\xi}^{-1} = -\boldsymbol{A}(\boldsymbol{x}_I)^{-1}\boldsymbol{A}(\boldsymbol{x}_I)_{,\xi}\boldsymbol{A}(\boldsymbol{x}_I)^{-1} \tag{4.37}$$

being,

$$\boldsymbol{A}(\boldsymbol{x}_I)_{,\xi} = \sum_{i=1}^{n} \frac{\partial W(\boldsymbol{x}_i - \boldsymbol{x}_I)}{\partial \xi}\boldsymbol{p}(\boldsymbol{x}_i)\boldsymbol{p}(\boldsymbol{x}_i)^{\mathrm{T}} \tag{4.38}$$

The partial derivative of the moment matrix $\boldsymbol{A}(\boldsymbol{x}_I)$ can be explicitly defined as,

$$\begin{aligned}\boldsymbol{A}(\boldsymbol{x}_I)_{,\xi} &= \sum_{i=1}^{n} \frac{\partial W(\boldsymbol{x}_i - \boldsymbol{x}_I)}{\partial \xi}\begin{bmatrix}1\\x_i\\y_i\\z_i\end{bmatrix}[1 \quad x_i \quad y_i \quad z_i] \\ &= \frac{\partial W(\boldsymbol{x}_1 - \boldsymbol{x}_I)}{\partial \xi}\begin{bmatrix}1 & x_1 & y_1 & z_1\\ x_1 & x_1^2 & x_1y_1 & x_1z_1\\ y_1 & y_1x_1 & y_1^2 & y_1z_1\\ z_1 & z_1x_1 & z_1y_1 & z_1^2\end{bmatrix} + \cdots \\ &+ \frac{\partial W(\boldsymbol{x}_n - \boldsymbol{x}_I)}{\partial \xi}\begin{bmatrix}1 & x_n & y_n & z_n\\ x_n & x_n^2 & x_ny_n & x_nz_n\\ y_n & y_nx_n & y_n^2 & y_nz_n\\ z_n & z_nx_n & z_ny_n & z_n^2\end{bmatrix}\end{aligned} \tag{4.39}$$

Following a similar methodology, it is possible to obtain the following second order partial derivatives of the approximation field function with respect to η (which is another generic variable representing x, y or z),

$$\frac{\partial^2 u^h(\boldsymbol{x}_I)}{\partial \xi \partial \eta} = \sum_{i=1}^{n} \frac{\partial^2 \varphi_i(\boldsymbol{x}_I)}{\partial \xi \partial \eta} u_i = \boldsymbol{\varphi}(\boldsymbol{x}_I)_{,\xi\eta}^{\mathrm{T}}\boldsymbol{u}_s \tag{4.40}$$

and the spatial second order partial derivatives of the MLS shape function with respect to ξ and η are obtained with,

$$
\begin{aligned}
\boldsymbol{\varphi}(\boldsymbol{x}_I)_{,\xi\eta} &= \left(\boldsymbol{p}(\boldsymbol{x}_I)_{,\xi}^{\mathrm{T}}\boldsymbol{A}(\boldsymbol{x}_I)^{-1}\boldsymbol{B}(\boldsymbol{x}_I) + \boldsymbol{p}(\boldsymbol{x}_I)^{\mathrm{T}}\boldsymbol{A}(\boldsymbol{x}_I)_{,\xi}^{-1}\boldsymbol{B}(\boldsymbol{x}_I) + \boldsymbol{p}(\boldsymbol{x}_I)^{\mathrm{T}}\boldsymbol{A}(\boldsymbol{x}_I)^{-1}\boldsymbol{B}(\boldsymbol{x}_I)_{,\xi}\right)_{,\eta} \\
&= \left(\boldsymbol{p}(\boldsymbol{x}_I)_{,\xi\eta}^{\mathrm{T}}\boldsymbol{A}(\boldsymbol{x}_I)^{-1}\boldsymbol{B}(\boldsymbol{x}_I) + \boldsymbol{p}(\boldsymbol{x}_I)_{,\xi}^{\mathrm{T}}\boldsymbol{A}(\boldsymbol{x}_I)_{,\eta}^{-1}\boldsymbol{B}(\boldsymbol{x}_I) + \boldsymbol{p}(\boldsymbol{x}_I)_{,\xi}^{\mathrm{T}}\boldsymbol{A}(\boldsymbol{x}_I)^{-1}\boldsymbol{B}(\boldsymbol{x}_I)_{,\eta}\right. \\
&\quad + \boldsymbol{p}(\boldsymbol{x}_I)_{,\eta}^{\mathrm{T}}\boldsymbol{A}(\boldsymbol{x}_I)_{,\xi}^{-1}\boldsymbol{B}(\boldsymbol{x}_I) + \boldsymbol{p}(\boldsymbol{x}_I)^{\mathrm{T}}\boldsymbol{A}(\boldsymbol{x}_I)_{,\xi\eta}^{-1}\boldsymbol{B}(\boldsymbol{x}_I) + \boldsymbol{p}(\boldsymbol{x}_I)^{\mathrm{T}}\boldsymbol{A}(\boldsymbol{x}_I)_{,\xi}^{-1}\boldsymbol{B}(\boldsymbol{x}_I)_{,\eta} \\
&\quad \left. + \boldsymbol{p}(\boldsymbol{x}_I)_{,\eta}^{\mathrm{T}}\boldsymbol{A}(\boldsymbol{x}_I)^{-1}\boldsymbol{B}(\boldsymbol{x}_I)_{,\xi} + \boldsymbol{p}(\boldsymbol{x}_I)^{\mathrm{T}}\boldsymbol{A}(\boldsymbol{x}_I)_{,\eta}^{-1}\boldsymbol{B}(\boldsymbol{x}_I)_{,\xi} + \boldsymbol{p}(\boldsymbol{x}_I)^{\mathrm{T}}\boldsymbol{A}(\boldsymbol{x}_I)^{-1}\boldsymbol{B}(\boldsymbol{x}_I)_{,\xi\eta}\right)
\end{aligned}
\tag{4.41}
$$

Being,

$$
\begin{aligned}
\boldsymbol{A}(\boldsymbol{x}_I)_{,\xi\eta}^{-1} &= \left(-\boldsymbol{A}(\boldsymbol{x}_I)^{-1}\boldsymbol{A}(\boldsymbol{x}_I)_{,\xi}\boldsymbol{A}(\boldsymbol{x}_I)^{-1}\right)_{,\eta} \\
&= \left(-\boldsymbol{A}(\boldsymbol{x}_I)_{,\eta}^{-1}\boldsymbol{A}(\boldsymbol{x}_I)_{,\xi}\boldsymbol{A}(\boldsymbol{x}_I)^{-1} - \boldsymbol{A}(\boldsymbol{x}_I)^{-1}\boldsymbol{A}(\boldsymbol{x}_I)_{,\xi\eta}\boldsymbol{A}(\boldsymbol{x}_I)^{-1} - \boldsymbol{A}(\boldsymbol{x}_I)^{-1}\boldsymbol{A}(\boldsymbol{x}_I)_{,\xi}\boldsymbol{A}(\boldsymbol{x}_I)_{,\eta}^{-1}\right)
\end{aligned}
\tag{4.42}
$$

and $\boldsymbol{A}(\boldsymbol{x}_I)_{,\xi n}$ obtained with,

$$
\boldsymbol{A}(\boldsymbol{x}_I)_{,\xi\eta} = \sum_{i=1}^{n} \frac{\partial^2 W(\boldsymbol{x}_i - \boldsymbol{x}_I)}{\partial\xi\partial\eta}\boldsymbol{p}(\boldsymbol{x}_i)\boldsymbol{p}(\boldsymbol{x}_i)^{\mathrm{T}} \tag{4.43}
$$

The second order partial derivative of matrix $\boldsymbol{B}(\boldsymbol{x}_I)$ with respect to ξ and η is obtained with,

$$
\boldsymbol{B}(\boldsymbol{x}_I)_{,\xi\eta} = \left[\frac{\partial^2 W(\boldsymbol{x}_1-\boldsymbol{x}_I)}{\partial\xi\partial\eta}\boldsymbol{p}(\boldsymbol{x}_1) \quad \frac{\partial^2 W(\boldsymbol{x}_2-\boldsymbol{x}_I)}{\partial\xi\partial\eta}\boldsymbol{p}(\boldsymbol{x}_2) \quad \cdots \quad \frac{\partial^2 W(\boldsymbol{x}_n-\boldsymbol{x}_I)}{\partial\xi\partial\eta}\boldsymbol{p}(\boldsymbol{x}_n)\right] \tag{4.44}
$$

Third, fourth and higher order partial derivatives of the MLS shape functions can be similarly obtained by repeated differentiations. It is perceptible that the continuity of the MLS shape function depends on the weight function continuity.

4.3.2 Weight Functions

In the construction of the MLS shape functions the weight function possesses a crucial importance. The weight function should be positive and bell-shaped, i.e., it should decrease in magnitude with the increase of the Euclidean norm between the interest point $\boldsymbol{x}_I$ and the ith node $\boldsymbol{x}_i$ inside the support-domain,

$$
d_i = \|\boldsymbol{x}_i - \boldsymbol{x}_I\| = \sqrt{(x_i - x_I)^2 + (y_i - y_I)^2 + (z_i - z_I)^2}, \quad \forall \boldsymbol{x}_i \in \boldsymbol{X}_I \tag{4.45}
$$

Being $\boldsymbol{X}_I = \{\boldsymbol{x}_1, \boldsymbol{x}_2, \ldots, \boldsymbol{x}_n\} \subset \boldsymbol{X}$ the support-domain of the interest point $\boldsymbol{x}_I$ and $\boldsymbol{X}$ the global nodal set discretizing the problem domain. Generally, the weight functions used to build the MSL approximation shape functions only depend on d_i,

$$W(\boldsymbol{x}_i - \boldsymbol{x}_I) = W(d_i) \tag{4.46}$$

and should satisfy the conditions:

i. $W(\boldsymbol{x}_i - \boldsymbol{x}_I) > 0, \ \forall \boldsymbol{x}_i \in \boldsymbol{X}_I$;
ii. $W(\boldsymbol{x}_i - \boldsymbol{x}_I) = 0, \ \forall \boldsymbol{x}_i \notin \boldsymbol{X}_I$;
iii. $W(\boldsymbol{x}_i - \boldsymbol{x}_I) > W(\boldsymbol{x}_j - \boldsymbol{x}_I), \ \forall \{\boldsymbol{x}_i, \boldsymbol{x}_j\} \in \boldsymbol{X}_I$, being $||\boldsymbol{x}_i - \boldsymbol{x}_I|| < ||\boldsymbol{x}_j - \boldsymbol{x}_I||$.

Notice that the first two conditions assures a compact support, leading to a banded system of equations. Additionally, the weight function should be bell-shaped to ensure the smooth inclusion and exclusion of nodes when the support-domain moves, conferring the compatibility property to the MLS shape function.

Equations (4.33) and (4.41) show that the continuity of the MLS shape function $\boldsymbol{\varphi}(\boldsymbol{x}_I)$ depends mainly the continuity of the weight function. Notice that if $W(\boldsymbol{x}_i - \boldsymbol{x}_I)$ is a C^{n_w}-function and $p(\boldsymbol{x})$ is a C^{n_p}-function, then the MLS shape function $\boldsymbol{\varphi}(\boldsymbol{x}_I)$ will be a C^{n_φ}-function, being $n_\varphi = \max\{n_w, n_p\}$. Usually, in the construction of the MLS shape function only linear or quadratic polynomial basis are used, therefore generally $n_w > n_p$. As long as the weight function is positive and continuous together with its derivatives up to the required degree, the choice of the weight functions is more or less arbitrary [12].

There are several weight functions available in the literature. In the first works on the EFGM [12] it were considered two types of weight functions. The exponential weight function,

$$W(d_i) = \begin{cases} \frac{e^{-(d_i/d_\alpha)^2} - e^{-(d_I/d_\alpha)^2}}{1 - e^{-(d_I/d_\alpha)^2}}, & d_i \leq d_I \\ 0, & d_i > d_I \end{cases} \tag{4.47}$$

with $d_\alpha = \alpha d_I$ and $1 \leq \alpha \leq 2$, and the conical weight function,

$$W(d_i) = W(s_i) = \begin{cases} 1 - s_i^2, & s_i \leq 1 \\ 0, & s_i > 1 \end{cases} \tag{4.48}$$

Being,

$$s_i = \frac{d_i}{d_I} = \frac{||\boldsymbol{x}_i - \boldsymbol{x}_I||}{d_I} \tag{4.49}$$

where d_I is the size of the influence-domain of the interest point $\boldsymbol{x}_I$, defined in a pre-processing phase or by the expression Eq. (4.50).

$$d_I = \max||\boldsymbol{x}_i - \boldsymbol{x}_I||, \quad \forall \boldsymbol{x}_i \in \boldsymbol{X}_I \tag{4.50}$$

However, since the exponential weight functions is computational demanding and the conical weight function presents a low continuity, other weight functions were proposed in several EFGM research works [13–16]. Presently the most popular weight functions used in the construction of the MLS shape function are the cubic spline weight function,

$$W(d_i) = W(s_i) = \begin{cases} \frac{2}{3} - 4s_i^2 + 4s_i^3, & s_i \leq 0.5 \\ \frac{4}{3} - 4s_i + 4s_i^2 - \frac{4}{3}s_i^3, & 0.5 < s_i \leq 1 \\ 0, & s_i > 1 \end{cases} \tag{4.51}$$

with a second order continuity, and the quartic spline weight function,

$$W(d_i) = W(s_i) = \begin{cases} 1 - 6s_i^2 + 8s_i^3 - 3s_i^4, & s_i \leq 1 \\ 0, & s_i > 1 \end{cases} \tag{4.52}$$

with a third order continuity. Notice that both weight functions are polynomials only dependent on the s_i radial variable. In the three-dimensional space s_i can be explicitly expressed as,

$$s_i = \frac{d_i}{d_I} = \frac{\left((x_i - x_I)^2 + (y_i - y_I)^2 + (z_i - z_I)^2\right)^{\frac{1}{2}}}{d_I} \tag{4.53}$$

The first order partial derivate with respect to a generic variable ξ, which can be x, y or z, of a s_i dependent polynomial term can be defined as,

$$\frac{\partial(a_n\, s_i^n)}{\partial \xi} = -\left(\frac{\xi_i - \xi_I}{d_i\, d_I}\right) n\, a_n\, s_i^{n-1} \tag{4.54}$$

Being $a_n \in \mathbb{R}$ a constant. Therefore, it is possible to present the first order partial derivate, with respect to ξ, of the cubic spline weight function,

$$\frac{\partial W(d_i)}{\partial \xi} = \frac{\partial W(s_i)}{\partial \xi} = -\left(\frac{\xi_i - \xi_I}{d_i d_I}\right) \begin{cases} -8s_i + 12s_i^2, & s_i \leq 0.5 \\ -4 + 8s_i - 4s_i^2, & 0.5 < s_i \leq 1 \\ 0, & s_i > 1 \end{cases} \tag{4.55}$$

In order to obtain the MLS shape function second order partial derivatives, Eq. (4.41), it is required to compute the weight function second order partial derivatives. Therefore, the second order partial derivative with respect to a standard variable ξ of a generic s_i dependent polynomial term of Eq. (4.55) can be obtained with,

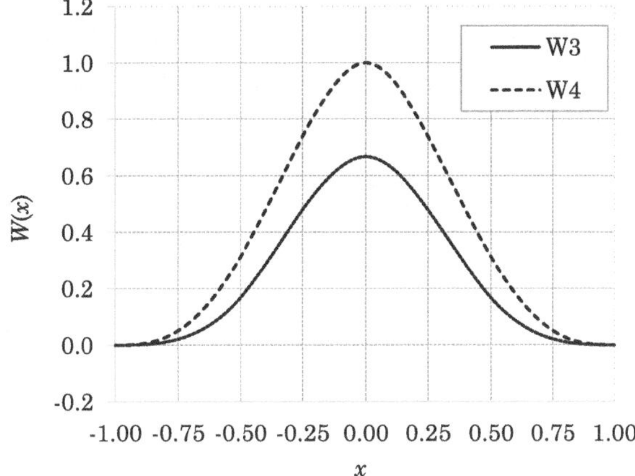

Fig. 4.4 1D weight functions. *W*3 cubic spline. *W*4 quartic spline

$$\frac{\partial^2\left(-\frac{\xi_i-\xi_I}{d_i\,d_I}a_n\,s_i^n\right)}{\partial\xi^2}=a_n\left(d_I^{-2}+(n-1)\left(\frac{\xi_i-\xi_I}{d_i\,d_I}\right)^2\right)s_i^{n-1} \tag{4.56}$$

The second order cross partial derivative with respect to variables ξ and η of the same s_i dependent polynomial term of Eq. (4.55) can be determined with,

$$\frac{\partial^2\left(-\frac{\xi_i-\xi_I}{d_i\,d_I}a_n\,s_i^n\right)}{\partial\xi\partial\eta}=a_n(n-1)\frac{(\xi_i-\xi_I)(\eta_i-\eta_I)}{d_i^2d_I^2}s_i^{n-1} \tag{4.57}$$

Both cubic and quartic spline weight functions are continuous in the entire support-domain, Fig. 4.4. However the cubic spline weight function does not present a smooth second order derivative and it is unable to provide a continuous third order derivative. The quartic spline weight function is numerically more efficient, it provides smooth and continuous first and second order derivatives. These features are presented in Figs. 4.5 and 4.6.

If a two-dimensional domain is considered the obtained cubic and quartic spline weight functions assume the bell-shaped surfaces indicated in Fig. 4.7. The respective first order derivatives are presented in Fig. 4.8 and the second order derivatives in Fig. 4.9. The second order derivative two-dimensional representation repeats the one-dimensional observation, the cubic spline weight function is not able to produce a smooth second order derivative surface, Fig. 4.10.

With the cubic and the quartic spline weight functions, the function itself and the respective first and the second order partial derivatives are exactly zero on the support-domain boundary. Therefore, both weight functions are capable to provide compatibility up to the second order. The quartic spline weight functions allow compatibility up to the third order.

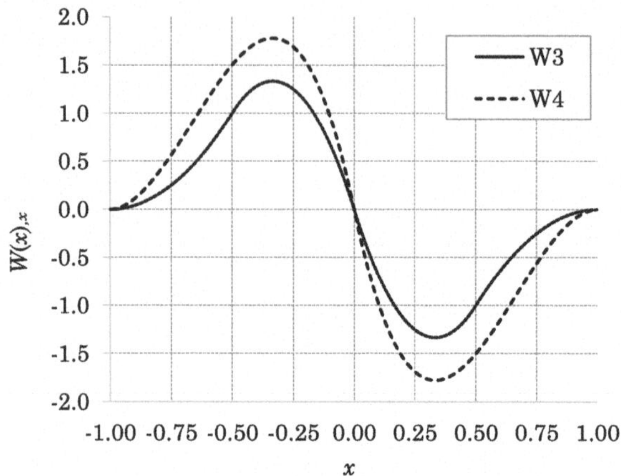

Fig. 4.5 1D weight functions first order derivatives. *W*3 cubic spline. *W*4 quartic spline

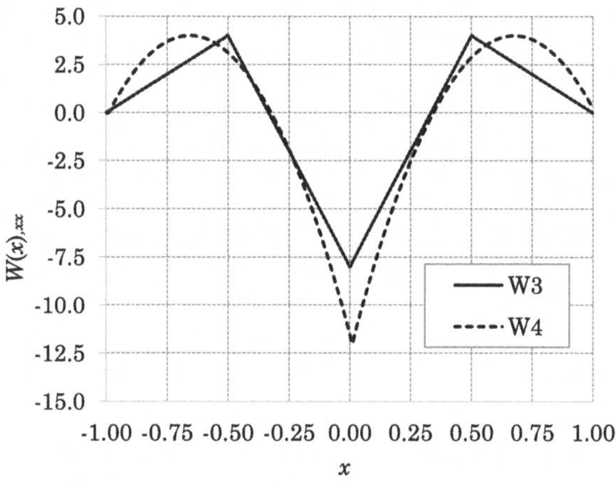

Fig. 4.6 1D weight functions second order derivatives. *W*3 cubic spline. *W*4 quartic spline

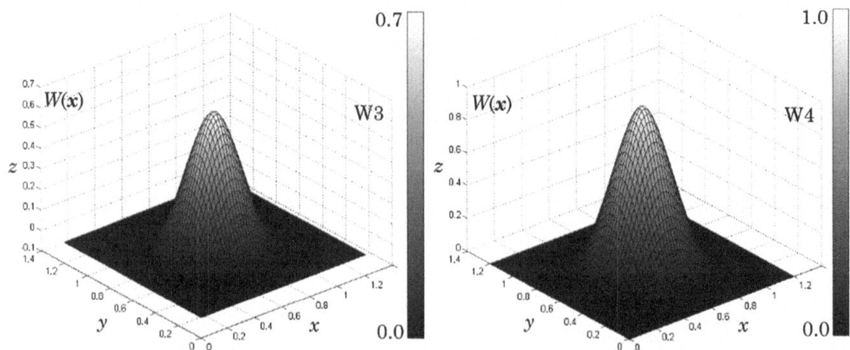

Fig. 4.7 2D weight functions. *W*3 cubic spline. *W*4 quartic spline

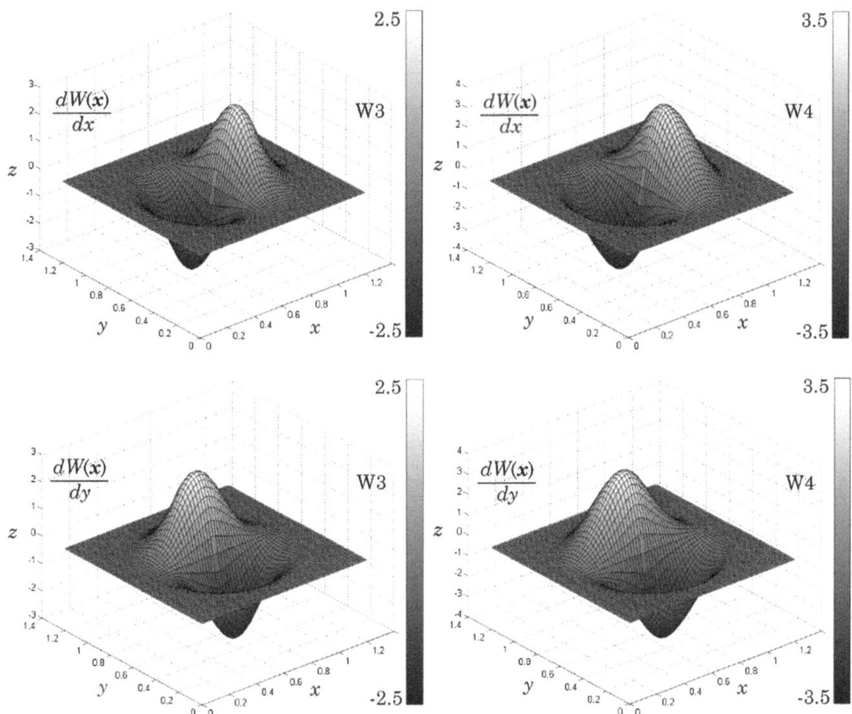

Fig. 4.8 2D weight functions first order partial derivatives. *W*3 cubic spline. *W*4 quartic spline

It is possible to construct a weight function with any desire order of continuity. A detailed description on efficient methodologies to construct weight functions can be found in the literature [5, 17].

4.3.3 MLS Shape Function Properties

4.3.3.1 Consistency

The shape function ability to reproduce the complete order of an unknown polynomial function is defined as consistency. The consistency of the MLS approximation depends on the complete order of the polynomial basis used in the Eq. (4.6), therefore a shape function possess C_m consistency when the used polynomial basis possess m monomials [18]. The MLS shape function consistency can be easily demonstrated [5, 18]. Consider a field defined by the following complete polynomial function,

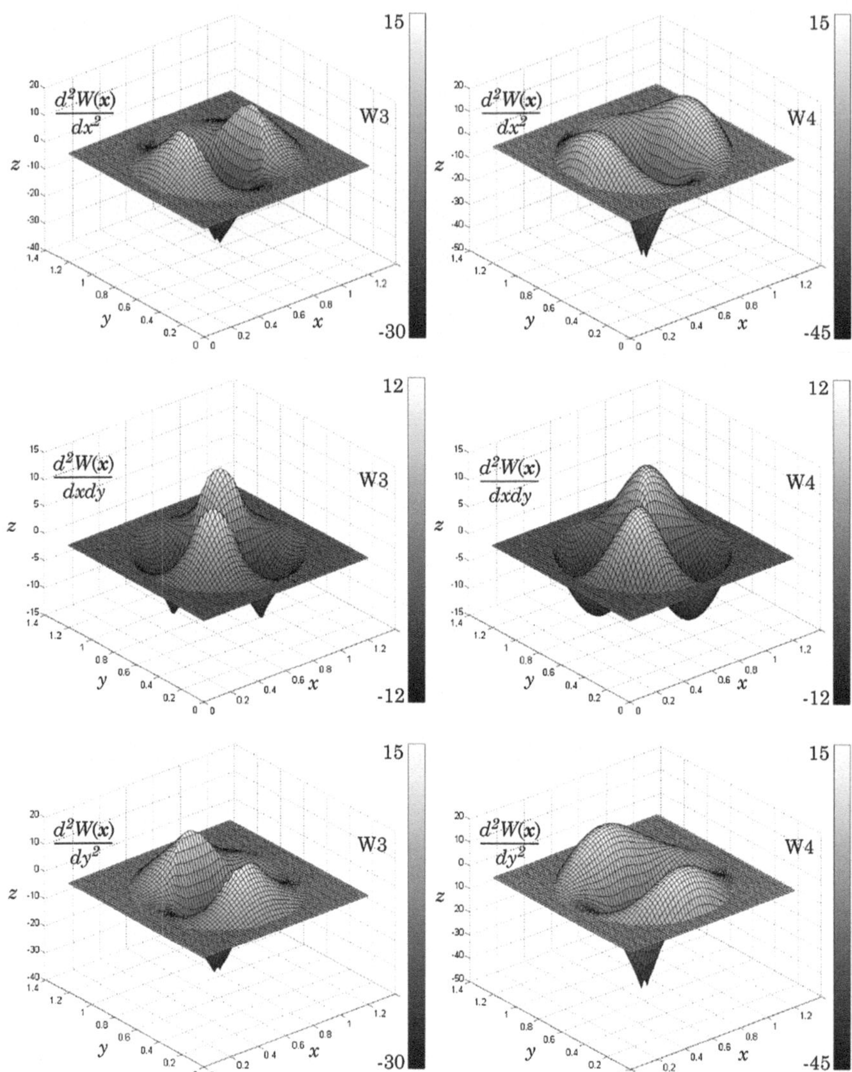

Fig. 4.9 2D weight functions second order partial derivatives. *W3* cubic spline. *W4* quartic spline

$$u(\boldsymbol{x}) = \sum_{i=1}^{k} p_i(\boldsymbol{x}) a_i(\boldsymbol{x}) = \boldsymbol{p}(\boldsymbol{x})^{\mathrm{T}} \boldsymbol{a}(\boldsymbol{x}) \tag{4.58}$$

being $k \leq m$. The approximation function $u^h(\boldsymbol{x})$ is defined with Eq. (4.6), therefore the complete polynomial function can be presented as,

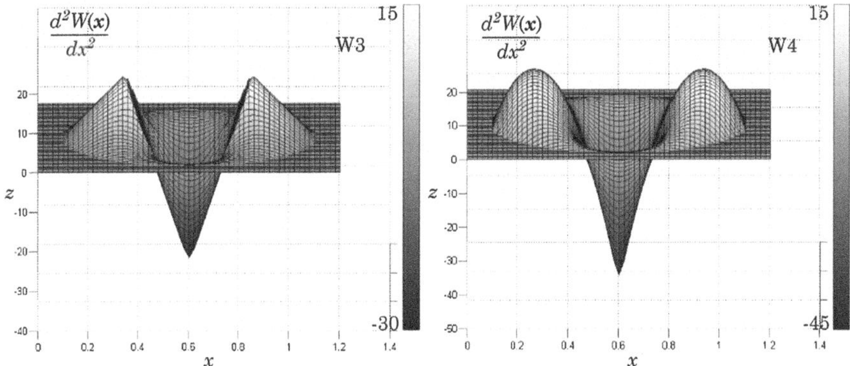

Fig. 4.10 Comparison between the cubic spline ($W3$) and the quartic spline ($W4$) weight functions second order partial derivatives

$$u(\boldsymbol{x}) = \sum_{i=1}^{k} p_i(\boldsymbol{x}) a_i(\boldsymbol{x}) + \sum_{j=k+1}^{m} p_j(\boldsymbol{x}) \cdot 0 \tag{4.59}$$

The non-constants coefficients $b_i(\boldsymbol{x})$ can be obtained with the minimization of the quadratic norm presented in Eq. (4.11). Then, substituting in Eq. (4.11) the approximation function, $u^h(\boldsymbol{x})$, from Eq. (4.6) and the field function, $u(\boldsymbol{x})$, defined by Eq. (4.59), it is possible to write,

$$J = \sum_{i=1}^{n} W(\boldsymbol{x}) \left[\left(\sum_{i=1}^{k} p_i(\boldsymbol{x}) b_i(\boldsymbol{x}) + \sum_{j=k+1}^{m} p_j(\boldsymbol{x}) b_j(\boldsymbol{x}) \right) - \left(\sum_{i=1}^{k} p_i(\boldsymbol{x}) a_i(\boldsymbol{x}) + \sum_{j=k+1}^{m} p_j(\boldsymbol{x}) \cdot 0 \right) \right]^2 \tag{4.60}$$

It is visible that $J = 0$ (and consequently $\partial J/\partial b = 0$) if $b_i(\boldsymbol{x}) = a_j(\boldsymbol{x}), i = \{1, 2, \ldots, k\}$ and $b_j(\boldsymbol{x}) = 0, j = \{k+1, \ldots, m\}$, leading to,

$$u^h(\boldsymbol{x}) = \sum_{i=1}^{k} p_i(\boldsymbol{x}) a_i(\boldsymbol{x}) = u(\boldsymbol{x}) \tag{4.61}$$

which proves the MLS approximation is capable to reproduce any set of monomials included in the polynomial basis of the MLS formulation.

4.3.3.2 Reproducibility

Since in meshless methods distinct types of basis functions can be used in the construction of the shape functions, within meshless methods the reproducibility property is not included in the consistency property. The reproducibility property is the ability to reproduce an unknown function that is included as a basis function in the shape functions construction [5]. In opposition to the consistency property, which focuses only on the reproducibility of complete polynomial functions, in reproducibility the unknown function may and may not be a polynomial function.

The reproducibility property can be proved with the following argument [18]. Consider a field defined by the following expression,

$$u(\boldsymbol{x}) = \sum_{j=1}^{k} f_j(\boldsymbol{x}) c_j(\boldsymbol{x}) = \boldsymbol{f}(\boldsymbol{x})^{\mathrm{T}} \boldsymbol{c}(\boldsymbol{x}) \tag{4.62}$$

where $\boldsymbol{f}(\boldsymbol{x})$ is defined in a functional space $f_k : \mathbb{R}^d \mapsto \mathbb{R}$ and $c_j(\boldsymbol{x})$ are arbitrary coefficients of $f_j(\boldsymbol{x})$. The approximation function is defined as,

$$u^h(\boldsymbol{x}) = \sum_{j=1}^{k} f_j(\boldsymbol{x}) a_j(\boldsymbol{x}) + \sum_{i=1}^{m} p_i(\boldsymbol{x}) b_i(\boldsymbol{x}) = \boldsymbol{f}(\boldsymbol{x})^{\mathrm{T}} \boldsymbol{a}(\boldsymbol{x}) + \boldsymbol{p}(\boldsymbol{x})^{\mathrm{T}} \boldsymbol{b}(\boldsymbol{x}) \tag{4.63}$$

It is possible to rewrite Eq. (4.63) as,

$$u^h(\boldsymbol{x}) = \left\{ \boldsymbol{f}(\boldsymbol{x})^{\mathrm{T}}, \boldsymbol{p}(\boldsymbol{x})^{\mathrm{T}} \right\} \left\{ \begin{matrix} \boldsymbol{a}(\boldsymbol{x}) \\ \boldsymbol{b}(\boldsymbol{x}) \end{matrix} \right\} = \boldsymbol{g}(\boldsymbol{x})^{\mathrm{T}} \boldsymbol{d}(\boldsymbol{x}) \tag{4.64}$$

Minimizing the quadratic norm presented in Eq. (4.11) it is possible to obtain a similar expression to Eq. (4.27),

$$u^h(\boldsymbol{x}) = \sum_{i=1}^{n} \boldsymbol{g}(\boldsymbol{x}) \boldsymbol{A}(\boldsymbol{x})^{-1} W(\boldsymbol{x}) \boldsymbol{g}(\boldsymbol{x})^{\mathrm{T}} u(\boldsymbol{x}) \tag{4.65}$$

The field functional defined in Eq. (4.62) can be written as,

$$u(\boldsymbol{x}) = \boldsymbol{f}(\boldsymbol{x})^{\mathrm{T}} \boldsymbol{c}(\boldsymbol{x}) = \{\boldsymbol{f}(\boldsymbol{x})^{\mathrm{T}}, \boldsymbol{p}(\boldsymbol{x})^{\mathrm{T}}\} \left\{ \begin{matrix} \boldsymbol{c}(\boldsymbol{x}) \\ \underset{[m \times 1]}{\boldsymbol{0}} \end{matrix} \right\} = \boldsymbol{g}(\boldsymbol{x})^{\mathrm{T}} \boldsymbol{e}(\boldsymbol{x}) \tag{4.66}$$

Substituting the field functional defined in Eq. (4.66) in the approximation functions, Eq. (4.65), it possible to obtain,

$$u^h(\boldsymbol{x}) = \sum_{i=1}^{n} \boldsymbol{g}(\boldsymbol{x})^{\mathrm{T}} \boldsymbol{A}(\boldsymbol{x})^{-1} W(\boldsymbol{x} - \boldsymbol{x}_i) \boldsymbol{g}(\boldsymbol{x}_i) \, \boldsymbol{g}(\boldsymbol{x}_i)^{\mathrm{T}} \boldsymbol{e}(\boldsymbol{x}) \tag{4.67}$$

Being the weighted moment matrix $\boldsymbol{A}(\boldsymbol{x})$ defined as in Eq. (4.21),

$$\boldsymbol{A}(\boldsymbol{x}) = \sum_{i=1}^{n} W(\boldsymbol{x} - \boldsymbol{x}_i) \boldsymbol{g}(\boldsymbol{x}_i) \, \boldsymbol{g}(\boldsymbol{x}_i)^{\mathrm{T}} \tag{4.68}$$

It possible to write Eq. (4.67) as,

$$u^h(\mathbf{x}) = \mathbf{g}(\mathbf{x})^{\mathrm{T}}\mathbf{A}(\mathbf{x})^{-1}\mathbf{A}(\mathbf{x})\mathbf{e}(\mathbf{x}) = \mathbf{g}(\mathbf{x})^{\mathrm{T}}\mathbf{e}(\mathbf{x}) = u(\mathbf{x}) \tag{4.69}$$

Proving that any functional appearing in the basis can be exactly reproduced by the MLS approximation. This property permits to construct MLS shape functions to deal with specific practical problems. The singular stress field at a crack tip can be obtained with the inclusion of singular functions in the basis combined with the polynomial basis [19, 20] as shown in Eq. (4.63). This technique permits to increase the solution accuracy, since using only a normal complete polynomial basis usually lead to significant errors in the analysis of crack tip problems. Nevertheless, if these enriched basis functions are included in the basis of the MLS approximation, it is important to certify the non-singularity of the weighted moment matrix $\mathbf{A}(\mathbf{x})$.

4.3.3.3 Partition of Unity

The MLS shape function $\varphi_i(\mathbf{x})$ satisfies the partition of unity,

$$\sum_{i=1}^{n} \varphi_i(\mathbf{x}) = 1 \tag{4.70}$$

if a constant is included in the basis. The argument used to prove the consistency property of the MLS shape function can be used to prove the partition of unity property. Consider the constant field $u(\mathbf{x}) = c$, with $c \in \mathbb{R}$, which can assume the same polynomial form as in Eq. (4.58). However in this case only the constant term of the polynomial exists, $k = 1$,

$$u(\mathbf{x}) = \sum_{i=1}^{k} p_i(\mathbf{x})a_i(\mathbf{x}) + \sum_{j=k+1}^{m} p_j(\mathbf{x}) \cdot 0 = a_1 + 0 = c \tag{4.71}$$

Once again, the non-constants coefficients $b_i(\mathbf{x})$ of the approximation function $u^h(\mathbf{x})$ defined with Eq. (4.6), can be obtained with the minimization of the quadratic norm presented in Eq. (4.11). Then, substituting in Eq. (4.11) the approximation function, $u^h(\mathbf{x})$, from Eq. (4.6), and the field function, $u(\mathbf{x})$, defined by Eq. (4.71), it is possible to write,

$$J = \sum_{i=1}^{n} W(\mathbf{x}) \left[\left(p_1(\mathbf{x})b_1(\mathbf{x}) + \sum_{j=2}^{m} p_j(\mathbf{x})b_j(\mathbf{x}) \right) - \left(p_1(\mathbf{x})a_1(\mathbf{x}) + \sum_{j=2}^{m} p_j(\mathbf{x}) \cdot 0 \right) \right]^2 \tag{4.72}$$

A minimum is obtained when $b_1(\boldsymbol{x}) = a_1(\boldsymbol{x}) = c$ and $b_j(\boldsymbol{x}) = 0,\ j = \{2, \ldots, m\}$. Therefore the approximation function can be presented as,

$$u^h(\boldsymbol{x}) = c \tag{4.73}$$

Since the field variable for an interest point $\boldsymbol{x}_I$ is approximated using the shape function values obtained at the nodes inside the support-domain of $\boldsymbol{x}_I$,

$$u^h(\boldsymbol{x}_I) = \sum_{i=1}^{n} \varphi_i(\boldsymbol{x}_I) u(\boldsymbol{x}_i) = \sum_{i=1}^{n} \varphi_i(\boldsymbol{x}_I) c = c \sum_{i=1}^{n} \varphi_i(\boldsymbol{x}_I) \tag{4.74}$$

and the approximation function is defined as $u^h(x) = c$, then,

$$c = c \sum_{i=1}^{n} \varphi_i(\boldsymbol{x}_I) \Leftrightarrow \sum_{i=1}^{n} \varphi_i(\boldsymbol{x}_I) = 1 \tag{4.75}$$

Proving that if the basis contains a constant term, then the MLS shape function is of the partition unity.

4.3.3.4 Kronecker Delta

In general, the approximation function obtained with MLS approximants is a smooth functional unable to pass through the nodal values. Hence, the MLS shape functions do not possess the Kronecker delta property,

$$\varphi_i(\boldsymbol{x}_j) \neq \delta_{ij} = \begin{cases} 1, & i = j \\ 0, & i \neq j \end{cases} \tag{4.76}$$

This property is demonstrated with the following one-dimensional example. Consider a one-dimensional domain discretized by a set of 5 nodes defined by $\boldsymbol{X} = \{\boldsymbol{x}_1, \boldsymbol{x}_2, \boldsymbol{x}_3, \boldsymbol{x}_4, \boldsymbol{x}_5\} \in \Omega \wedge \boldsymbol{x}_i \in \mathbb{R}^1$. The mesh density parameter of $\boldsymbol{X}$ is identified by Eq. (4.4), being h considered constant in this example. The domain is represented in Fig. 4.11. Consider now an interest point $\boldsymbol{x}_I \in \Omega$ coincident with $\boldsymbol{x}_3$ possessing an influence-domain containing all the nodes of the domain, $\boldsymbol{X}$. If the MLS approximation is able to construct interpolation functions, then the MLS shape function for the interest point $\boldsymbol{x}_I \in \Omega$ should resembles the shape function of node $\boldsymbol{x}_3$, $\varphi_I(\boldsymbol{x}) = \varphi_3(\boldsymbol{x})$ presented in Fig. 4.11,

$$\varphi_I(\boldsymbol{x}) = \varphi_3(\boldsymbol{x}) = \{0 \quad 0 \quad 1 \quad 0 \quad 0\}^{\mathrm{T}} \tag{4.77}$$

The construction of the MLS shape function respect Eq. (4.31). To simplify the present demonstration consider the following spatial coordinates for each node of the one-dimensional domain,

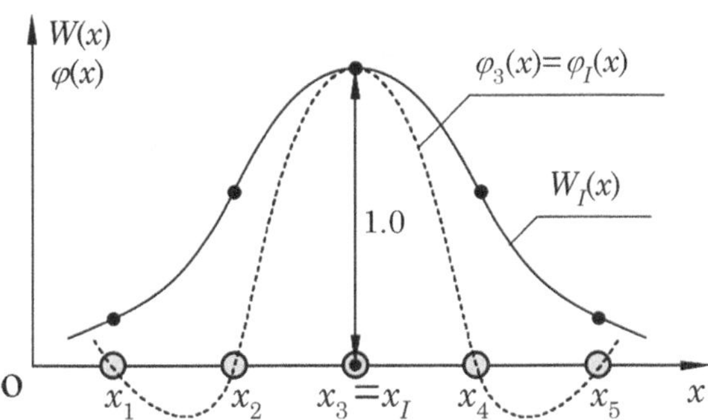

Fig. 4.11 Representation of the one-dimensional domain, the interest point weight function and potential shape function

$$\boldsymbol{X} = \{\boldsymbol{x}_1 \quad \boldsymbol{x}_2 \quad \boldsymbol{x}_3 \quad \boldsymbol{x}_4 \quad \boldsymbol{x}_5\} = \{-2 \quad -1 \quad 0 \quad 1 \quad 2\} \in \mathbb{R}^1 \tag{4.78}$$

A generic bell-shaped weight function $\boldsymbol{W}_I(\boldsymbol{x})$ is considered, Fig. 4.11. Being $\boldsymbol{x}_I \equiv \boldsymbol{x}_3$, the weight function is centred in the problem domain. Therefore,

$$\begin{cases} \boldsymbol{W}_I(\boldsymbol{x}_1) = \boldsymbol{W}_I(\boldsymbol{x}_5) = w_1 \\ \boldsymbol{W}_I(\boldsymbol{x}_2) = \boldsymbol{W}_I(\boldsymbol{x}_4) = w_2 \\ \boldsymbol{W}_I(\boldsymbol{x}_3) = w_3 = 1 \end{cases} \tag{4.79}$$

It is assumed a linear polynomial basis, $\boldsymbol{p}(\boldsymbol{x}) = \{1 \quad \boldsymbol{x}\}$, therefore for the interest point $\boldsymbol{x}_I$ it is obtained,

$$\boldsymbol{p}(\boldsymbol{x}_I) = \boldsymbol{p}(\boldsymbol{x}_3) = \{1 \quad \boldsymbol{x}_I\} = \{1 \quad 0\} \tag{4.80}$$

The weighted polynomial matrix, $\boldsymbol{B}(\boldsymbol{x})$, defined in Eq. (4.23), can be obtained with,

$$\boldsymbol{B}(\boldsymbol{x}) = \left[w_1 \begin{bmatrix} 1 \\ x_1 \end{bmatrix} \quad w_2 \begin{bmatrix} 1 \\ x_2 \end{bmatrix} \quad w_3 \begin{bmatrix} 1 \\ x_3 \end{bmatrix} \quad w_2 \begin{bmatrix} 1 \\ x_4 \end{bmatrix} \quad w_1 \begin{bmatrix} 1 \\ x_5 \end{bmatrix} \right] \tag{4.81}$$

and after the substitution with the nodal values,

$$\boldsymbol{B}(\boldsymbol{x}) = \begin{bmatrix} w_1 & w_2 & 1 & w_2 & w_1 \\ -2w_1 & -w_2 & 0 & w_2 & 2w_1 \end{bmatrix} \tag{4.82}$$

In this demonstration the coefficients of $\boldsymbol{A}(\boldsymbol{x})^{-1}$, which can be obtained inverting the weighted moment matrix $\boldsymbol{A}(\boldsymbol{x})$ defined in Eq. (4.21), are considered as unknowns,

$$\boldsymbol{A}(\boldsymbol{x})^{-1} = \begin{bmatrix} a_{11} & a_{12} \\ a_{12} & a_{22} \end{bmatrix} \tag{4.83}$$

Therefore, if the MLS approximation is able to construct interpolation functions, then Eq. (4.30) must be verified,

$$\begin{Bmatrix} 0 \\ 0 \\ 1 \\ 0 \\ 0 \end{Bmatrix} = \{1 \quad x_I\} \begin{bmatrix} a_{11} & a_{12} \\ a_{12} & a_{22} \end{bmatrix} \begin{bmatrix} w_1 & w_2 & 1 & w_2 & w_1 \\ -2w_1 & -w_2 & 0 & w_2 & 2w_1 \end{bmatrix} \tag{4.84}$$

Which can be developed to obtain five equations,

$$\begin{Bmatrix} 0 \\ 0 \\ 1 \\ 0 \\ 0 \end{Bmatrix} = \begin{Bmatrix} w_1(a_{11} + a_{12}x_I) - 2w_1(a_{12} + a_{22}x_I) \\ w_2(a_{11} + a_{12}x_I) - w_2(a_{12} + a_{22}x_I) \\ a_{11} + a_{12}x_I \\ w_2(a_{11} + a_{12}x_I) + w_2(a_{12} + a_{22}x_I) \\ w_1(a_{11} + a_{12}x_I) + 2w_1(a_{12} + a_{22}x_I) \end{Bmatrix} \tag{4.85}$$

Adding the first equation to the fifth equation,

$$a_{11} = -a_{12}x_I \tag{4.86}$$

Substituting in third equation gives $1 = 0$, indicating that equation system presented in Eq. (4.85) is not true. It is not possible to define a $\boldsymbol{A}(\boldsymbol{x})^{-1}$ matrix capable to produce the interpolation function presented in Eq. (4.77). Therefore, the MLS shape functions do not possess the Kronecker delta property.

4.3.3.5 Compact Support

The MLS shape functions is obtained considering only the nodes inside an initially defined compact support-domain. Since the value of $\varphi(\boldsymbol{x}_I)$ outside the support-domain is zero, the MLS shape functions possess compact support. This property permits to create sparse and banded discretized systems of equations.

4.3.3.6 Compatibility

The MLS shape functions are compatible in a local support-domain because the support-domain movement does not perturb the continuity of the approximated field function. The bell-shaped weight function, used to construct the MSL shape function, permit to smoothly update the nodes entering and leaving the support-domain.

Consider the one-dimensional domain described in Fig. 4.12, being $X = \{\boldsymbol{x}_1, \boldsymbol{x}_2, \ldots, \boldsymbol{x}_{12}\} \in \Omega \wedge \boldsymbol{x}_i \in \mathbb{R}^1$ the set of nodes discretizing the problem domain. In this example the average nodal spacing is considered equal to the mesh density

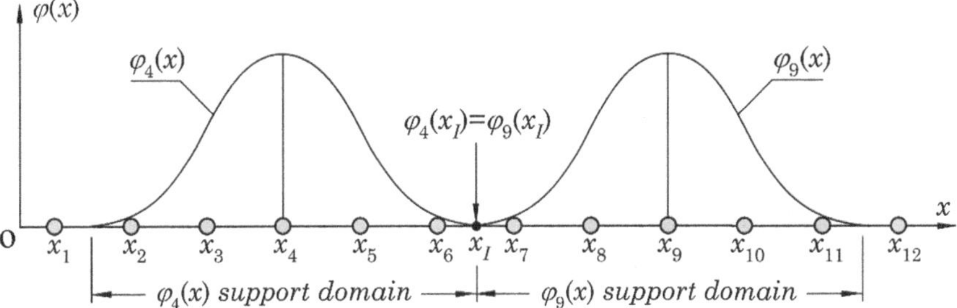

Fig. 4.12 MLS shape functions compatibility representation

parameter, h, which is considered constant. Thus, the support-domain of the MLS shape functions is defined as $d_s = 2.5\ h$. Consider now an interest point $\boldsymbol{x}_I \in \mathbb{R} \wedge \boldsymbol{x}_I \subset \Omega$ defined simultaneously in the boundary of the support-domains of node $\boldsymbol{x}_4$ and node $\boldsymbol{x}_9$. As Fig. 4.12 shows, $\varphi_4(\boldsymbol{x}_I) = \varphi_9(\boldsymbol{x}_I) = 0$, indicating that the constructed MLS shape functions do not end abruptly and possess first order compatibility. If the weight function is selected accordingly, it is possible to obtain higher orders of compatibility. With the cubic spline weight function, Eq. (4.51), and quartic spline weight function, Eq. (4.52), it is possible to obtain a compatibility up to the third order, which would permit in the present example to obtain: $\varphi_4(\boldsymbol{x}_I)_{,x} = \varphi_9(\boldsymbol{x}_I)_{,x}$ and $\varphi_4(\boldsymbol{x}_I)_{,xx} = \varphi_9(\boldsymbol{x}_I)_{,xx}$.

4.3.4 MLS Shape Functions Examples

The main objective of this subsection is to show the effect of the weight function on the obtained MLS shape function. Thus, it is considered a two-dimensional domain $\boldsymbol{x} \in \mathbb{R}^2 : x \in [0, 10], y \in [0, 10]$ discretized in a regular mesh with 7×7 nodes uniformly distributed along both dimensions. It is assumed a constant square support-domain, $d_s = 2.5\ h$, being h the mesh density parameter, which in this case is equal to the average nodal spacing, d_a, and can be defined as $h = 6/10$. For the MLS shape function construction it was considered a linear polynomial basis and two weight functions: the cubic spline (W3), Eq. (4.51), and the quartic spline (W4), Eq. (4.52). The obtained MLS shape functions for the middle node, $\boldsymbol{x} = \{5, 5\}$, are presented in Fig. 4.13. The first order and the second order partial derivatives of the MLS shape functions are presented in Figs. 4.14 and 4.15, respectively. Notice that the silhouette of the obtained MLS shape functions resembles with the shape of the respective weight function. In Fig. 4.16 another view is used to present the second order partial derivatives of the MLS shape function. It is visible that when the cubic spline (W3) weight function is considered, the second order derivative of the MLS shape functions shows the same lack

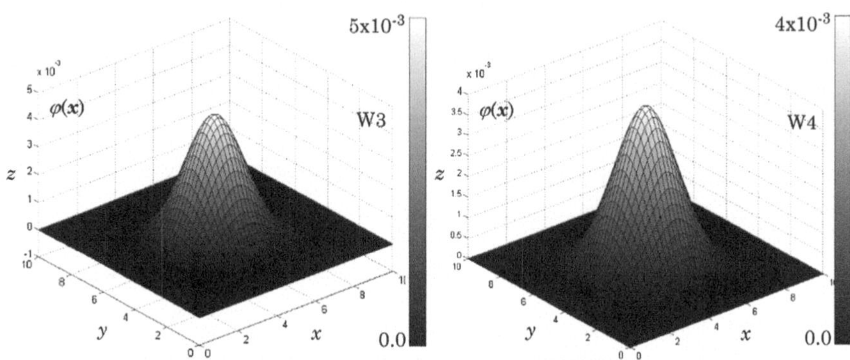

Fig. 4.13 MLS shape functions obtained with the cubic spline (*W*3) and the quartic spline (*W*4)

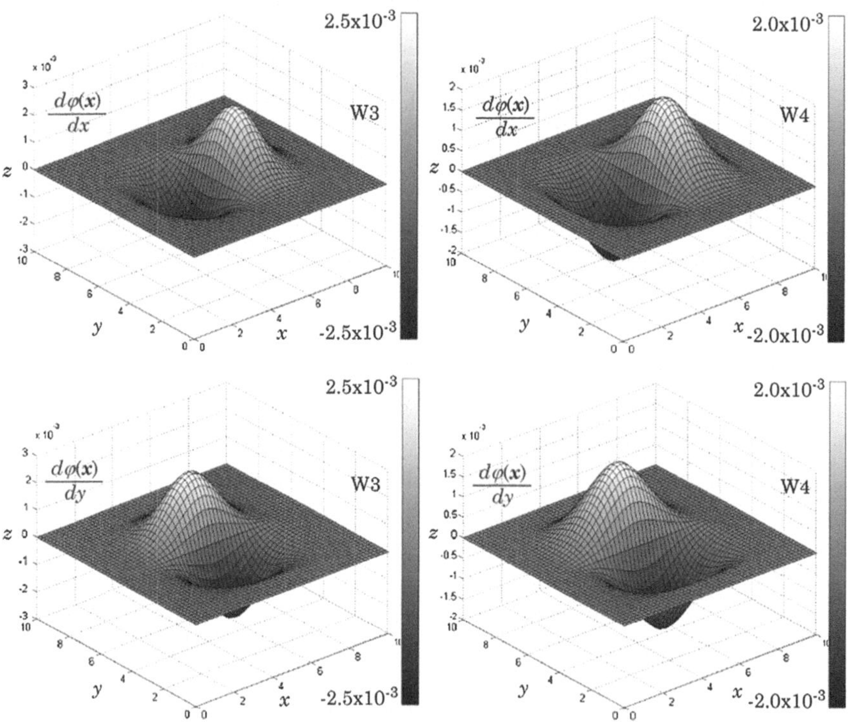

Fig. 4.14 First partial derivatives of the MLS shape functions obtained with the cubic spline (*W*3) and the quartic spline (*W*4)

of smoothness of the cubic spline weight function, Fig. 4.10. This observation proves that the MLS shape functions inherits the continuity of the considered weight function.

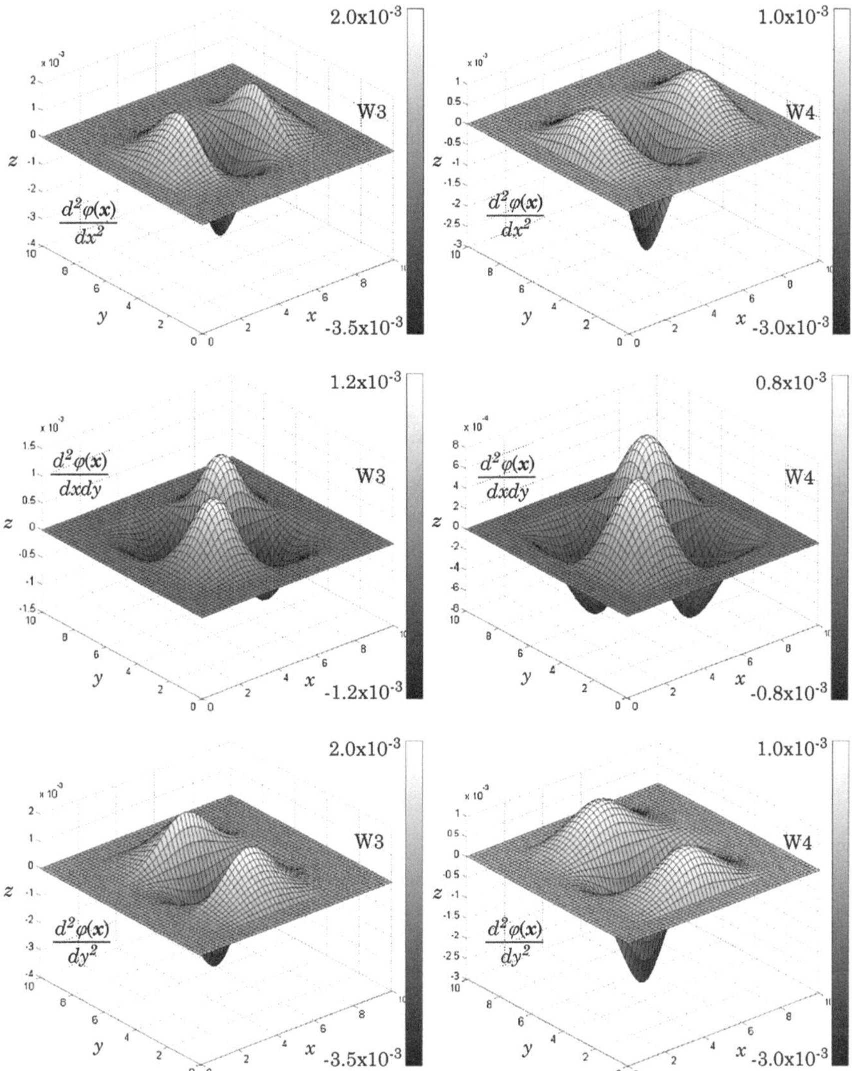

Fig. 4.15 Second partial derivatives of the MLS shape functions obtained with the cubic spline ($W3$) and the quartic spline ($W4$)

4.3.5 MLS Shape Functions Calculation

This section provides a simple and practical flow-chart to compute the MLS shape functions. The MLS shape function, and the respective partial derivatives, of an interest point $\boldsymbol{x}_I$, with an influence-domain defined by $\boldsymbol{X}_I = \{\boldsymbol{x}_1, \boldsymbol{x}_2, \ldots, \boldsymbol{x}_n\} \in \mathbb{R}^d$, can be calculated with the following procedure:

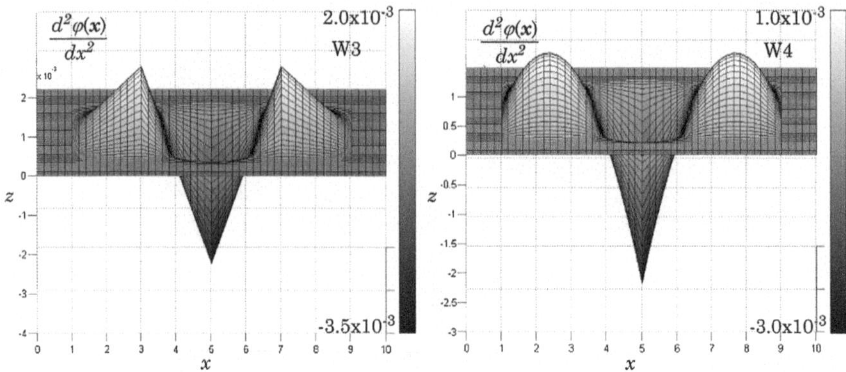

Fig. 4.16 Comparison between the second order partial derivatives of the MLS shape functions obtained with the cubic spline ($W3$) and the quartic spline ($W4$)

1. The MLS shape function support-domain is defined based on the influence-domain of the interest point $\boldsymbol{x}_I$, $d_s = \max\|\boldsymbol{x}_i - \boldsymbol{x}_I\|, \forall \boldsymbol{x}_i \in \boldsymbol{X}_I$.
2. The polynomial vector for interest point $\boldsymbol{x}_I$ is defined: $\boldsymbol{p}(\boldsymbol{x}_I)$. The dimension of $\boldsymbol{p}(\boldsymbol{x}_I)$ depends on the monomials, $[m \times 1]$. For example, in a three-dimensional space, the linear polynomial vector for interest point $\boldsymbol{x}_I$ is defined as: $\boldsymbol{p}(\boldsymbol{x}_I) = \{1 \quad x_I \quad y_I \quad z_I\}^{\mathrm{T}}$.
3. Next, the weight of each node can be determined using a weight function, $\boldsymbol{W}(\boldsymbol{x}_I) = \{W(\boldsymbol{x}_1 - \boldsymbol{x}_I)W(\boldsymbol{x}_2 - \boldsymbol{x}_I)\cdots W(\boldsymbol{x}_n - \boldsymbol{x}_I)\}$.
4. With the nodal weight determined it is possible to construct the weighted polynomial matrix $\boldsymbol{B}(\boldsymbol{x}_I)$ as indicated in Eq. (4.24). This matrix has a dimension $[m \times n]$.
5. The momentum matrix $\boldsymbol{A}(\boldsymbol{x}_I)$ is constructed as in Eq. (4.22) and then $\boldsymbol{A}(\boldsymbol{x}_I)^{-1}$ is computed. The momentum matrix is a square matrix with $[m \times m]$.
6. Finally, the interpolation function $\boldsymbol{\varphi}(\boldsymbol{x}_I)$ is obtained with Eq. (4.31).

The MLS shape function first order partial derivatives can be obtained applying Eq. (4.33) and the second order partial derivatives are obtained with Eq. (4.41).

4.3.6 Influence of the Size of the Support-Domain

In this subsection the importance of the support-domain size, which was referred in Sect. 4.2, is shown with a simple example.

Consider a one-dimensional domain $\Omega \subset \mathbb{R}$ with $\boldsymbol{x} \in \mathbb{R} : x \in [0, 1]$, being $X = \{\boldsymbol{x}_1, \boldsymbol{x}_2, \ldots, \boldsymbol{x}_{11}\} \in \Omega \wedge \boldsymbol{x}_i \in \mathbb{R}^1$ the set of nodes discretizing the problem domain. Two distinct nodal distributions are considered in this example: a uniform nodal distribution and an irregular nodal distribution. The objective is to adjust an approximation function $u^h(\boldsymbol{x})$ to the N discrete nodal values $u(\boldsymbol{x}_i)$. The spatial

Table 4.1 Spatial location of each node discretizing the problem domain and the respective nodal value

Regular distribution											
x	0.00	0.10	0.20	0.30	0.40	0.50	0.60	0.70	0.80	0.90	1.00
$u(x)$	0.00	0.05	0.10	0.12	0.09	0.08	0.11	0.13	0.09	0.04	0.00
Irregular distribution											
x	0.00	0.12	0.19	0.31	0.43	0.49	0.58	0.71	0.79	0.92	1.00
$u(x)$	0.00	0.05	0.10	0.12	0.09	0.08	0.11	0.13	0.09	0.04	0.00

location of each node $\boldsymbol{x}_i$ discretizing the problem domain and the respective nodal values $u(\boldsymbol{x}_i)$ are presents in Table 4.1.

To obtain the approximation function $u^h(\boldsymbol{x})$ the following procedure is performed:

1. It is created a background mesh of interest points covering the problem domain, $\boldsymbol{Q} = \{\boldsymbol{q}_1, \boldsymbol{q}_2, \ldots, \boldsymbol{q}_{101}\} \in \Omega \wedge \boldsymbol{q}_i \in \mathbb{R}^1$. This background mesh is equivalent to the integration mesh that will be required to numerically integrate the integro-differential equations governing the studied physical phenomenon.
2. The size of the support-domain of the shape functions is defined as: $d_s = s \cdot h$. In this case it is considered $h = 1/10$.
3. Based on the support-domain it is defined the size of the influence-domain of each interest point $\boldsymbol{q}_I$, $d_I = d_s$. Then, each interest point $\boldsymbol{q}_I$ searches for the n field nodes within the radial distance d_I, establishing the individual influence-domains of interest points $\boldsymbol{q}_I$.
4. The MLS shape functions are constructed for each interest point $\boldsymbol{q}_I$ following the procedure indicated in Sect. 4.3.5. In this example it is used the cubic spline weight function and the linear polynomial basis.
5. For each interest point $\boldsymbol{q}_I$ it is obtained the approximation field value with: $u^h(\boldsymbol{q}_I) = \sum_{i=1}^{n} \varphi_i(\boldsymbol{q}_I) u(\boldsymbol{x}_i)$

The previously described procedure is performed for five distinct support-domains, considering $s = \{\, 1.0 \quad 2.0 \quad 3.0 \quad 4.0 \quad 5.0 \,\}$.

For the regular nodal distribution, the results of each analysis are presented in Fig. 4.17a. In Fig. 4.17b it is possible to observe the effect of s on the weight function shape. Notice that the use of support-domains with reduce size, $s = 1$, permitted to interpolate the data. However this effect it is not universal, it was only possible because the nodal distribution is uniform and the number of nodes inside each influence-domain fulfil the minimum to create a non-singular momentum matrix. The increase of the size of the support-domain leads to a smoother approximated solution $u^h(\boldsymbol{x})$.

The results regarding the irregular nodal distribution are presented in Fig. 4.18a. The effect of the support-domain size on the weight function shape is presented in Fig. 4.18b. It is visible that the support-domain with reduce size, $s = 1$, does not provide an acceptable approximation solution. In this case the size of the support-domain is not enough to guarantee the construction of non-singular

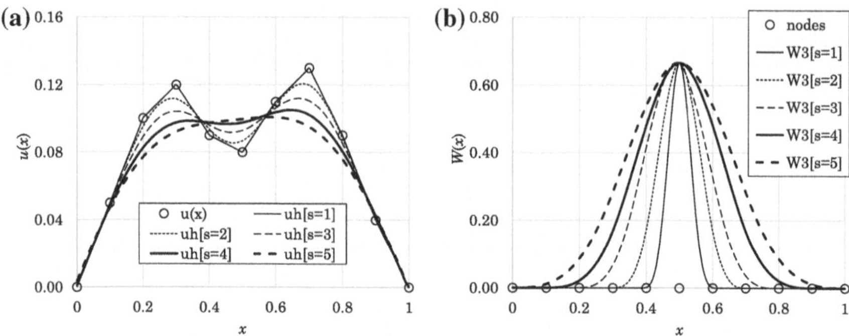

Fig. 4.17 Regular nodal distribution. **a** Obtained approximation solutions. **b** Shape of the used weight functions

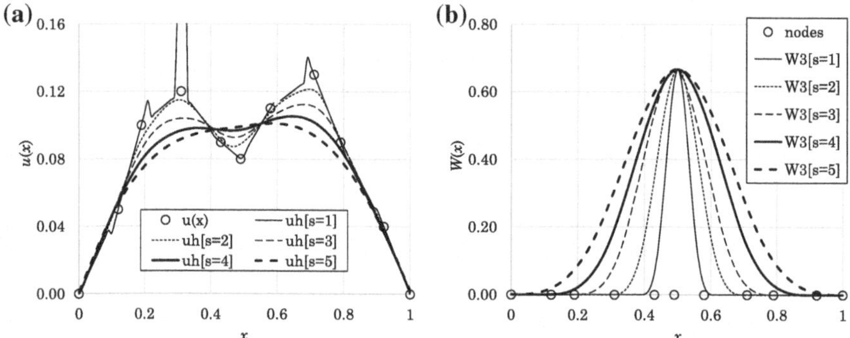

Fig. 4.18 Irregular nodal distribution. **a** Obtained approximation solutions. **b** Shape of the used weight functions

momentum matrices for all interest points and consequently it is not possible to obtain an approximation value in all interest points.

In Figs. 4.17a and 4.18a it is possible to observe that with the increase of the support-domain size the approximated solution $u^h(\boldsymbol{x})$ starts to deviate from the nodal values $u(\boldsymbol{x}_i)$. It are recommended support-domain sizes between $d_s = 2.0 \cdot d_a$ and $d_s = 3.0 \cdot d_a$ since this interval assures the existent of non-singular and stable momentum matrices and permits to construct approximation functions close enough to the discrete data.

4.4 Radial Point Interpolators

The Radial Point Interpolator (RPI) is a numerical technique belonging to the Point Interpolation Methods (PIM), which combines polynomial basis functions with radial basis functions. The PIM [21] is an efficient numerical tool capable to construct interpolation shape functions for meshless methods. The PIM [21] was originally developed using uniquely polynomial basis functions. The PIM formulation permits to easily construct shape functions and allows to obtain solutions with high accuracy [21]. However this polynomial version of the PIM presents a heavy disadvantage, the construction of the PIM shape function it is not always possible. In some cases, the perfect alignment of the nodes lead to singular moment matrices precluding to obtain the shape function.

In order to solve this PIM drawback, radial basis functions were included in the PIM formulation [22, 23]. This PIM version is entitled Radial Point Interpolation Method (RPIM) and it permits to obtain always a non-singular moment matrix and consequently it allows to construct consistently the interpolation shape function. Comparing with the polynomial PIM formulation, the radial point interpolation (RPI) formulation is more complex however it permits to obtain numerical solutions with higher accuracy.

4.4.1 PIM Generic Shape Functions

To understand the generic PIM construction, consider a function space T on $\Omega \subset \mathbb{R}^d$. It is possible to define the finite dimensional function space $T_H \subset T$ discretizing the domain Ω with: $T_H := f_k(\boldsymbol{x})$, being $f_k : \mathbb{R}^d \mapsto \mathbb{R}$ defined in the functional space. It is assumed that the d-dimensional spatial domain is discretized in N nodes: $\boldsymbol{X} = \{\boldsymbol{x}_1, \boldsymbol{x}_2, \ldots, \boldsymbol{x}_N\} \in \Omega \wedge \boldsymbol{x}_i \in \mathbb{R}^d$.

Considering a continuous scalar function $u(\boldsymbol{x})$, being $u \in T$, it is possible to define for an interest point $\boldsymbol{x}_I \in \mathbb{R}^d$, not necessarily coincident with $\boldsymbol{X}$, the PIM interpolation function of $u(\boldsymbol{x}_I)$ as,

$$u^h(\boldsymbol{x}_I) = \sum_{i=1}^{m} f_i(\boldsymbol{x}_I) b_i(\boldsymbol{x}_I) = \boldsymbol{f}(\boldsymbol{x}_I)^{\mathrm{T}} \boldsymbol{b}(\boldsymbol{x}_I) \tag{4.87}$$

Being $b_i(\boldsymbol{x}_I)$ the non-constant coefficients of $f_i(\boldsymbol{x}_I)$ and m the number of functions $f_i(\boldsymbol{x}_I)$ used as basis. Notice that if the m functions, $f_i(\boldsymbol{x}_I)$, of Eq. (4.87) are substituted by m monomial terms $p_i(\boldsymbol{x}_I)$ of a complete polynomial basis obtained from the triangle of Pascal, Fig. 4.2, Eq. (4.87) equalizes Eq. (4.6), which leads to the classic polynomial PIM formulation [21].

In opposition to the MLS approximation function, the PIM interpolation function $u^h(\boldsymbol{x})$ match the continuous scalar function $u(\boldsymbol{x})$, Fig. 4.19, therefore $u^h(\boldsymbol{x}) = u(\boldsymbol{x})$. It is possible to obtain $u^h(\boldsymbol{x}) = u(\boldsymbol{x})$ because within the PIM interpolation the number of nodes n on the support-domain of the interest point $\boldsymbol{x}_I$ is equal to m, the number of unknowns coefficients of $\boldsymbol{b}(\boldsymbol{x}_I)$.

The non-constant coefficients $\boldsymbol{b}(\boldsymbol{x}_I)$ can be obtained enforcing $u^h(\boldsymbol{x}_I)$ to pass through all the n nodal values on the support-domain of $\boldsymbol{x}_I$. Thus, Eq. (4.87) must

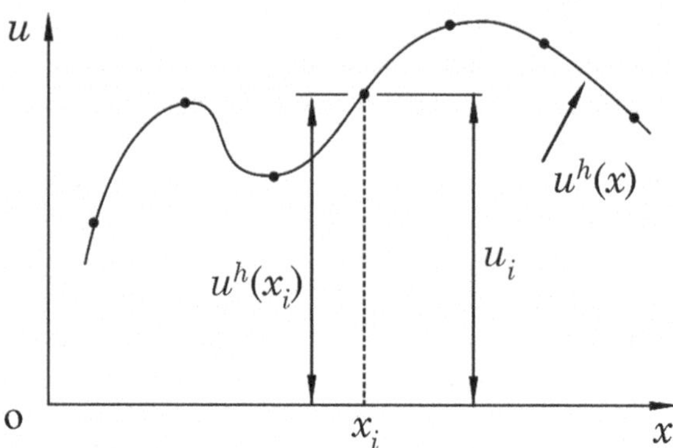

Fig. 4.19 The discrete nodal parameters $u(\boldsymbol{x}_i)$ and the PIM interpolation function $u^h(\boldsymbol{x})$

be satisfied to all n nodes on the support-domain of the interest point $\boldsymbol{x}_I$, yielding n equations with m unknowns,

$$\begin{cases} \sum_{i=1}^{m} f_i(\boldsymbol{x}_1)b_i(\boldsymbol{x}_I) = u(\boldsymbol{x}_1) = f_1(\boldsymbol{x}_1)b_1(\boldsymbol{x}_I) + f_2(\boldsymbol{x}_1)b_2(\boldsymbol{x}_I) + \cdots + f_m(\boldsymbol{x}_1)b_m(\boldsymbol{x}_I) \\ \sum_{i=1}^{m} f_i(\boldsymbol{x}_2)b_i(\boldsymbol{x}_2) = u(\boldsymbol{x}_2) = f_1(\boldsymbol{x}_2)b_1(\boldsymbol{x}_2) + f_2(\boldsymbol{x}_2)b_2(\boldsymbol{x}_2) + \cdots + f_m(\boldsymbol{x}_2)b_m(\boldsymbol{x}_2) \\ \vdots \\ \sum_{i=1}^{m} f_i(\boldsymbol{x}_n)b_i(\boldsymbol{x}_n) = u(\boldsymbol{x}_n) = f_1(\boldsymbol{x}_n)b_1(\boldsymbol{x}_n) + f_2(\boldsymbol{x}_n)b_2(\boldsymbol{x}_n) + \cdots + f_m(\boldsymbol{x}_n)b_m(\boldsymbol{x}_n) \end{cases} \tag{4.88}$$

Being by imposition $m = n$, if all the equations are independent, it is possible to determine uniquely all the $b_i(\boldsymbol{x}_I)$ unknowns. The equation system in Eq. (4.88) can be presented in the following matrix form,

$$\boldsymbol{F} \cdot \boldsymbol{b}(\boldsymbol{x}_I) = \boldsymbol{u}_s \tag{4.89}$$

where $\boldsymbol{b}(\boldsymbol{x}_I)$ and $\boldsymbol{u}_s$ are defined respectively by Eqs. (4.7) and (4.20). The moment matrix $\boldsymbol{F}$ with dimensions $n \times m$ is in fact square, because $m = n$. Additionally, $\boldsymbol{F}$ does not depend on the interest point spatial position $\boldsymbol{x}_I$, it only depends on the spatial position of the n nodes inside the support-domain of the interest point $\boldsymbol{x}_I$. Therefore, the components of moment matrix $\boldsymbol{F}$ are constants. The moment matrix $\boldsymbol{F}$ is defined as,

$$\boldsymbol{F} = \begin{bmatrix} f_1(\boldsymbol{x}_1) & f_2(\boldsymbol{x}_1) & \cdots & f_m(\boldsymbol{x}_1) \\ f_1(\boldsymbol{x}_2) & f_2(\boldsymbol{x}_2) & \cdots & f_m(\boldsymbol{x}_2) \\ \vdots & \vdots & \ddots & \vdots \\ f_1(\boldsymbol{x}_n) & f_2(\boldsymbol{x}_n) & \cdots & f_m(\boldsymbol{x}_n) \end{bmatrix} \tag{4.90}$$

Solving Eq. (4.89) it is possible to obtain the coefficients $\boldsymbol{b}(\boldsymbol{x}_I)$,

$$\boldsymbol{b}(\boldsymbol{x}_I) = \boldsymbol{F}^{-1}\boldsymbol{u}_s \tag{4.91}$$

The obtained coefficients $\boldsymbol{b}(\boldsymbol{x}_I)$ are in fact constant as long as the same n nodes inside the support-domain of the interest point $\boldsymbol{x}_I$ are maintained. The PIM shape functions can be obtained substituting Eq. (4.91) in (4.87),

$$u^h(\boldsymbol{x}_I) = \boldsymbol{f}(\boldsymbol{x}_I)^{\mathrm{T}}\boldsymbol{F}^{-1}\boldsymbol{u}_s = \sum_{i=1}^{n} \varphi_i(\boldsymbol{x}_I)\, u(\boldsymbol{x}_i) = \boldsymbol{\varphi}(\boldsymbol{x}_I)^{\mathrm{T}}\boldsymbol{u}_s = u(\boldsymbol{x}_I) \tag{4.92}$$

being $\varphi_i(\boldsymbol{x}_I)$ the shape function value of interest point $\boldsymbol{x}_I$ on the ith node, obtained considering the nodes inside the support-domain of interest point $\boldsymbol{x}_I$. The PIM shape function vector for the n nodes inside the support-domain of $\boldsymbol{x}_I$ is defined as,

$$\boldsymbol{\varphi}(\boldsymbol{x}_I)^{\mathrm{T}} = \{\varphi_1(\boldsymbol{x}_I) \quad \varphi_2(\boldsymbol{x}_I) \quad \ldots \quad \varphi_n(\boldsymbol{x}_I)\} = \boldsymbol{f}(\boldsymbol{x}_I)^{\mathrm{T}}\boldsymbol{F}^{-1} \tag{4.93}$$

Notice that in opposition to the MLS shape functions, in which the moment matrix $\boldsymbol{A}(\boldsymbol{x}_I)$ and the weighted polynomial matrix $\boldsymbol{B}(\boldsymbol{x}_I)$, used to construct the approximation function $u^h(\boldsymbol{x})$ in Eq. (4.26), have to be defined for a specific interest point $\boldsymbol{x}_I$ possessing a particular support-domain with n nodes, in the PIM shape functions construction the moment matrix $\boldsymbol{F}$ does not depend on the interest point $\boldsymbol{x}_I$ spatial position, therefore $\boldsymbol{F}$ is valid for other interest points possessing the same support-domain.

It is necessary to determine the PIM shape functions partial derivatives in order to obtain the partial derivatives of the interpolated field function, Eq. (4.92). Compared with the MLS shape functions, the PIM shape functions partial derivatives are much more simple to obtain. The first order partial derivatives of interpolated field function, with respect to a generic variable ξ, can be obtained with,

$$\frac{\partial u^h(\boldsymbol{x}_I)}{\partial \xi} = \sum_{i=1}^{n} \frac{\partial \varphi_i(\boldsymbol{x}_I)}{\partial \xi} u_i = \boldsymbol{\varphi}(\boldsymbol{x}_I)_{,\xi}^{\mathrm{T}}\boldsymbol{u}_s \tag{4.94}$$

being the first order partial derivative with respect to ξ of the PIM shape function defined as,

$$\boldsymbol{\varphi}(\boldsymbol{x}_I)_{,\xi} = \left(\boldsymbol{f}(\boldsymbol{x}_I)^{\mathrm{T}}\boldsymbol{F}^{-1}\right)_{,\xi} = \frac{\partial \boldsymbol{f}(\boldsymbol{x}_I)^{\mathrm{T}}}{\partial \xi}\boldsymbol{F}^{-1} + \boldsymbol{f}(\boldsymbol{x}_I)^{\mathrm{T}}\underbrace{\frac{\partial \boldsymbol{F}^{-1}}{\partial \xi}}_{0} \tag{4.95}$$

Notice that, since the moment matrix $\boldsymbol{F}$ does not depend on the interest point $\boldsymbol{x}_I$ spatial position, it is only require to obtain the partial derivative of the functional $\boldsymbol{f}(\boldsymbol{x}_I)$. The second order partial derivative of the interpolated field function, with respect to the generic variables ξ and η, is obtained with,

$$\frac{\partial^2 u^h(\boldsymbol{x}_I)}{\partial \xi \partial \eta} = \sum_{i=1}^{n} \frac{\partial^2 \varphi_i(\boldsymbol{x}_I)}{\partial \xi \partial \eta} u_i = \boldsymbol{\varphi}(\boldsymbol{x}_I)_{,\xi\eta}^{\mathrm{T}} \boldsymbol{u}_s \tag{4.96}$$

As expected, the spatial second order partial derivative of the PIM shape function with respect to ξ and η, is defined as,

$$\boldsymbol{\varphi}(\boldsymbol{x}_I)_{,\xi\eta} = \left(\boldsymbol{f}(\boldsymbol{x}_I)^{\mathrm{T}}\boldsymbol{F}^{-1}\right)_{,\xi\eta} = \frac{\partial^2 \boldsymbol{f}(\boldsymbol{x}_I)^{\mathrm{T}}}{\partial \xi \partial \eta}\boldsymbol{F}^{-1} \tag{4.97}$$

If the functional $\boldsymbol{f}(\boldsymbol{x}_I)$ is defined by a complete polynomial, being $f_i(\boldsymbol{x}_I)$ monomial terms of a complete polynomial basis, it is possible that $\boldsymbol{F}$ becomes singular or ill-conditioned, leading to the PIM failure. The most common reason for the inexistence of $\boldsymbol{F}^{-1}$ is the spatial collinearity of field nodes belonging to the same support-domain, which is recurrent in uniformly distributed nodal meshes or linear domain boundaries.

This drawback can be shown with the following simple example. Consider an interest point $\boldsymbol{x}_I \in \Omega$ with a support-domain containing the following six nodes defined in the two-dimensional space,

$$\boldsymbol{X}_I = \{\boldsymbol{x}_1 \quad \boldsymbol{x}_2 \quad \boldsymbol{x}_3 \quad \boldsymbol{x}_4 \quad \boldsymbol{x}_5 \quad \boldsymbol{x}_6\} = \begin{Bmatrix} 0 & 0 & 0 & 1 & 1 & 1 \\ 0 & 1 & 2 & 0 & 1 & 2 \end{Bmatrix} \tag{4.98}$$

It is possible to observe that nodes $\boldsymbol{x}_1$, $\boldsymbol{x}_2$ and $\boldsymbol{x}_3$ are collinear, as well as nodes $\boldsymbol{x}_4$, $\boldsymbol{x}_5$ and $\boldsymbol{x}_6$. Since the support-domain has six nodes, $n = 6$, in order to construct a polynomial PIM shape function it is required a moment matrix $\boldsymbol{F}$ defined by a complete polynomial basis with $m = 6$,

$$f(\boldsymbol{x}) = \{p_1(\boldsymbol{x}) \quad p_2(\boldsymbol{x}) \quad \ldots \quad p_6(\boldsymbol{x})\} = \{1 \quad x \quad y \quad x^2 \quad xy \quad y^2\} \tag{4.99}$$

Therefore the moment matrix can be written as,

$$F = \begin{bmatrix} p_1(\boldsymbol{x}_1) & p_2(\boldsymbol{x}_1) & \cdots & p_m(\boldsymbol{x}_1) \\ p_1(\boldsymbol{x}_2) & p_2(\boldsymbol{x}_2) & \cdots & p_m(\boldsymbol{x}_2) \\ \vdots & ; \vdots & \ddots & \vdots \\ p_1(\boldsymbol{x}_n) & p_2(\boldsymbol{x}_n) & \cdots & p_m(\boldsymbol{x}_n) \end{bmatrix} = \begin{bmatrix} 1 & 0 & 0 & 0 & 0 & 0 \\ 1 & 0 & 1 & 0 & 0 & 1 \\ 1 & 0 & 2 & 0 & 0 & 4 \\ 1 & 1 & 0 & 1 & 0 & 0 \\ 1 & 1 & 1 & 1 & 1 & 1 \\ 1 & 1 & 2 & 1 & 2 & 4 \end{bmatrix} \tag{4.100}$$

It is clear that the obtained moment matrix singular. It is not possible to obtain a PIM shape function.

Several techniques are available to avoid the singularity of $\boldsymbol{F}$ [4, 21], such as moving the field nodes, transformation of the coordinate system, algorithms for matrix triangulation and the inclusion of radial basis functions in the PIM formulation.

4.4.2 RPI Shape Functions

This section presents the Radial Point Interpolators (RPI). Consider again the function space T on $\Omega \subset \mathbb{R}^d$. The finite dimensional space $T_H \subset T$ which discretizes the domain Ω is defined by,

$$T_H := \langle r(\boldsymbol{x}_i - \boldsymbol{x}) : i \in \mathbb{N} \wedge i \leq N \rangle + p_k(\boldsymbol{x}) \tag{4.101}$$

Being $r : \mathbb{R}^d \mapsto \mathbb{R}$ at least a C^1-function. The polynomial function $p_k : \mathbb{R}^d \mapsto \mathbb{R}$ is defined in the space of polynomials of degree less than k. The d-dimensional spatial domain is discretized in a set of N nodes, whose coordinates are defined as $\boldsymbol{X} = \{\boldsymbol{x}_1, \boldsymbol{x}_2, \ldots, \boldsymbol{x}_N\} \in \Omega \wedge \boldsymbol{x}_i \in \mathbb{R}^d$. The density of the nodal distributions is defined by h, Eq. (4.4).

Considering a continuous scalar function $u(\boldsymbol{x})$, with $u \in T$, it is possible to define for an interest point $\boldsymbol{x}_I \in \mathbb{R}^d$, not necessarily coincident with $\boldsymbol{X}$, the RPIM interpolation function of $u(\boldsymbol{x}_I)$ as,

$$u^h(\boldsymbol{x}_I) = \sum_{i=1}^{n} r_i(\boldsymbol{x}_i - \boldsymbol{x}_I) a_i(\boldsymbol{x}_I) + \sum_{j=1}^{m} p_j(\boldsymbol{x}_I) b_j(\boldsymbol{x}_I) = \boldsymbol{r}(\boldsymbol{x}_I)^{\mathrm{T}} \boldsymbol{a}(\boldsymbol{x}_I) + \boldsymbol{p}(\boldsymbol{x}_I)^{\mathrm{T}} \boldsymbol{b}(\boldsymbol{x}_I) \tag{4.102}$$

Being $a_i(\boldsymbol{x}_I)$ and $b_j(\boldsymbol{x}_I)$ the non-constant coefficients of $r(\boldsymbol{x}_I)$ and $\boldsymbol{p}(\boldsymbol{x}_I)$ respectively, which can be defined as,

$$\boldsymbol{a}(\boldsymbol{x}_I) = \{a_1(\boldsymbol{x}_I) \quad a_2(\boldsymbol{x}_I) \quad \ldots \quad a_n(\boldsymbol{x}_I)\}^T \tag{4.103}$$

$$\boldsymbol{b}(\boldsymbol{x}_I) = \{\, b_1(\boldsymbol{x}_I) \quad b_2(\boldsymbol{x}_I) \quad \ldots \quad b_m(\boldsymbol{x}_I)\,\}^{\mathrm{T}} \tag{4.104}$$

where n is the number of field nodes inside the support-domain of interest point $\boldsymbol{x}_I$ and m is the number of monomials of the complete polynomial basis $p_j(\boldsymbol{x}_I)$, which can be defined by the triangle of Pascal, Fig. 4.2, presenting the following vector form,

$$\boldsymbol{p}(\boldsymbol{x}_I) = \{p_1(\boldsymbol{x}_I) \quad p_2(\boldsymbol{x}_I) \quad \ldots \quad p_m(\boldsymbol{x}_I)\}^{\mathrm{T}} \tag{4.105}$$

The Radial Basis Function (RBF) can be defined as,

$$\begin{aligned}\boldsymbol{r}(\boldsymbol{x}_I) &= \{r_1(\boldsymbol{x}_I) \quad r_2(\boldsymbol{x}_I) \quad \cdots \quad r_n(\boldsymbol{x}_I)\}^{\mathrm{T}} \\ &= \{r(\boldsymbol{x}_1 - \boldsymbol{x}_I) \quad r(\boldsymbol{x}_2 - \boldsymbol{x}_I) \quad \cdots \quad r(\boldsymbol{x}_n - \boldsymbol{x}_I)\}^{\mathrm{T}}\end{aligned} \tag{4.106}$$

The only variable in the RBF is the Euclidean norm between the field nodes and the interest point, d_{iI}, which can be defined for a three-dimensional space, $\boldsymbol{x} = \{x, y, z\}$, as,

$$d_{iI} = \sqrt{(x_i - x_I)^2 + (y_i - y_I)^2 + (z_i - z_I)^2} \tag{4.107}$$

In the literature it is possible to find several appropriate RBF to incorporate the RPI formulation [22–25]. As indicated in [26], within the meshless RPI methods the most frequently used globally supported RBFs are the multi-quadrics (MQ) function,

$$r_i(\boldsymbol{x}_I) = \left(d_{iI}^2 + (\gamma d_a)^2\right)^p \tag{4.108}$$

the Gaussian function,

$$r_i(\boldsymbol{x}_I) = e^{\left(-\gamma\left(\frac{d_{iI}}{d_a}\right)^2\right)} \tag{4.109}$$

and the thin plate spline function,

$$r_i(\boldsymbol{x}_I) = d_{iI}^p \tag{4.110}$$

being γ and p the RBF shape parameters. The coefficient d_a is a size coefficient, indicating the influence size of the interest $\boldsymbol{x}_I$. For meshless methods using the classical influence-domain concept, such as the EFGM and the RPIM, d_a is the average nodal spacing of the n nodes inside the support-domain of $\boldsymbol{x}_I$ defined by Eq. (4.2). However, for the NNRPIM, which uses the influence-cell concept, the coefficient d_a can be considered as the size of the Voronoï cell of interest point $\boldsymbol{x}_I$. For the MQ-RBF, if $\boldsymbol{x}_I$ is an integration point, then the coefficient d_a can be considered $d_a = \widehat{w}_I$, being $\widehat{w}_I$ integration weight of $\boldsymbol{x}_I$. The main drawback of MQ-RBF is that the shape parameters γ and p need to be determined and optimize to obtain accurate results.

Also other RBF have been developed, the compactly supported RBFs [27, 28]. These RBF compactly supported are strictly positive definite for all $d_{iI} \leq d_s$, being $d_s = \max\|\boldsymbol{x}_i - \boldsymbol{x}_I\|, \forall \boldsymbol{x}_i \in \boldsymbol{X}_I$ and can be constructed to have the desire amount of smoothness [26].

Recently, Belinha et al. [29–31] proposed the Natural Radial Element Method (NREM), a new meshless method requiring reduced size support-domains and using a new RPI approach, in which the MQ-RBF is substituted by the square of the Euclidean norm.

The RPI formulation used in this book only uses the MQ-RBF [32]. In an initial phase several other RBF were used with the NNRPIM, however the best results were obtain with the MQ-RBF.

Regarding the MQ-RBF, it was found that for the RPIM formulation [22, 23] the optimal shape parameters are: $\gamma = 1.42 d_a^{-1}$ and $p = 1.03$. However in the work developed by Belinha et al. [33–35], other optimal values for the shape parameters were obtained. These optimization studies are presented in the next chapter.

The preliminary studies on RPI have found that the absence of a polynomial basis in the RPI formulation leads to the failure of the standard patch tests, i.e., the pure RBF RPI formulation it is unable to reconstruct exactly a linear polynomial field. The addition of a linear polynomial basis is sufficient to assure the require C^1-consistency to pass the standard patch test. It was observed in the meshless methods using the weak formulation approach that the addition of a polynomial basis enhance the stability of the RPI shape functions, reducing the condition number of the moment matrix.

Analogous with the generic PIM formulation, the non-constant coefficients $\boldsymbol{a}(\boldsymbol{x}_I)$ and $\boldsymbol{b}(\boldsymbol{x}_I)$ can be obtained by imposing $u^h(\boldsymbol{x}_I)$, defined in Eq. (4.102), to pass through all the n nodal values on the support-domain of $\boldsymbol{x}_I$. Therefore, satisfying Eq. (4.102) for all n nodes on the support-domain of the interest point $\boldsymbol{x}_I$, a system of equations with n equations with $n + m$ unknowns is obtained,

$$\begin{cases} u^h(\boldsymbol{x}_1) = \sum_{i=1}^{n} r_i(\boldsymbol{x}_i - \boldsymbol{x}_1) a_i(\boldsymbol{x}_I) + \sum_{j=1}^{m} p_j(\boldsymbol{x}_1) b_j(\boldsymbol{x}_I) = u(\boldsymbol{x}_1) \\ u^h(\boldsymbol{x}_2) = \sum_{i=1}^{n} r_i(\boldsymbol{x}_i - \boldsymbol{x}_2) a_i(\boldsymbol{x}_I) + \sum_{j=1}^{m} p_j(\boldsymbol{x}_2) b_j(\boldsymbol{x}_I) = u(\boldsymbol{x}_2) \\ \vdots \\ u^h(\boldsymbol{x}_n) = \sum_{i=1}^{n} r_i(\boldsymbol{x}_i - \boldsymbol{x}_n) a_i(\boldsymbol{x}_I) + \sum_{j=1}^{m} p_j(\boldsymbol{x}_n) b_j(\boldsymbol{x}_I) = u(\boldsymbol{x}_n) \end{cases} \tag{4.111}$$

Which can be expressed in the matrix form as,

$$\boldsymbol{R a}(\boldsymbol{x}_I)+\boldsymbol{P b}(\boldsymbol{x}_I)=\boldsymbol{u}_s \tag{4.112}$$

where $\boldsymbol{u}_s$ is the vector containing the nodal parameters of the field function for each node inside the support-domain of the RPI shape function defined in Eq. (4.20). The generic moment matrix of the RBF is defined as,

$$\boldsymbol{R}=\begin{bmatrix} r_1(\boldsymbol{x}_1) & r_2(\boldsymbol{x}_1) & \cdots & r_n(\boldsymbol{x}_1) \\ r_1(\boldsymbol{x}_2) & r_2(\boldsymbol{x}_2) & \cdots & r_n(\boldsymbol{x}_2) \\ \vdots & \vdots & \ddots & \vdots \\ r_1(\boldsymbol{x}_n) & r_2(\boldsymbol{x}_n) & \cdots & r_n(\boldsymbol{x}_n) \end{bmatrix} \tag{4.113}$$

Specifically for the MQ-RBF the radial moment matrix is defined as,

$$\boldsymbol{R}=\begin{bmatrix} \left(d_{11}^2+(\gamma d_c)^2\right)^p & \left(d_{12}^2+(\gamma d_c)^2\right)^p & \cdots & \left(d_{1n}^2+(\gamma d_c)^2\right)^p \\ \left(d_{21}^2+(\gamma d_c)^2\right)^p & \left(d_{22}^2+(\gamma d_c)^2\right)^p & \cdots & \left(d_{2n}^2+(\gamma d_c)^2\right)^p \\ \vdots & \vdots & \ddots & \vdots \\ \left(d_{n1}^2+(\gamma d_c)^2\right)^p & \left(d_{n2}^2+(\gamma d_c)^2\right)^p & \cdots & \left(d_{nn}^2+(\gamma d_c)^2\right)^p \end{bmatrix} \tag{4.114}$$

The polynomial moment matrix is defined as,

$$\boldsymbol{P}=\begin{bmatrix} p_1(\boldsymbol{x}_1) & p_2(\boldsymbol{x}_1) & \cdots & p_m(\boldsymbol{x}_1) \\ p_1(\boldsymbol{x}_2) & p_2(\boldsymbol{x}_2) & \cdots & p_m(\boldsymbol{x}_2) \\ \vdots & \vdots & \ddots & \vdots \\ p_1(\boldsymbol{x}_n) & p_2(\boldsymbol{x}_n) & \cdots & p_m(\boldsymbol{x}_n) \end{bmatrix} \tag{4.115}$$

Considering a linear polynomial basis, $p(\boldsymbol{x})^T=\{1 \quad x \quad y \quad z\}$ with $m=4$, defined in the three-dimensional space, $\boldsymbol{x}=\{x \quad y \quad z\}$, the polynomial moment matrix $\boldsymbol{P}$ can be explicitly defined as,

$$\boldsymbol{P}=\begin{bmatrix} 1 & x_1 & y_1 & z_1 \\ 1 & x_2 & y_2 & z_2 \\ \vdots & \vdots & \ddots & \vdots \\ 1 & x_n & y_n & z_n \end{bmatrix} \tag{4.116}$$

Notice that Eq. (4.111) has $n+m$ unknowns. To obtain an unique solution an extra set of equations has to be considered [36], as a consequence of a theorem of Duchon [37]. Therefore, the following supplementary m equations can be added to the initial equation system,

$$\begin{cases} \sum_{i=1}^{n} p_1(\boldsymbol{x}_i)a_i(\boldsymbol{x}_I) = p_1(\boldsymbol{x}_1)a_1(\boldsymbol{x}_I) + p_1(\boldsymbol{x}_2)a_2(\boldsymbol{x}_I) + \cdots + p_1(\boldsymbol{x}_n)a_n(\boldsymbol{x}_I) = 0 \\ \sum_{i=1}^{n} p_2(\boldsymbol{x}_i)a_i(\boldsymbol{x}_I) = p_2(\boldsymbol{x}_1)a_1(\boldsymbol{x}_I) + p_2(\boldsymbol{x}_2)a_2(\boldsymbol{x}_I) + \cdots + p_2(\boldsymbol{x}_n)a_n(\boldsymbol{x}_I) = 0 \\ \vdots \\ \sum_{i=1}^{n} p_m(\boldsymbol{x}_i)a_i(\boldsymbol{x}_I) = p_m(\boldsymbol{x}_1)a_1(\boldsymbol{x}_I) + p_m(\boldsymbol{x}_2)a_2(\boldsymbol{x}_I) + \cdots + p_m(\boldsymbol{x}_n)a_n(\boldsymbol{x}_I) = 0 \end{cases} \tag{4.117}$$

Which can be presented in the matrix form as,

$$\boldsymbol{P}^{\mathrm{T}}\boldsymbol{a}(\boldsymbol{x}_I) = 0 \tag{4.118}$$

Combining Eq. (4.112) with (4.118) it is obtained the following set of equations written in matrix form,

$$\begin{bmatrix} \boldsymbol{R} & \boldsymbol{P} \\ \boldsymbol{P}^{\mathrm{T}} & \boldsymbol{Z} \end{bmatrix} \begin{Bmatrix} \boldsymbol{a}(\boldsymbol{x}_I) \\ \boldsymbol{b}(\boldsymbol{x}_I) \end{Bmatrix} = \boldsymbol{M}_T \begin{Bmatrix} \boldsymbol{a}(\boldsymbol{x}_I) \\ \boldsymbol{b}(\boldsymbol{x}_I) \end{Bmatrix} = \begin{Bmatrix} \boldsymbol{u}_s \\ z \end{Bmatrix} \tag{4.119}$$

Being $Z_{ij} = 0$ for $\{i, j\} = 1, 2, \ldots, m$ and $z_i = 0$ for $i = 1, 2, \ldots, m$. Since the radial moment matrix $\boldsymbol{R}$ is symmetric, the total moment matrix $\boldsymbol{M}_T$ will also be symmetric. With Eq. (4.119) it is possible to obtain the non-constant coefficients $\boldsymbol{a}(\boldsymbol{x}_I)$ and $\boldsymbol{b}(\boldsymbol{x}_I)$,

$$\begin{Bmatrix} \boldsymbol{a}(\boldsymbol{x}_I) \\ \boldsymbol{b}(\boldsymbol{x}_I) \end{Bmatrix} = \boldsymbol{M}_T^{-1} \begin{Bmatrix} \boldsymbol{u}_s \\ z \end{Bmatrix} \tag{4.120}$$

Within Eq. (4.102) the polynomial basis vector $\boldsymbol{p}(\boldsymbol{x}_I)$ and the RBF vector $\boldsymbol{r}(\boldsymbol{x}_I)$ obtained for the interest point $\boldsymbol{x}_I$ can be combined in a single line vector $\{\boldsymbol{r}(\boldsymbol{x}_I)^{\mathrm{T}}\boldsymbol{p}(\boldsymbol{x}_I)^{\mathrm{T}}\}$. Thus, substituting the solution from Eq. (4.120) back in Eq. (4.102), it possible to re-write Eq. (4.102) as,

$$u^h(\boldsymbol{x}_I) = \{\boldsymbol{r}(\boldsymbol{x}_I)^{\mathrm{T}} \quad \boldsymbol{p}(\boldsymbol{x}_I)^{\mathrm{T}}\} \begin{Bmatrix} \boldsymbol{a}(\boldsymbol{x}_I) \\ \boldsymbol{b}(\boldsymbol{x}_I) \end{Bmatrix} = \{\boldsymbol{r}(\boldsymbol{x}_I)^{\mathrm{T}} \quad \boldsymbol{p}(\boldsymbol{x}_I)^{\mathrm{T}}\}\boldsymbol{M}_T^{-1} \begin{Bmatrix} \boldsymbol{u}_s \\ z \end{Bmatrix} \tag{4.121}$$

Recall that the field variable value for an interest point $\boldsymbol{x}_I$ is interpolated using the shape function values obtained at the nodes inside the support-domain of $\boldsymbol{x}_I$,

$$u^h(\boldsymbol{x}_I) = \sum_{i=1}^{n} \varphi_i(\boldsymbol{x}_I)u_i = \boldsymbol{\varphi}(\boldsymbol{x}_I)^{\mathrm{T}}\boldsymbol{u}_s \tag{4.122}$$

Therefore, it is possible to identify the interpolation function vector $\boldsymbol{\varphi}(\boldsymbol{x}_I)$ on Eq. (4.121),

$$u^h(\boldsymbol{x}_I) = \left\{ \boldsymbol{r}(\boldsymbol{x}_I)^{\mathrm{T}} \quad \boldsymbol{p}(\boldsymbol{x}_I)^{\mathrm{T}} \right\} \boldsymbol{M}_T^{-1} \left\{ \begin{matrix} \boldsymbol{u}_s \\ \boldsymbol{z} \end{matrix} \right\} = \left\{ \boldsymbol{\varphi}(\boldsymbol{x}_I)^{\mathrm{T}} \quad \boldsymbol{\psi}(\boldsymbol{x}_I)^{\mathrm{T}} \right\} \left\{ \begin{matrix} \boldsymbol{u}_s \\ \boldsymbol{z} \end{matrix} \right\} \tag{4.123}$$

The interpolation function vector $\boldsymbol{\varphi}(\boldsymbol{x}_I)$ and the byproduct vector $\boldsymbol{\psi}(\boldsymbol{x}_I)$ are defined as,

$$\boldsymbol{\varphi}(\boldsymbol{x}_I)^{\mathrm{T}} = \{ \varphi_1(\boldsymbol{x}_I) \quad \varphi_2(\boldsymbol{x}_I) \quad \ldots \quad \varphi_n(\boldsymbol{x}_I) \} \tag{4.124}$$

$$\boldsymbol{\psi}(\boldsymbol{x}_I)^{\mathrm{T}} = \{ \psi_1(\boldsymbol{x}_I) \quad \psi_2(\boldsymbol{x}_I) \quad \ldots \quad \psi_m(\boldsymbol{x}_I) \} \tag{4.125}$$

Notice that the byproduct vector $\boldsymbol{\psi}(\boldsymbol{x}_I)$ only exists if a polynomial basis is considered, otherwise it does not appears. The components of vector $\boldsymbol{\psi}(\boldsymbol{x}_I)$ do not possess any relevant physical meaning. Additionally, to obtain the interpolation field variable, the byproduct vector $\boldsymbol{\psi}(\boldsymbol{x}_I)$ is multiplied to the null vector z, therefore $\boldsymbol{\psi}(\boldsymbol{x}_I)$ can be completely neglected.

$$u^h(\boldsymbol{x}_I) = \sum_{i=1}^{n} \varphi_i(\boldsymbol{x}_I) u_i + \underbrace{\sum_{i=1}^{m} \psi_i(\boldsymbol{x}_I) z_i}_{0} = \sum_{i=1}^{n} \varphi_i(\boldsymbol{x}_I) u_i \tag{4.126}$$

being $\boldsymbol{\varphi}_i(\boldsymbol{x}_I)$ the shape function value of the interest point $\boldsymbol{x}_I$ for the ith node obtained considering the nodes n inside the support-domain of interest point $\boldsymbol{x}_I$. Consider now two distinct interest points $\{\boldsymbol{x}_I, \boldsymbol{x}_J\} \in \Omega \wedge \boldsymbol{x}_i \in \mathbb{R}^d$, being $\boldsymbol{x}_J \neq \boldsymbol{x}_I$. The shape functions of both interest points possess the exactly same support-domain nodal set $\boldsymbol{N}_I = \boldsymbol{N}_J = \{n_1, n_2, \ldots, n_n\} \subset \boldsymbol{N}$, being $\boldsymbol{N}$ the complete nodal set discretizing the problem domain. Notice that for the interest point $\boldsymbol{x}_I$, the total moment matrix $\boldsymbol{M}_T^I$ constructed using the $\boldsymbol{N}_I$ nodal set will be equal to the total moment matrix $\boldsymbol{M}_T^J$ of interest point $\boldsymbol{x}_J$ obtained for the $\boldsymbol{N}_J$ nodal set, leading to $a_i(\boldsymbol{x}_I) = a_i(\boldsymbol{x}_J)$ and $b_i(\boldsymbol{x}_I) = b_i(\boldsymbol{x}_J)$. Therefore the obtained coefficients $a_i(\boldsymbol{x}_I)$ and $b_j(\boldsymbol{x}_I)$ are in fact constant as long as the same n nodes inside the support-domain of the interest point $\boldsymbol{x}_I$ are maintained. The most important conclusion is that the total moment matrix, and also the RBF and the polynomial moment matrices, are not directly dependent on the spatial position of the interest point $\boldsymbol{x}_I$. Consequently,

$$\left\{ \frac{\partial \boldsymbol{R}}{\partial \boldsymbol{x}} \quad \frac{\partial \boldsymbol{P}}{\partial \boldsymbol{x}} \quad \frac{\partial \boldsymbol{M}_T}{\partial \boldsymbol{x}} \right\} = \{0 \quad 0 \quad 0\} \tag{4.127}$$

In order to compute the partial derivatives of the interpolated field function, Eq. (4.123), it is necessary to obtain the respective RPI shape functions partial derivatives.

$$\frac{\partial u^h(\boldsymbol{x}_I)}{\partial \xi} = \sum_{i=1}^{n} \frac{\partial \varphi_i(\boldsymbol{x}_I)}{\partial \xi} u_i + \underbrace{\sum_{i=1}^{m} \frac{\partial \psi_i(\boldsymbol{x}_I)}{\partial \xi} z_i}_{0} = \boldsymbol{\varphi}(\boldsymbol{x}_I)^{\mathrm{T}}_{,\xi} \boldsymbol{u}_s \tag{4.128}$$

Thus, with respect to a generic variable ξ, the first order partial derivatives of interpolated field function can be determined by,

$$\left(\left\{ \boldsymbol{\varphi}(\boldsymbol{x}_I)^{\mathrm{T}} \quad \boldsymbol{\psi}(\boldsymbol{x}_I)^{\mathrm{T}} \right\}\right)_{,\xi} = \left(\left\{ \boldsymbol{r}(\boldsymbol{x}_I)^{\mathrm{T}} \quad \boldsymbol{p}(\boldsymbol{x}_I)^{\mathrm{T}} \right\} \boldsymbol{M}_T^{-1}\right)_{,\xi} \tag{4.129}$$

and considering Eq. (4.127),

$$\left(\left\{ \boldsymbol{r}(\boldsymbol{x}_I)^{\mathrm{T}} \quad \boldsymbol{p}(\boldsymbol{x}_I)^{\mathrm{T}} \right\} \boldsymbol{M}_T^{-1}\right)_{,\xi} = \left(\left\{ \boldsymbol{r}(\boldsymbol{x}_I)^{\mathrm{T}} \quad \boldsymbol{p}(\boldsymbol{x}_I)^{\mathrm{T}} \right\}\right)_{,\xi} \boldsymbol{M}_T^{-1} + \left\{ \boldsymbol{r}(\boldsymbol{x}_I)^{\mathrm{T}} \quad \boldsymbol{p}(\boldsymbol{x}_I)^{\mathrm{T}} \right\} \underbrace{\boldsymbol{M}_{T,\xi}^{-1}}_{0} \tag{4.130}$$

Equation (4.129) can be presented as,

$$\left\{ \boldsymbol{\varphi}(\boldsymbol{x}_I)^{\mathrm{T}}_{,\xi} \quad \boldsymbol{\psi}(\boldsymbol{x}_I)^{\mathrm{T}}_{,\xi} \right\} = \left\{ \boldsymbol{r}(\boldsymbol{x}_I)^{\mathrm{T}}_{,\xi} \quad \boldsymbol{p}(\boldsymbol{x}_I)^{\mathrm{T}}_{,\xi} \right\} \boldsymbol{M}_T^{-1} \tag{4.131}$$

The first order partial derivative of the RBF vector with respect to the same generic variable ξ is defined as,

$$\boldsymbol{r}(\boldsymbol{x}_I)_{,\xi} = \left\{ r_1(\boldsymbol{x}_I)_{,\xi} \quad r_2(\boldsymbol{x}_I)_{,\xi} \quad \cdots \quad r_n(\boldsymbol{x}_I)_{,\xi} \right\}^{\mathrm{T}} = \left\{ \frac{\partial r_1(\boldsymbol{x}_I)}{\partial \xi} \quad \frac{\partial r_2(\boldsymbol{x}_I)}{\partial \xi} \quad \cdots \quad \frac{\partial r_n(\boldsymbol{x}_I)}{\partial \xi} \right\}^{\mathrm{T}} \tag{4.132}$$

Being for the MQ-RBF,

$$\frac{\partial r_i(\boldsymbol{x}_I)}{\partial \xi} = -2p(\xi_i - \xi_I)\left(d_{iI}^2 + (\gamma d_c)^2\right)^{p-1} \tag{4.133}$$

For the same generic variable ξ, the first order partial derivative of the polynomial basis vector is obtained with,

$$\boldsymbol{p}(\boldsymbol{x}_I)_{,\xi} = \left\{ p_1(\boldsymbol{x}_I)_{,\xi} \quad p_2(\boldsymbol{x}_I)_{,\xi} \quad \cdots \quad p_m(\boldsymbol{x}_I)_{,\xi} \right\}^{\mathrm{T}} = \left\{ \frac{\partial p_1(\boldsymbol{x}_I)}{\partial \xi} \quad \frac{\partial p_2(\boldsymbol{x}_I)}{\partial \xi} \quad \cdots \quad \frac{\partial p_n(\boldsymbol{x}_I)}{\partial \xi} \right\}^{\mathrm{T}} \tag{4.134}$$

It is possible to obtain the second order partial derivative of the interpolated field function, with respect to the generic variables ξ and η, with the following expression,

$$\frac{\partial^2 u^h(\boldsymbol{x}_I)}{\partial\xi\partial\eta} = \sum_{i=1}^{n}\frac{\partial^2 \varphi_i(\boldsymbol{x}_I)}{\partial\xi\partial\eta}u_i + \underbrace{\sum_{i=1}^{m}\frac{\partial^2 \psi_i(\boldsymbol{x}_I)}{\partial\xi\partial\eta}z_i}_{0} = \boldsymbol{\varphi}(\boldsymbol{x}_I)^{\mathrm{T}}_{,\xi\eta}\boldsymbol{u}_s \tag{4.135}$$

Following the same argument used to obtain Eq. (4.131), the spatial second order partial derivative of the RPI shape function with respect to ξ and η, is defined as,

$$\left\{ \boldsymbol{\varphi}(\boldsymbol{x}_I)^{\mathrm{T}}_{,\xi\eta} \quad \boldsymbol{\psi}(\boldsymbol{x}_I)^{\mathrm{T}}_{,\xi\eta} \right\} = \left\{ \boldsymbol{r}(\boldsymbol{x}_I)^{\mathrm{T}}_{,\xi\eta} \quad \boldsymbol{p}(\boldsymbol{x}_I)^{\mathrm{T}}_{,\xi\eta} \right\}\boldsymbol{M}_T^{-1} \tag{4.136}$$

in which the second order cross partial derivative of the RBF vector is defined as,

$$\begin{aligned}\boldsymbol{r}(\boldsymbol{x}_I)_{,\xi\eta} &= \left\{ r_1(\boldsymbol{x}_I)_{,\xi\eta} \quad r_2(\boldsymbol{x}_I)_{,\xi\eta} \quad \cdots \quad r_n(\boldsymbol{x}_I)_{,\xi\eta} \right\}^{\mathrm{T}} \\ &= \left\{ \frac{\partial^2 r_1(\boldsymbol{x}_I)}{\partial\xi\partial\eta} \quad \frac{\partial^2 r_2(\boldsymbol{x}_I)}{\partial\xi\partial\eta} \quad \cdots \quad \frac{\partial^2 r_n(\boldsymbol{x}_I)}{\partial\xi\partial\eta} \right\}^{\mathrm{T}}\end{aligned} \tag{4.137}$$

being,

$$\frac{\partial^2 r_i(\boldsymbol{x}_I)}{\partial\xi\partial\eta} = 4p(p-1)(\xi_i - \xi_I)(\eta_i - \eta_I)\left(d_{iI}^2 + (\gamma d_c)^2\right)^{p-2} \tag{4.138}$$

The respective second order cross partial derivative of the polynomial basis vector is obtained with,

$$\begin{aligned}\boldsymbol{p}(\boldsymbol{x}_I)_{,\xi\eta} &= \left\{ p_1(\boldsymbol{x}_I)_{,\xi\eta} \quad p_2(\boldsymbol{x}_I)_{,\xi\eta} \quad \cdots \quad p_m(\boldsymbol{x}_I)_{,\xi\eta} \right\}^{\mathrm{T}} \\ &= \left\{ \frac{\partial^2 p_1(\boldsymbol{x}_I)}{\partial\xi\partial\eta} \quad \frac{\partial^2 p_2(\boldsymbol{x}_I)}{\partial\xi\partial\eta} \quad \cdots \quad \frac{\partial^2 p_n(\boldsymbol{x}_I)}{\partial\xi\partial\eta} \right\}^{\mathrm{T}}\end{aligned} \tag{4.139}$$

The second order partial derivative of the interpolated field function with respect to the generic variable ξ can be determined with,

$$\frac{\partial^2 u^h(\boldsymbol{x}_I)}{\partial\xi^2} = \sum_{i=1}^{n}\frac{\partial^2 \varphi_i(\boldsymbol{x}_I)}{\partial\xi^2}u_i + \underbrace{\sum_{i=1}^{m}\frac{\partial^2 \psi_i(\boldsymbol{x}_I)}{\partial\xi^2}z_i}_{0} = \boldsymbol{\varphi}(\boldsymbol{x}_I)^{\mathrm{T}}_{,\xi\xi}\boldsymbol{u}_s \tag{4.140}$$

Following the previous argument, the spatial second order partial derivative of the PIM shape function with respect to ξ is obtained.

$$\left\{ \boldsymbol{\varphi}(\boldsymbol{x}_I)^{\mathrm{T}}_{,\xi\xi} \quad \boldsymbol{\psi}(\boldsymbol{x}_I)^{\mathrm{T}}_{,\xi\xi} \right\} = \left\{ \boldsymbol{r}(\boldsymbol{x}_I)^{\mathrm{T}}_{,\xi\xi} \quad \boldsymbol{p}(\boldsymbol{x}_I)^{\mathrm{T}}_{,\xi\xi} \right\}\boldsymbol{M}_T^{-1} \tag{4.141}$$

being the second order partial derivative of the RBF vector defined as,

$$\begin{aligned} \boldsymbol{r}(\boldsymbol{x}_I)_{,\xi\xi} &= \left\{ r_1(\boldsymbol{x}_I)_{,\xi\xi} \quad r_2(\boldsymbol{x}_I)_{,\xi\xi} \quad \ldots \quad r_n(\boldsymbol{x}_I)_{,\xi\xi} \right\}^{\mathrm{T}} \\ &= \left\{ \frac{\partial^2 r_1(\boldsymbol{x}_I)}{\partial \xi^2} \quad \frac{\partial^2 r_2(\boldsymbol{x}_I)}{\partial \xi^2} \quad \ldots \quad \frac{\partial^2 r_n(\boldsymbol{x}_I)}{\partial \xi^2} \right\}^{\mathrm{T}} \end{aligned} \tag{4.142}$$

where,

$$\frac{\partial^2 r_i(\boldsymbol{x}_I)}{\partial \xi^2} = 2p\left(d_{iI}^2 + (\gamma d_c)^2\right)^{p-1} + 4p(p-1)(\xi_i - \xi_I)^2\left(d_{iI}^2 + (\gamma d_c)^2\right)^{p-2} \tag{4.143}$$

and the second order partial derivative of the polynomial basis vector defined as,

$$\begin{aligned} \boldsymbol{p}(\boldsymbol{x}_I)_{,\xi\xi} &= \left\{ p_1(\boldsymbol{x}_I)_{,\xi\xi} \quad p_2(\boldsymbol{x}_I)_{,\xi\xi} \quad \ldots \quad p_m(\boldsymbol{x}_I)_{,\xi\xi} \right\}^{\mathrm{T}} \\ &= \left\{ \frac{\partial^2 p_1(\boldsymbol{x}_I)}{\partial \xi^2} \quad \frac{\partial^2 p_2(\boldsymbol{x}_I)}{\partial \xi^2} \quad \ldots \quad \frac{\partial^2 p_n(\boldsymbol{x}_I)}{\partial \xi^2} \right\}^{\mathrm{T}} \end{aligned} \tag{4.144}$$

Generally, for a random nodal distribution, the total moment matrix $\boldsymbol{M}_T$ is not singular [28], however the number of nodes inside the support-domain should be much larger than the number of polynomial terms of the polynomial basis function, $n \gg m$. In this book only low order polynomials are used, therefore the $n \gg m$ rule is always respected. However it should be mentioned that support-domains containing a large number of nodes can lead to ill-conditioned RBF moment matrix, $\boldsymbol{R}$. Within the RPI shape function construction, the inclusion of the RBF as a functional basis permits to avoid the singularity problem of the polynomial PIM.

However compared with the polynomial PIM, the use of RBF increases significantly the computational cost. One of the most significantly advantage of using the RBF in the PIM formulation is the virtually infinity continuity of the RBF, permitting to construct shape functions possessing high continuity.

4.4.3 RPI Shape Functions Calculation

Comparing with the MLS shape function construction procedure, the RPI shape function construction is much simple and easy to program. Consider an interest point $\boldsymbol{x}_I$ with an influence-domain defined by $\boldsymbol{X}_I = \{\boldsymbol{x}_1, \boldsymbol{x}_2, \ldots, \boldsymbol{x}_n\} \in \mathbb{R}^d$. The objective is to construct a RPI shape function using a polynomial basis function combined with the MQ-RBF, which require two shape parameters γ and p that should be defined in a pre-processing phase. The RPI shape function, and the respective partial derivatives, are obtained following the presented algorithm:

1. The MQ-RBF expression requires the definition of a size parameter d_a, Eq. (4.108). Within the RPIM d_a can be assumed as the RPI shape function support-domain, d_s, which can be defined based on the influence-domain of the

interest point $\boldsymbol{x}_I$, $d_s = \max\|\boldsymbol{x}_i - \boldsymbol{x}_I\|, \forall \boldsymbol{x}_i \in \boldsymbol{X}_I$. Since the NNRPIM uses the influence-cell concept, the coefficient d_a can be considered as the size of the Voronoï cell of interest point $\boldsymbol{x}_I$; or, if $\boldsymbol{x}_I$ is an integration point, then $d_a = \widehat{w}_I$, being $\widehat{w}_I$ integration weight of $\boldsymbol{x}_I$.

2. Next, the polynomial vector for interest point $\boldsymbol{x}_I$ is defined: $\boldsymbol{p}(\boldsymbol{x}_I)$. The dimension of $\boldsymbol{p}(\boldsymbol{x}_I)$ depends on the monomials, $[m \times 1]$. For example, in a two-dimensional space, the quadratic polynomial vector for interest point $\boldsymbol{x}_I$ is defined as: $\boldsymbol{p}(\boldsymbol{x}_I) = \left\{1 \quad x_I \quad y_I \quad x_I^2 \quad x_I y_I \quad y_I^2\right\}^T$.
3. Then, the MQ-RBF vector $\boldsymbol{r}(\boldsymbol{x}_I)$ is constructed, Eq. (4.106). The distances between the interest point $\boldsymbol{x}_I$ and all the nodes inside the influence-domain of $\boldsymbol{x}_I$ are determined, d_{Ii}. Afterwards the component of the MQ-RBF vector can be determined using Eq. (4.108). The dimension of the MQ-RBF vector $\boldsymbol{r}(\boldsymbol{x}_I)$ is $[n \times 1]$.
4. The $[n \times n]$ radial moment matrix $\boldsymbol{R}$ is determined using Eq. (4.114).
5. The $[n \times m]$ polynomial moment matrix $\boldsymbol{P}$ is constructed with Eq. (4.115).
6. The total moment matrix $\boldsymbol{M}$ is defined using the radial moment matrix $\boldsymbol{R}$ and the polynomial moment matrix $\boldsymbol{P}$, Eq. (4.119). Then $\boldsymbol{M}^{-1}$ is determined. The total momentum matrix is a square matrix with $[(n + m) \times (n + m)]$ size.
7. The MQ-RBF vector $\boldsymbol{r}(\boldsymbol{x}_I)$ and the polynomial vector $\boldsymbol{p}(\boldsymbol{x}_I)$ are assembled in one single line vector: $\left\{\boldsymbol{r}(\boldsymbol{x}_I)^{\mathrm{T}}\boldsymbol{p}(\boldsymbol{x}_I)^{\mathrm{T}}\right\}$ with size $[1 \times (n + m)]$.
8. Finally, the interpolation function $\boldsymbol{\varphi}(\boldsymbol{x}_I)$, Eq. (4.124), is obtained with Eq. (4.123).

The RPI shape function first order partial derivatives can be obtained applying Eq. (4.131) and the second order partial derivatives are obtained with Eqs. (4.136) and (4.141).

4.4.4 Radial Basis Functions

Although in the RPI approach the most commonly used RBFs are the multi-quadrics RBF, Eq. (4.108), the Gaussian RBF, Eq. (4.109), and the thin plate spline RBF, Eq. (4.110), there other RBF that can be used in the RPI formulation [27, 38]. The major disadvantage of using non-compactly supported RBFs, such are the MQ-RBF, the Gaussian-RBF and the thin-plate spline RBF, is the existence of shape parameters that require a previous calibration and optimization study.

In opposition, the compactly supported RBFs [27, 28] do not require shape parameters and are constructed using uniquely as variable the normalized radial distance. Five example of compactly supported RBFs are presented from Eq. (4.145) to (4.149). In Eqs. (4.145) and (4.146) are presented compactly supported RBFs proposed by Wu [27] and in Eqs. (4.147), (4.148) and (4.149) are presented compactly supported RBFs proposed by Wendland [28].

$$r_i(\boldsymbol{x}_I) = (1 - \bar{d}_{iI})^5 \left(8 + 40\bar{d}_{iI} + 48\bar{d}_{iI}^2 + 25\bar{d}_{iI}^3 + 5\bar{d}_{iI}^4\right), \text{ [RBF - Wu1]} \quad (4.145)$$

$$r_i(\boldsymbol{x}_I) = (1 - \bar{d}_{iI})^6 \left(6 + 36\bar{d}_{iI} + 82\bar{d}_{iI}^2 + 72\bar{d}_{iI}^3 + 30\bar{d}_{iI}^4 + 5\bar{d}_{iI}^5\right), \text{ [RBF - Wu2]} \quad (4.146)$$

$$r_i(\boldsymbol{x}_I) = (1 - \bar{d}_{iI})^4 (3 + 4\bar{d}_{iI}), \text{ [RBF - We1]} \quad (4.147)$$

$$r_i(\boldsymbol{x}_I) = (1 - \bar{d}_{iI})^6 \left(3 + 18\bar{d}_{iI} + 35\bar{d}_{iI}^2\right), \text{ [RBF - We2]} \quad (4.148)$$

$$r_i(\boldsymbol{x}_I) = (1 - \bar{d}_{iI})^8 \left(1 + 8\bar{d}_{iI} + 25\bar{d}_{iI}^2 + 32\bar{d}_{iI}^3\right), \text{ [RBF - We3]} \quad (4.149)$$

Being $\bar{d}_{iI} = d_{iI}/d_s$. The compactly supported RBF proposed by Wu and Wendland are strictly positive definite for all $d_{iI} \leq d_s$, being d_s the shape function support-domain. When $\bar{d}_{iI} > 1$ the RBF assume a null value. In the literature [5] it is possible to find works assuring that, in comparison with non-compactly supported RBF, there is no clear advantage of using compactly supported RBF to solve solid mechanical problems or surface fitting.

The RBFs referred in this text are presented in Fig. 4.20. In all examples the RBF application domain is $x = [-0.5, 0.5] \wedge x \in \mathbb{R}$ and the centre of the RBF is $x = 0$ with a support-domain $d_s = 0.5$. The MQ-RBFs presented in Fig. 4.20a were obtained considering a fixed value for the shape parameter $p = 1.001$ and varying the shape parameter γ, $\gamma = \{0.001, 1.001, 2.001\}$. In Fig. 4.20b are presented Gaussian-RBFs curves obtained considering $\gamma = \{0.001, 1.001, 10.001\}$ and in Fig. 4.20c are presented thin plate spline RBFs constructed using $p = \{0.001, 1.001, 2.001\}$.

It is possible to observe in Fig. 4.20a, b, c that the constructed RBFs are not confined to the support-domain and that the variation of the shape parameters permit to create distinct curves within the same RBF.

In Fig. 4.20d are presented the compactly supported RBFs from Eq. (4.145) to (4.149). Notice that these RBFs curves are bell shaped and assume a null value on the support-domain limit. Since it is the first term of Eq. (4.145) to (4.149) that imposes the RBF continuity, these compactly supported RBFs can be constructed with the desire amount of smoothness.

In this work only the MQ-RBF is used. With the purpose of showing the importance of the RBF shape parameter value on the constructed RPI shape function, consider a one-dimensional domain $\Omega \subset \mathbb{R}$ discretized by the nodal set $\boldsymbol{X} = \{\boldsymbol{x}_1, \boldsymbol{x}_2, \ldots, \boldsymbol{x}_{11}\} \in \Omega$, being $\boldsymbol{x} \in \mathbb{R} : x \in [0, 1]$. Two distinct nodal distributions are considered in this example: a uniform nodal distribution and an irregular nodal distribution. As in Sect. 4.3.6, an approximation function $u^h(\boldsymbol{x})$ will be adjust to the N discrete nodal values $u(\boldsymbol{x}_i)$. In Table 4.1 are presented the spatial location of each node $\boldsymbol{x}_i$ discretizing the problem domain and the respective nodal values $u(\boldsymbol{x}_i)$. The described procedure is followed to obtain the approximation function $u^h(\boldsymbol{x})$:

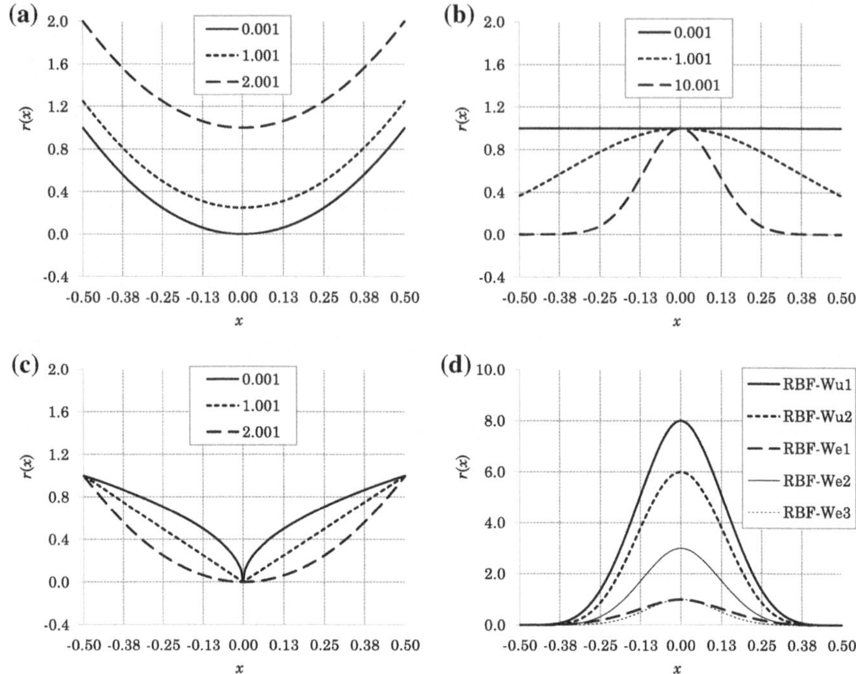

Fig. 4.20 **a** MQ-RBF obtained considering $p = 1.001$ and $\gamma = \{0.001, 1.001, 2.001\}$. **b** Gaussian-RBF obtained considering $\gamma = \{0.001, 1.001, 10.001\}$. **c** Thin-plate-RBF obtained considering $p = \{0.001, 1.001, 2.001\}$. **d** Compactly supported RBFs

1. A background mesh of interest points, covering the problem domain, is constructed, $\boldsymbol{Q} = \{\boldsymbol{q}_1, \boldsymbol{q}_2, \ldots, \boldsymbol{q}_{101}\} \in \Omega \wedge \boldsymbol{q}_i \in \mathbb{R}^1$.
2. The size of the support-domain of the shape functions is defined as: $d_s = 3.0001 \cdot h$. In this case it is considered $h = 1/10$.
3. To simplify the analysis, the size of the influence-domain of each interest point $\boldsymbol{q}_I$ is defined as $d_I = d_s$. Next, each interest point $\boldsymbol{q}_I$ searches for the n field nodes within the radial distance d_I, establishing the individual influence-domains of interest points $\boldsymbol{q}_I$.
4. The RPI shape functions are constructed for each interest point $\boldsymbol{q}_I$ following the procedure indicated in Sect. 4.4.3. In this example it is used the MQ-RBF and the linear polynomial basis.
5. For each interest point $\boldsymbol{q}_I$ it is obtained the approximation field value with: $u^h(\boldsymbol{q}_I) = \sum_{i=1}^{n} \varphi_i(\boldsymbol{q}_I) u(\boldsymbol{x}_i)$.

The previously described procedure is performed considering a permanent shape parameter $p = 1.001$ and five distinct values for the shape parameter γ, being $c = \gamma d_s$, $c = \{0.001 \quad 0.251 \quad 0.501 \quad 1.001 \quad 1.501\}$.

The results of each analysis are presented in Fig. 4.21a, b, respectively for the regular nodal distribution and for the irregular nodal distribution. It seems that for

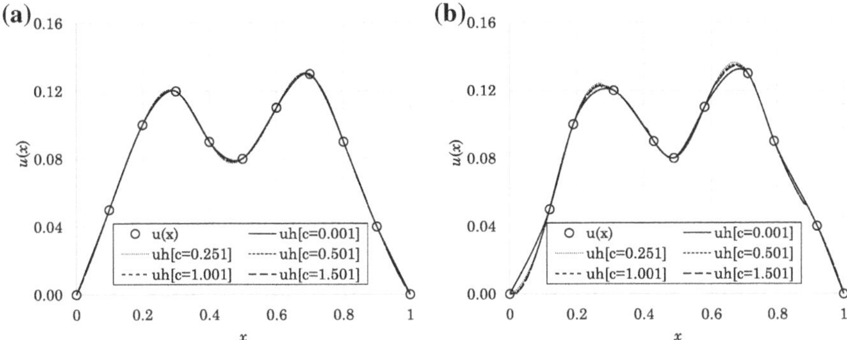

Fig. 4.21 Obtained approximation solutions for the **a** regular nodal distribution and the **b** irregular nodal distribution

Table 4.2 Error obtained in each node using the regular nodal distribution

Node	Shape parameter				
	c = 0.001	c = 0.251	c = 0.501	c = 1.001	c = 1.501
x_1	–	–	–	–	–
x_2	0.00E+00	0.00E+00	0.00E+00	−5.80E−07	−1.11E−05
x_3	0.00E+00	0.00E+00	1.00E−08	3.05E−06	1.22E−04
x_4	0.00E+00	0.00E+00	0.00E+00	1.12E−05	4.07E−04
x_5	0.00E+00	0.00E+00	−1.11E−08	2.29E−05	−6.97E−03
x_6	0.00E+00	0.00E+00	−2.50E−08	−1.67E−06	−2.72E−03
x_7	0.00E+00	0.00E+00	0.00E+00	2.77E−05	4.44E−03
x_8	0.00E+00	0.00E+00	0.00E+00	−1.42E−05	9.47E−03
x_9	0.00E+00	0.00E+00	1.11E−08	−4.22E−07	1.49E−04
x_{10}	0.00E+00	0.00E+00	0.00E+00	−4.00E−07	−4.43E−05
x_{11}	–	–	–	–	–

the regular nodal distribution the value of the shape parameter γ does not influences significantly the approximation. It is also perceptible that regardless the considered shape parameter γ value the approximation function interpolates perfectly the nodal values. However these observations are not true.

In Tables 4.2 and 4.3 are presented the local errors for each node discretizing the problem domain. The error is obtained with: $error = (u(\boldsymbol{x}_i) - u^h(\boldsymbol{x}_i))/(u(\boldsymbol{x}_i))$. For both the regular and irregular nodal distributions, Tables 4.2 and 4.3 show that in fact the RPI approximation is not capable to interpolate accurately the nodal values when $c > 0.5$. Increasing the shape parameter value leads to RPI shape functions without the Kronecker delta property.

In Fig. 4.22a it is presented the RPI shape function of the central node $\boldsymbol{x}_6$ obtained with the regular nodal distribution. The RPI shape function obtained with the irregular nodal distribution is presented in Fig. 4.22b. These figures permit to visualize directly the effects of the variation of the shape parameter. Using low

Table 4.3 Error obtained in each node using the irregular nodal distribution

Node	Shape parameter				
	c = 0.001	c = 0.251	c = 0.501	c = 1.001	c = 1.501
x_1	–	–	–	–	–
x_2	0.00E+00	0.00E+00	0.00E+00	2.00E−08	0.00E+00
x_3	0.00E+00	0.00E+00	1.00E−08	−5.30E−06	2.69E−04
x_4	0.00E+00	0.00E+00	0.00E+00	1.83E–06	−6.31E−04
x_5	0.00E+00	0.00E+00	0.00E+00	1.39E−05	3.92E−04
x_6	0.00E+00	0.00E+00	−3.75E−08	1.49E−04	1.22E−03
x_7	0.00E+00	0.00E+00	0.00E+00	1.73E−06	−3.88E−04
x_8	0.00E+00	0.00E+00	−7.69E−08	−3.41E−05	1.31E−03
x_9	0.00E+00	0.00E+00	−1.11E−08	−8.44E−07	−5.43E−05
x_{10}	0.00E+00	0.00E+00	0.00E+00	−2.50E−08	−1.50E−07
x_{11}	–	–	–	–	–

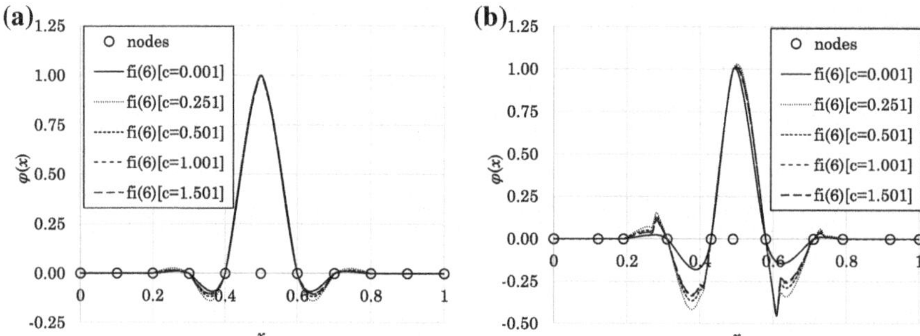

Fig. 4.22 Obtained RPI shape function for node 6 using the **a** regular nodal distribution and the **b** irregular nodal distribution

values for the shape parameter γ allow to obtain smooth RPI shape functions possessing the Kronecker delta property. In Table 4.4 are presented the obtained values for node 8, of the RPI shape function constructed in node 6. It is visible that the $\varphi_6(x_8)$ values are very close to zero when $c < 0.5$. This simple example demonstrates that the RPI shape function, constructed using the MQ-RBF, require a low shape parameter γ in order to assure the Kronecker delta property.

The use of the correct shape parameter γ assumes greater importance in higher dimensional spaces. Consider a two-dimensional domain $\Omega \subset \mathbb{R}^2$ discretized by the eight nodes represented in Fig. 4.23a. In order to constructed the RPI shape functions of the central node, $\boldsymbol{x}_1$, using the other seven nodes, it was considered the MQ-RBF and a linear polynomial basis. Three values for the shape parameter γ were assumed $\gamma = \{0.01, 0.50, 2.00\}$ and a permanent exponential shape parameter $p = 1.001$ was considered. The obtained two-dimensional RPI shape functions are presented in Fig. 4.23b, c, d. As it is possible to observe the shape parameter γ affects significantly the RPI shape function silhouette.

Table 4.4 Obtained values for node 8, of the RPI shape function constructed in node 6

	Shape parameter				
	c = 0.001	c = 0.251	c = 0.501	c = 1.001	c = 1.501
$\varphi_6(x_8)$ [*regular*]	−3.98E−13	2.91E−11	7.45E−09	1.53E−05	−9.77E−04
$\varphi_6(x_8)$ [*irregular*]	−1.25E−12	−1.16E−10	1.49E−08	1.53E−05	−3.91E−03

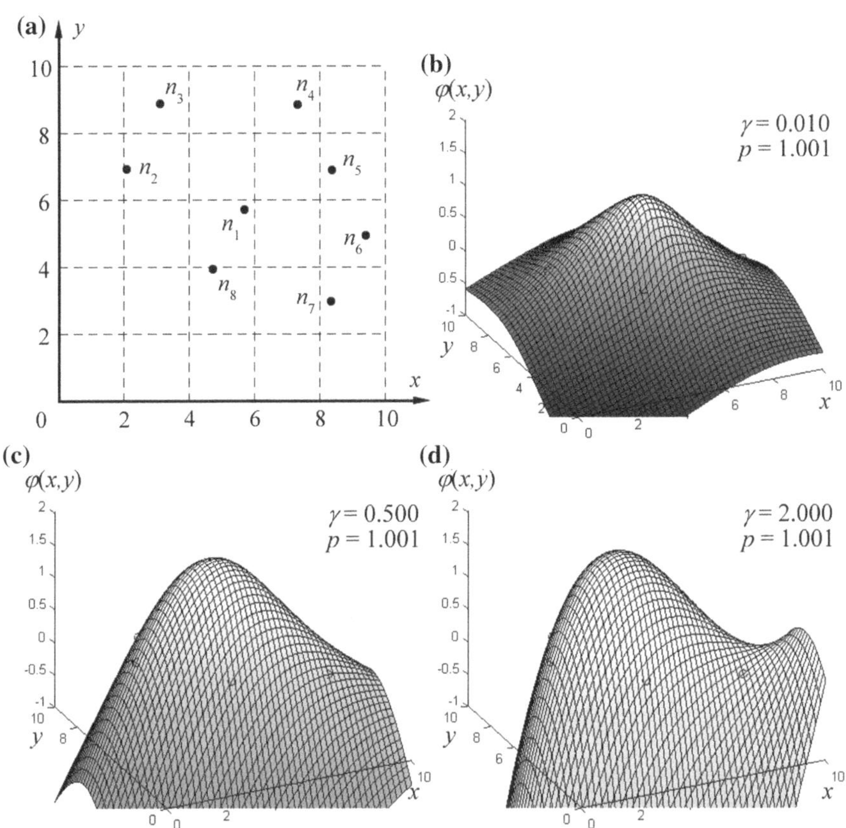

Fig. 4.23 **a** Nodal discretization. RPI shape function obtained a linear polynomial basis and a MQ-RBF considering. **b** $\gamma = 0.01$. **c** $\gamma = 0.50$. **d** $\gamma = 2.00$

4.4.5 RPI Shape Functions Properties

4.4.5.1 Consistency

As mentioned in Sect. 4.3.3.1, consistency is the capacity of a shape function to reproduce the complete order of a polynomial functional. The consistency of the RPI approximation depends on the complete order of the polynomial basis used to construct the approximation, Eq. (4.102). The RPI shape functions possess the consistency property if the polynomial basis is included as basis. Thus, being m the complete monomial order of the polynomial basis function, then the RPI shape function possess C_m consistency. A field functional defined by a complete polynomial functional, in which $t \leq m$, can be expressed by,

$$u(\boldsymbol{x}) = \sum_{j=1}^{t} p_j(\boldsymbol{x}) c_j(\boldsymbol{x}) = \boldsymbol{p}(\boldsymbol{x})^{\mathrm{T}} \boldsymbol{c}(\boldsymbol{x}) \tag{4.150}$$

Consider the approximation function $u^h(\boldsymbol{x})$ defined with Eq. (4.102). Thus, the complete polynomial field functional can be conveniently presented as,

$$u(\boldsymbol{x}) = \sum_{i=1}^{n} r_i(\boldsymbol{x}_i - \boldsymbol{x}_I) \cdot 0 + \sum_{j=1}^{t} p_j(\boldsymbol{x}) c_j(\boldsymbol{x}) + \sum_{k=t+1}^{m} p_k(\boldsymbol{x}) \cdot 0 \tag{4.151}$$

The n nodal values on the support-domain of $\boldsymbol{x}_I$ are obtained with,

$$\begin{cases} u(\boldsymbol{x}_1) = \sum_{i=1}^{n} r_i(\boldsymbol{x}_i - \boldsymbol{x}_1) \cdot 0 + \sum_{j=1}^{t} p_j(\boldsymbol{x}_1) c_j(\boldsymbol{x}_I) + \sum_{k=t+1}^{m} p_k(\boldsymbol{x}_1) \cdot 0 \\ u(\boldsymbol{x}_2) = \sum_{i=1}^{n} r_i(\boldsymbol{x}_i - \boldsymbol{x}_2) \cdot 0 + \sum_{j=1}^{t} p_j(\boldsymbol{x}_2) c_j(\boldsymbol{x}_I) + \sum_{k=t+1}^{m} p_k(\boldsymbol{x}_2) \cdot 0 \\ \vdots \\ u(\boldsymbol{x}_n) = \sum_{i=1}^{n} r_i(\boldsymbol{x}_i - \boldsymbol{x}_n) \cdot 0 + \sum_{j=1}^{t} p_j(\boldsymbol{x}_n) c_j(\boldsymbol{x}_I) + \sum_{k=t+1}^{m} p_k(\boldsymbol{x}_n) \cdot 0 \end{cases} \tag{4.152}$$

Using Eq. (4.102) it is possible to obtain the coefficients $\boldsymbol{a}(\boldsymbol{x}_I)$ and $\boldsymbol{b}(\boldsymbol{x}_I)$ by imposing $u^h(\boldsymbol{x}_i) = u(\boldsymbol{x}_i)$. Therefore, satisfying Eq. (4.102) for all n nodes on the support-domain of the interest point $\boldsymbol{x}_I$, a system of n equations with $n + m$ unknowns is obtained,

$$\begin{cases} u^h(\boldsymbol{x}_1) = \sum_{i=1}^{n} r_i(\boldsymbol{x}_i - \boldsymbol{x}_1) a_i(\boldsymbol{x}_I) + \sum_{j=1}^{t} p_j(\boldsymbol{x}_1) b_j(\boldsymbol{x}_I) + \sum_{k=1+t}^{m} p_k(\boldsymbol{x}_1) b_k(\boldsymbol{x}_I) = u(\boldsymbol{x}_1) \\ u^h(\boldsymbol{x}_2) = \sum_{i=1}^{n} r_i(\boldsymbol{x}_i - \boldsymbol{x}_2) a_i(\boldsymbol{x}_I) + \sum_{j=1}^{t} p_j(\boldsymbol{x}_2) b_j(\boldsymbol{x}_I) + \sum_{k=1+t}^{m} p_k(\boldsymbol{x}_2) b_k(\boldsymbol{x}_I) = u(\boldsymbol{x}_2) \\ \vdots \\ u^h(\boldsymbol{x}_n) = \sum_{i=1}^{n} r_i(\boldsymbol{x}_i - \boldsymbol{x}_n) a_i(\boldsymbol{x}_I) + \sum_{j=1}^{t} p_j(\boldsymbol{x}_n) b_j(\boldsymbol{x}_I) + \sum_{k=1+t}^{m} p_k(\boldsymbol{x}_n) b_k(\boldsymbol{x}_I) = u(\boldsymbol{x}_n) \end{cases} \tag{4.153}$$

Then, substituting the equations on Eq. (4.152) on (4.153) it is possible to obtain the coefficients $\boldsymbol{a}(\boldsymbol{x}_I)$ and $\boldsymbol{b}(\boldsymbol{x}_I)$. Equation (4.153) is true only if $\boldsymbol{a}_i(\boldsymbol{x}_I) = 0$, with $i = \{1, 2, \ldots, n\}$, and $b_j(\boldsymbol{x}_I) = c_j(\boldsymbol{x}_I)$, with $j = \{1, 2, \ldots, t\}$, and $b_k(\boldsymbol{x}_I) = 0$, with $k = \{t + 1, t + 2, \ldots, m\}$. Thus,

$$u^h(\boldsymbol{x}) = \sum_{j=1}^{t} p_j(\boldsymbol{x}) c_j(\boldsymbol{x}) = \boldsymbol{p}(\boldsymbol{x})^{\mathrm{T}} \boldsymbol{c}(\boldsymbol{x}) = u(\boldsymbol{x}) \tag{4.154}$$

Proving that the RPI approximation is capable to reproduce any set of monomials included in the polynomial basis of the RPI formulation.

4.4.5.2 Reproducibility

As mentioned before, in meshless methods the reproducibility property is not included in the consistency property because in meshless methods it is possible to include in the shape function construction procedure distinct types of basis functions. If the meshless shape function is capable to reproduce an unknown function which is included as a basis function in the meshless shape functions formulation, then the meshless shape function possess the reproducibility property. While the consistency property focuses only on the reproducibility of complete polynomial functions, the reproducibility property is concern with any kind of functional besides polynomial functions.

The argument used to prove the consistency can be used to demonstrate the reproducibility properties of the RPI shape function. Consider a field defined by the following function,

$$u(\boldsymbol{x}) = \sum_{j=1}^{k} f_j(\boldsymbol{x}) c_j(\boldsymbol{x}) = \boldsymbol{f}(\boldsymbol{x})^{\mathrm{T}} \boldsymbol{c}(\boldsymbol{x}) \tag{4.155}$$

Being $c_j(\boldsymbol{x})$ arbitrary coefficients of $f_j(\boldsymbol{x})$, which is a function defined in a functional space $f_k : \mathbb{R}^d \mapsto \mathbb{R}$. The RPI approximation function can be defined as,

$$u^h(\boldsymbol{x}_I) = \sum_{i=1}^{n} r_i(\boldsymbol{x}_i - \boldsymbol{x}_I) a_i(\boldsymbol{x}_I) + \sum_{j=1}^{m} p_j(\boldsymbol{x}_I) b_j(\boldsymbol{x}_I) + \sum_{k=1}^{t} f_k(\boldsymbol{x}_I) d_k(\boldsymbol{x}_I) \tag{4.156}$$

and in the matrix form as,

$$u^h(\boldsymbol{x}) = \boldsymbol{r}(\boldsymbol{x})^{\mathrm{T}} \boldsymbol{a}(\boldsymbol{x}) + \boldsymbol{p}(\boldsymbol{x})^{\mathrm{T}} \boldsymbol{b}(\boldsymbol{x}) + \boldsymbol{f}(\boldsymbol{x})^{\mathrm{T}} \boldsymbol{d}(\boldsymbol{x}) = \left\{ \boldsymbol{r}(\boldsymbol{x})^{\mathrm{T}} \quad \boldsymbol{p}(\boldsymbol{x})^{\mathrm{T}} \quad \boldsymbol{f}(\boldsymbol{x})^{\mathrm{T}} \right\} \left\{ \begin{array}{c} \boldsymbol{a}(\boldsymbol{x}) \\ \boldsymbol{b}(\boldsymbol{x}) \\ \boldsymbol{d}(\boldsymbol{x}) \end{array} \right\} \tag{4.157}$$

It is possible to construct a system of n equations with $n + m + t$ unknowns satisfying Eq. (4.156) for all n nodes on the support-domain of the interest point $\boldsymbol{x}_I$,

$$\begin{cases} u^h(\boldsymbol{x}_1) = \sum_{i=1}^{n} r_i(\boldsymbol{x}_i - \boldsymbol{x}_1)a_i(\boldsymbol{x}_I) + \sum_{j=1}^{m} p_j(\boldsymbol{x}_1)b_j(\boldsymbol{x}_I) + \sum_{k=1}^{t} f_k(\boldsymbol{x}_1)d_k(\boldsymbol{x}_I) \\ u^h(\boldsymbol{x}_2) = \sum_{i=1}^{n} r_i(\boldsymbol{x}_i - \boldsymbol{x}_2)a_i(\boldsymbol{x}_I) + \sum_{j=1}^{m} p_j(\boldsymbol{x}_2)b_j(\boldsymbol{x}_I) + \sum_{k=1}^{t} f_k(\boldsymbol{x}_2)d_k(\boldsymbol{x}_I) \\ \vdots \\ u^h(\boldsymbol{x}_n) = \sum_{i=1}^{n} r_i(\boldsymbol{x}_i - \boldsymbol{x}_n)a_i(\boldsymbol{x}_I) + \sum_{j=1}^{m} p_j(\boldsymbol{x}_n)b_j(\boldsymbol{x}_I) + \sum_{k=1}^{t} f_k(\boldsymbol{x}_n)d_k(\boldsymbol{x}_I) \end{cases} \tag{4.158}$$

Since Eq. (4.158) has $n + m + t$ unknowns it is require to consider extra sets of equations in order to obtain an unique solution [36]. Therefore, the following supplementary equations can be can be added to the initial equation system,

$$\begin{cases} \sum_{i=1}^{n} p_1(\boldsymbol{x}_i)a_i(\boldsymbol{x}_I) = 0 \\ \sum_{i=1}^{n} p_2(\boldsymbol{x}_i)a_i(\boldsymbol{x}_I) = 0 \\ \vdots \\ \sum_{i=1}^{n} p_m(\boldsymbol{x}_i)a_i(\boldsymbol{x}_I) = 0 \end{cases} \tag{4.159}$$

and,

$$\begin{cases} \sum_{i=1}^{n} f_1(\boldsymbol{x}_i)a_i(\boldsymbol{x}_I) = 0 \\ \sum_{i=1}^{n} f_2(\boldsymbol{x}_i)a_i(\boldsymbol{x}_I) = 0 \\ \vdots \\ \sum_{i=1}^{n} f_t(\boldsymbol{x}_i)a_i(\boldsymbol{x}_I) = 0 \end{cases} \tag{4.160}$$

The extra system of equations can be presented in the following matrix form,

$$\boldsymbol{P}^{\mathrm{T}} \boldsymbol{a}(\boldsymbol{x}_I) = 0 \tag{4.161}$$

and

$$\boldsymbol{F}^{\mathrm{T}} \boldsymbol{a}(\boldsymbol{x}_I) = 0 \tag{4.162}$$

Combining Eq. (4.158) with (4.159) and (4.160) it is obtained the following system of equations,

$$\begin{Bmatrix} u^h(\boldsymbol{x}) \\ z_P \\ z_F \end{Bmatrix} = \begin{bmatrix} \boldsymbol{R} & \boldsymbol{P} & \boldsymbol{F} \\ \boldsymbol{P}^{\mathrm{T}} & \boldsymbol{Z}_1 & \boldsymbol{Z}_2 \\ \boldsymbol{F}^{\mathrm{T}} & \boldsymbol{Z}_2^T & \boldsymbol{Z}_3 \end{bmatrix} \begin{Bmatrix} \boldsymbol{a}(\boldsymbol{x}_I) \\ \boldsymbol{b}(\boldsymbol{x}_I) \\ \boldsymbol{d}(\boldsymbol{x}_I) \end{Bmatrix} = \boldsymbol{M}_T \begin{Bmatrix} \boldsymbol{a}(\boldsymbol{x}_I) \\ \boldsymbol{b}(\boldsymbol{x}_I) \\ \boldsymbol{d}(\boldsymbol{x}_I) \end{Bmatrix} \tag{4.163}$$

Being $(Z_1)_{ij} = 0$ for $\{i, j\} = 1, 2, \ldots, m$, $(Z_2)_{ij} = 0$ for $i = 1, 2, \ldots, m \wedge j = 1, 2, \ldots, t$ and $(Z_3)_{ij} = 0$ for $\{i, j\} = 1, 2, \ldots, t$. The null vectors are defined as $(z_F)_i = 0$ with $i = \{1, 2, \ldots, t\}$ and $(z_P)_i = 0$ with $i = \{1, 2, \ldots, m\}$. As in previous subsection, the field functional $u(\boldsymbol{x})$ from Eq. (4.155) can also be presented as,

$$u(\boldsymbol{x}) = \sum_{i=1}^{n} r_i(\boldsymbol{x}_i - \boldsymbol{x}_I) \cdot 0 + \sum_{j=1}^{m} p_j(\boldsymbol{x}) \cdot 0 + \sum_{k=1}^{t} f_k(\boldsymbol{x}) c_k(\boldsymbol{x}) \tag{4.164}$$

and in the matrix form as,

$$u(\boldsymbol{x}) = \boldsymbol{r}(\boldsymbol{x})^{\mathrm{T}} z_R + \boldsymbol{p}(\boldsymbol{x})^{\mathrm{T}} z_P + \boldsymbol{f}(\boldsymbol{x})^{\mathrm{T}} \boldsymbol{c}(\boldsymbol{x}) = \left\{ \boldsymbol{r}(\boldsymbol{x})^{\mathrm{T}} \quad \boldsymbol{p}(\boldsymbol{x})^{\mathrm{T}} \quad \boldsymbol{f}(\boldsymbol{x})^{\mathrm{T}} \right\} \begin{Bmatrix} z_R \\ z_P \\ \boldsymbol{c}(\boldsymbol{x}) \end{Bmatrix} \tag{4.165}$$

being $(z_R)_i = 0$ with $i = \{1, 2, \ldots, n\}$ and $(z_P)_i = 0$ with $i = \{1, 2, \ldots, m\}$. Thus, it is possible to obtain the n nodal values on the support-domain of $\boldsymbol{x}_I$ with the following system of equations,

$$\begin{cases} u(\boldsymbol{x}_1) = \sum_{i=1}^{n} r_i(\boldsymbol{x}_i - \boldsymbol{x}_1) \cdot 0 + \sum_{j=1}^{m} p_j(\boldsymbol{x}_1) \cdot 0 + \sum_{k=1}^{t} f_k(\boldsymbol{x}_1) c_k(\boldsymbol{x}_I) \\ u(\boldsymbol{x}_2) = \sum_{i=1}^{n} r_i(\boldsymbol{x}_i - \boldsymbol{x}_2) \cdot 0 + \sum_{j=1}^{m} p_j(\boldsymbol{x}_2) \cdot 0 + \sum_{k=1}^{t} f_k(\boldsymbol{x}_2) c_k(\boldsymbol{x}_I) \\ \vdots \\ u(\boldsymbol{x}_n) = \sum_{i=1}^{n} r_i(\boldsymbol{x}_i - \boldsymbol{x}_n) \cdot 0 + \sum_{j=1}^{m} p_j(\boldsymbol{x}_n) \cdot 0 + \sum_{k=1}^{t} f_k(\boldsymbol{x}_n) c_k(\boldsymbol{x}_I) \end{cases} \tag{4.166}$$

Supplementary trivial equations can be can be added to the initial equation system, Eq. (4.166), in order to obtain a system of equations similar with Eq. (4.163).

$$\boldsymbol{P}^{\mathrm{T}} z_R = 0 \tag{4.167}$$

and

$$\boldsymbol{F}^{\mathrm{T}} z_R = 0 \tag{4.168}$$

Combining again Eq. (4.165) with (4.167) and (4.168) the following set of equations are obtained,

$$\begin{Bmatrix} \boldsymbol{u}(\boldsymbol{x}) \\ z_P \\ z_F \end{Bmatrix} = \begin{bmatrix} \boldsymbol{R} & \boldsymbol{P} & \boldsymbol{F} \\ \boldsymbol{P}^{\mathrm{T}} & \boldsymbol{Z}_1 & \boldsymbol{Z}_2 \\ \boldsymbol{F}^{\mathrm{T}} & \boldsymbol{Z}_2^T & \boldsymbol{Z}_3 \end{bmatrix} \begin{Bmatrix} z_R \\ z_P \\ \boldsymbol{c}(\boldsymbol{x}_I) \end{Bmatrix} = \boldsymbol{M}_T \begin{Bmatrix} z_R \\ z_P \\ \boldsymbol{c}(\boldsymbol{x}_I) \end{Bmatrix} \tag{4.169}$$

Imposing the interpolation: $u^h(\boldsymbol{x}_i) = u(\boldsymbol{x}_i)$, from Eq. (4.163) and Eq. (4.169) it is possible to conclude that $\boldsymbol{d}(\boldsymbol{x}_I) = \boldsymbol{c}(\boldsymbol{x}_I)$,

$$\begin{Bmatrix} \boldsymbol{u}^h(\boldsymbol{x}) \\ z_P \\ z_F \end{Bmatrix} = \begin{Bmatrix} \boldsymbol{u}(\boldsymbol{x}) \\ z_P \\ z_F \end{Bmatrix} \Rightarrow \boldsymbol{M}_T \begin{Bmatrix} z_R \\ z_P \\ \boldsymbol{d}(\boldsymbol{x}_I) \end{Bmatrix} = \boldsymbol{M}_T \begin{Bmatrix} z_R \\ z_P \\ \boldsymbol{c}(\boldsymbol{x}_I) \end{Bmatrix} \Rightarrow \boldsymbol{d}(\boldsymbol{x}_I) = \boldsymbol{c}(\boldsymbol{x}_I) \tag{4.170}$$

Showing that any functional included in the basis of the RPI shape function can be exactly reproduced by the RPI approximation. This important property permits to construct RPI shape functions capable of dealing with discontinuous domains, such as cracks and abrupt material variation. The conventional RPI formulation can be enhanced with the use of suitable trigonometric basis functions, reflecting the properties of the stress filed around the crack tip [39]. It was proven [39] that comparing with the classical RPI formulation the enriched RPI formulation presents a similar accuracy to fit a polynomial surface; a much better accuracy to fit a trigonometric surface; and a similar interpolation stability without increase of the condition number of the RBF interpolation matrix.

Nevertheless, as in the MLS formulation, it is important to certify the non-singularity of the total moment matrix $\boldsymbol{M}_T$ if these enriched basis functions are included in the basis of the RPI approximation.

4.4.5.3 Partition of Unity

The partition of unity property of the RPI shape function can be proved using the same argument used to prove the consistency property. The partition of unity is satisfied by the RPI shape function $\varphi_i(\boldsymbol{x})$,

$$\sum_{i=1}^{n} \varphi_i(\boldsymbol{x}) = 1 \quad (4.171)$$

if at least a constant is included in the basis. Consider a constant field $u(\boldsymbol{x}) = c$, being $c \in \mathbb{R}$. The polynomial form presented in Eq. (4.151) can be used to represent the constant field, however in this example only the constant term of the polynomial is different from zero, $t = 1$,

$$u(\boldsymbol{x}) = \sum_{i=1}^{n} r_i(\boldsymbol{x}_i - \boldsymbol{x}_I) \cdot 0 + \sum_{j=1}^{t} p_j(\boldsymbol{x}) c_j(\boldsymbol{x}) + \sum_{k=t+1}^{m} p_k(\boldsymbol{x}) \cdot 0 = c_1 + 0 = c \quad (4.172)$$

Once again, the non-constants coefficients $b_i(\boldsymbol{x})$ of the approximation function $u^h(\boldsymbol{x})$ defined with Eq. (4.102), can be obtained imposing the interpolation, $u^h(\boldsymbol{x}_i) = u(\boldsymbol{x}_i)$. Thus, the substitution of the n nodal values $u(\boldsymbol{x}_i)$, obtained with Eq. (4.152), back in the equation system of Eq. (4.153) allow to obtain the coefficients $\boldsymbol{a}(\boldsymbol{x}_I)$ and $\boldsymbol{b}(\boldsymbol{x}_I)$. It is perceptible that Eq. (4.153) is true only if $\boldsymbol{a}_i(\boldsymbol{x}_I) = 0$ for $i = \{1, 2, \ldots, n\}$, and $b_1(\boldsymbol{x}_I) = c_1(\boldsymbol{x}_I)$, and $b_k(\boldsymbol{x}_I) = 0$ for $k = \{2, 3, \ldots, m\}$. Therefore,

$$u^h(\boldsymbol{x}) = b_1(\boldsymbol{x}_I) \cdot 1 = c_1(\boldsymbol{x}_I) \cdot 1 = c \quad (4.173)$$

The approximated field variable value of an interest point $\boldsymbol{x}_I$ is determined using the shape function values obtained at the nodes within the support-domain of $\boldsymbol{x}_I$,

$$u^h(\boldsymbol{x}_I) = \sum_{i=1}^{n} \varphi_i(\boldsymbol{x}_I) u(\boldsymbol{x}_i) = \sum_{i=1}^{n} \varphi_i(\boldsymbol{x}_I) c = c \sum_{i=1}^{n} \varphi_i(\boldsymbol{x}_I) \quad (4.174)$$

being the approximation function defined as $u^h(\boldsymbol{x}) = c$ it is possible to write,

$$c = c \sum_{i=1}^{n} \varphi_i(\boldsymbol{x}_I) \Leftrightarrow \sum_{i=1}^{n} \varphi_i(\boldsymbol{x}_I) = 1 \quad (4.175)$$

Therefore, if a functional basis contains a constant term, then the RPI shape function is of the partition unit. Notice that in the case of absence of the polynomial basis, if another functional basis is used to construct the RPI shape function and it contains a constant term, the previous demonstration is still valid and the RPI shape function will continue to possess the partition unit property.

If in the RPI formulation the only basis function used is the RBF, the partition of unit depends on the RBF. If compactly supported RBFs are used, the constructed RPI shape functions will possess the partition of unit property, since all the presented compactly supported RBFs, Eq. (4.145) to (4.149), clearly present a constant term. However some non-compactly supported RBFs, such as the MQ-

RBF, Eq. (4.108), the Gaussian RBF, Eq. (4.109), and the thin plate spline RBF, Eq. (4.110), do not show explicitly constant terms.

Regarding the MQ-RBF, it is possible to find the constant term of the MQ-RBF with the following procedure. Consider the MQ-RBF defined by Eq. (4.108) for the one-dimensional space,

$$r(x) = \left(d_i^2 + (\gamma d_a)^2\right)^p = \left((x_i - x)^2 + c^2\right)^p = \left((x_i^2 + c^2) - 2x_i x + x^2\right)^p \quad (4.176)$$

Notice that both x_i and c are scalar values. This infinite continuous functions can be represented in $x = 0$ by an infinite Taylor series expansion polynomial $f(x)$,

$$f(x) = \sum_{k=0}^{\infty} \left[\frac{r^{(k)}(0)}{k!} x^k\right] = r(0) + r^{(1)}(0)x + \frac{r^{(2)}(0)}{2!} x^2 + \cdots \quad (4.177)$$

Being $r^{(k)}(x)$ the kth derivative of the continuous function $r(x)$. The first term of the infinite polynomial series is clearly a constant term, $r(0) = (x_i^2 + c^2)^p$. The existence of this constant term in the MQ-RBF permits to construct RPI shape functions possessing the partition of unit using only the MQ-RBF as a functional basis.

4.4.5.4 Kronecker Delta

One of the most important properties displayed by a shape function is certainly the Kronecker delta property. The RPI shape functions have interpolating properties, which has been numerically shown in Sect. 4.4.4. Nevertheless, it is possible to demonstrate with the following simple example that the RPI shape function, constructed using the MQ-RBF, possess the Kronecker delta property. For simplicity sake consider $r : \mathbb{R}^d \mapsto \mathbb{R}$ and the absence of a polynomial base. Regardless the chosen spatial dimension of the function space T, in which the domain Ω is discretized, in this example it is considered a very small support-domain, containing only $n = 3$ nodes. The total moment matrix $\boldsymbol{M}_T$ is then defined using Eq. (4.119), however in this particular case the total moment matrix is built using only the radial moment matrix,

$$\boldsymbol{M}_T = \begin{bmatrix} r_{11} & r_{12} & r_{13} \\ r_{21} & r_{22} & r_{23} \\ r_{31} & r_{32} & r_{33} \end{bmatrix} = \begin{bmatrix} 0 & r_{12} & r_{13} \\ r_{12} & 0 & r_{23} \\ r_{13} & r_{23} & 0 \end{bmatrix} \quad (4.178)$$

The r_{ij} components of the total moment matrix $\boldsymbol{M}_T$ are defined using Eq. (4.108) and the diagonal terms are null because $\gamma d_a \cong 0$. The total moment matrix $\boldsymbol{M}_T$ is always invertible, as long as $r_{ij} \neq 0 : i \neq j$, i.e., there are no coincident nodes inside the support-domain.

$$\boldsymbol{M}_T^{-1} = \frac{1}{2r_{12}r_{13}r_{23}} \begin{bmatrix} -r_{23}^2 & r_{13}r_{23} & r_{12}r_{23} \\ r_{13}r_{23} & -r_{13}^2 & r_{12}r_{13} \\ r_{12}r_{23} & r_{12}r_{13} & -r_{12}^2 \end{bmatrix} \tag{4.179}$$

If an interest point $\boldsymbol{x}_I$ is considered coincident with one of the nodes inside the influence-domain, lets say node 1, $\boldsymbol{x}_I = \boldsymbol{x}_1$, then with Eq. (4.123) is possible to obtain,

$$\boldsymbol{\varphi}(\boldsymbol{x}_I) = \boldsymbol{r}(\boldsymbol{x}_I)^{\mathrm{T}} \boldsymbol{M}_T^{-1} = \{0 \quad r_{12} \quad r_{13}\} \boldsymbol{M}_T^{-1} = \{1 \quad 0 \quad 0\}^T \tag{4.180}$$

showing the Kronecker delta property. This demonstration can be extended to a support-domain containing n and also for RPI shape functions containing polynomial basis functions. As has been shown in Sect. 4.4.4, the RPI shape functions constructed with the MQ-RBF only possess the Kronecker delta property if $\gamma d_a \cong 0$.

To determine if the total moment matrix $\boldsymbol{M}_T$ is well-conditioned, the condition number of $\boldsymbol{M}_T$ must be determined. The condition number is obtained with $Cond(\boldsymbol{M}_T) = |||\boldsymbol{M}_T||| \cdot |||\boldsymbol{M}_T^{-1}|||$, being the matrix norm defined by $|||\boldsymbol{M}_T||| = \max\limits_{1 \le j \le n} \sum_{i=1}^{n} |(M_T)_{ij}|$. The nodal spatial disposition minimizing $Cond(\boldsymbol{M}_T)$ is an equidistant nodal distribution, for which, considering again Eq. (4.178), the total moment matrix $\boldsymbol{M}_T$ components should be defined with $g_{12} = g_{13} = g_{23}$. However in meshless methods, nodes can be arbitrarily distributed, therefore $Cond(\boldsymbol{M}_T)$ is maximized when two nodes are extremely close to each other when compared with a third node. In this case the total moment matrix $\boldsymbol{M}_T$ components would be defined with $g_{13} = g_{23} \gg g_{12}$. Therefore,

$$|||\boldsymbol{G}||| = max\left\langle \begin{array}{c} |g_{12}| + |g_{13}| \\ |g_{12}| + |g_{23}| \\ |g_{13}| + |g_{23}| \end{array} \right\rangle = 2|g_{13}| \tag{4.181}$$

and,

$$|||\boldsymbol{G}^{-1}||| = \frac{1}{2g_{12}g_{13}g_{23}} max\left\langle \begin{array}{c} \left|-g_{23}^2\right| + |g_{13}g_{23}| + |g_{12}g_{23}| \\ |g_{13}g_{23}| + \left|-g_{13}^2\right| + |g_{12}g_{13}| \\ |g_{12}g_{23}| + |g_{12}g_{13}| + \left|-g_{12}^2\right| \end{array} \right\rangle = \frac{1}{g_{12}} + \frac{1}{2g_{13}} \tag{4.182}$$

the total moment matrix $\boldsymbol{M}_T$ condition number is defined with $Cond(\boldsymbol{M}_T) = 1 + 2g_{13}/g_{12}$.

In this simple example, the lowest condition number is achieved when all nodes are equidistant, $Cond(\boldsymbol{M}_T) = 3$, which indicates a well-conditioned matrix.

Notice that $Cond(\boldsymbol{M}_T)$ is proportional to the relation between the radial distance determined for the nodes of the support-domain, Eq. (4.108). The most important conclusion from this example is that if the radial distance between two arbitrary nodes inside the support-domain is much lower than the average nodal distance of the support-domain nodal set, then an ill-conditioned total moment matrix $\boldsymbol{M}_T$ can occur.

4.4.5.5 Compact Support

Recall that within the RPIM formulation the support-domain is defined based on a radial search and for the NNRPIM formulation the support-domain is directly obtained by the natural neighbours. Nevertheless, regardless the procedure to defined the support-domain, in order to construct the RPI shape functions, only the nodes within the defined compact support-domain are considered. The RPI shape functions possess compact support because the value of $\varphi(\boldsymbol{x}_I)$ outside the support-domain is zero. As already mentioned for the MLS shape functions, this property is very important since it permits to create to sparse and banded discretized systems of equations, increasing the computational efficiency.

4.4.5.6 Compatibility

The RPI shape functions are not compatible in a local support-domain because the abrupt change of the field nodes used to construct the RPI shape function, caused by the spatial movement of the support-domain, perturb the continuity of the approximated field function.

The lack of compatibility can be understand with the following example. Consider the one-dimensional domain described in Fig. 4.24, being the set of nodes discretizing the problem domain defined by $\boldsymbol{X} = \{\boldsymbol{x}_1, \boldsymbol{x}_2, \ldots, \boldsymbol{x}_{12}\} \in \Omega \wedge \boldsymbol{x}_i \in \mathbb{R}^1$ and irregulary distributed. The average nodal spacing is considered equal to the average mesh density parameter, h, which permits to define the support-domain of the RPI shape functions as $d_s = 2.5\ h$. Consider an interest point $\boldsymbol{x}_I \in \mathbb{R} \wedge \boldsymbol{x}_I \subset \Omega$ defined simultaneously in the boundary of the support-domains of node $\boldsymbol{x}_4$ and node $\boldsymbol{x}_9$. As Fig. 4.24 shows, $\varphi_4(x_I) \neq \varphi_9(x_I)$, indicating that the constructed RPI shape functions ends abruptly and do not possess first order compatibility. The RPI shape function is not capable to assure any kind of compatibility order, since it is not constructed using weight functions as the MLS shape function.

Nevertheless, the compatibility of the RPI shape functions can be achieved using the conforming RPI formulation [40]. Studies on the conforming RPI formulation and classic RPI formulation have concluded that conforming RPI can exactly pass the standard patch tests and the classic RPI cannot [4, 40, 41]. However, for the problems considered in the studies, the classic RPI formulation was also convergent and permitted to obtain satisfactory results. Additionally, the classic RPI formulation is simpler and much more efficient than the conforming RPI formulation [40, 42].

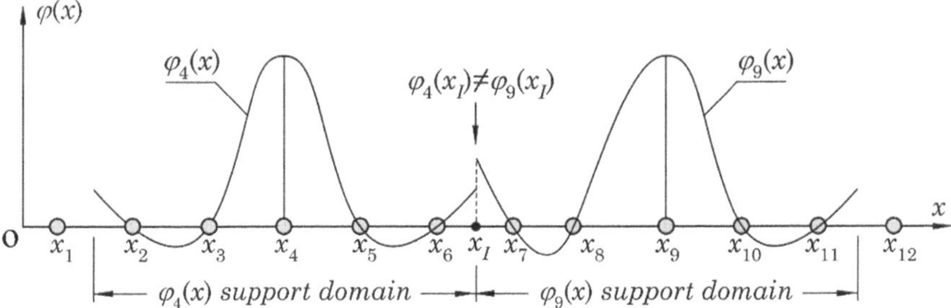

Fig. 4.24 RPI shape functions compatibility representation

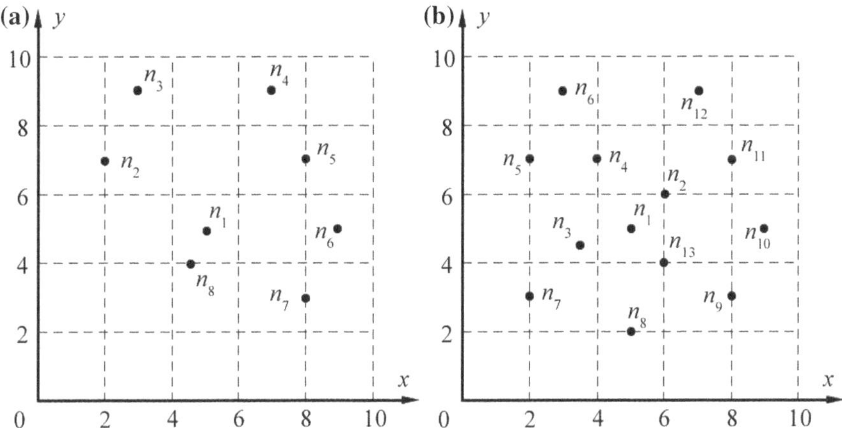

Fig. 4.25 **a** Nodal discretization X_1. **b** Nodal discretization X_2

4.4.5.7 RPI Shape Functions Examples

The RPI approach is capable to produce smooth shape functions with interpolating properties. In the present subsection are shown two distinct RPI shape functions, both constructed in the two-dimensional space. One of the RPI shape functions is obtained using a reduced size support-domain and the other RPI shape function is constructed considering a large support-domain.

Thus, consider a two-dimensional domain $\Omega \subset \mathbb{R}^2$ discretized by two distinct nodal sets: $\boldsymbol{X}_1 = \{\boldsymbol{x}_1, \boldsymbol{x}_2, \ldots, \boldsymbol{x}_8\} \in \mathbb{R}^2$, Fig. 4.25a, and $\boldsymbol{X}_2 = \{\boldsymbol{x}_1, \boldsymbol{x}_2, \ldots, \boldsymbol{x}_{13}\} \in \mathbb{R}^2$, Fig. 4.25b. The nodes of the studied nodal sets are irregularly distributed in the two-dimensional space.

Both RPI shape functions were constructed for the central node, $\boldsymbol{x}_1$, following the procedure indicated in Sect. 4.4.3. In the RPI formulation it was considered the linear polynomial basis and the MQ-RBF with $\gamma = 0.1$ and $p = 0.9$. The size parameter d_a of Eq. (4.108) was obtained with: $d_a = \max\|\boldsymbol{x}_i - \boldsymbol{x}_1\|, \forall \boldsymbol{x}_i \in \boldsymbol{X}_J$.

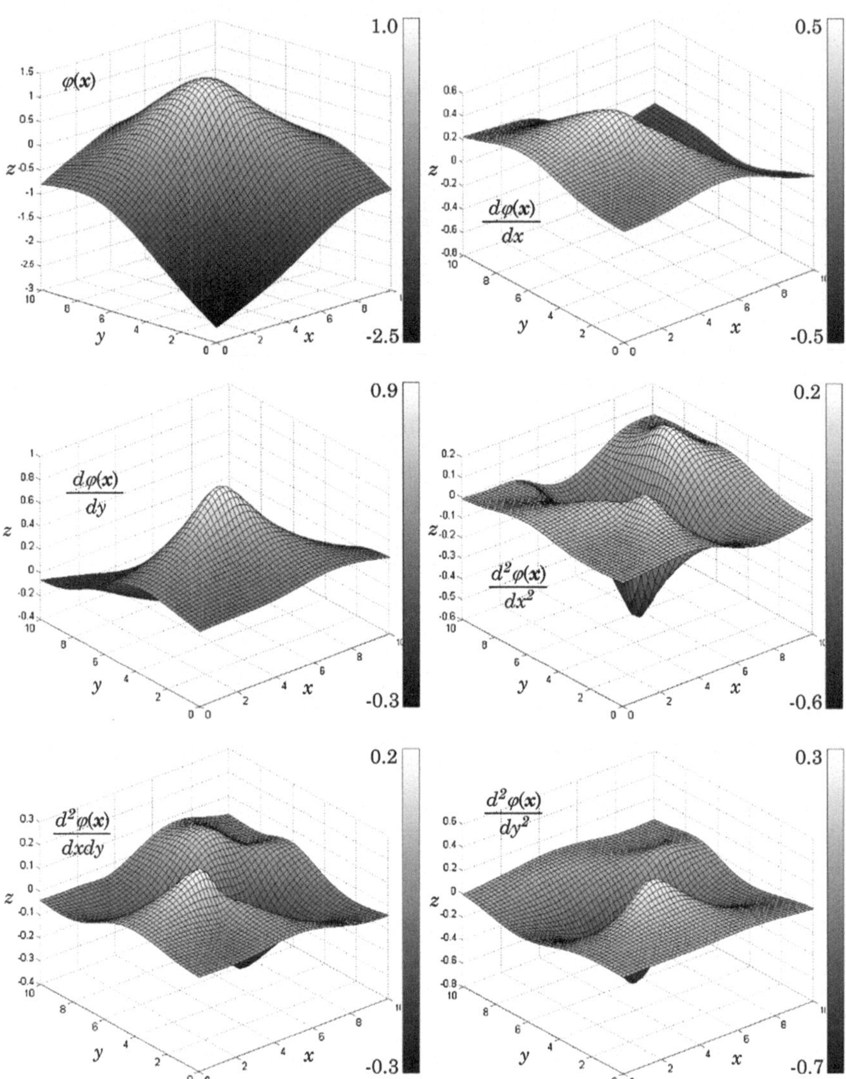

Fig. 4.26 Node $\boldsymbol{x}_1$ RPI shape function obtained for the nodal discretization $\boldsymbol{X}_1$

The RPI shape function obtained for the middle node, $\boldsymbol{x}_1 = \{5, 5\}$, using the nodal discretization $\boldsymbol{X}_1$ is presented in Fig. 4.26. The first order and the second order partial derivatives of the RPI shape functions are similarly presented in Fig. 4.26. Notice that the shape of the obtained RPI shape functions is smooth and continuous.

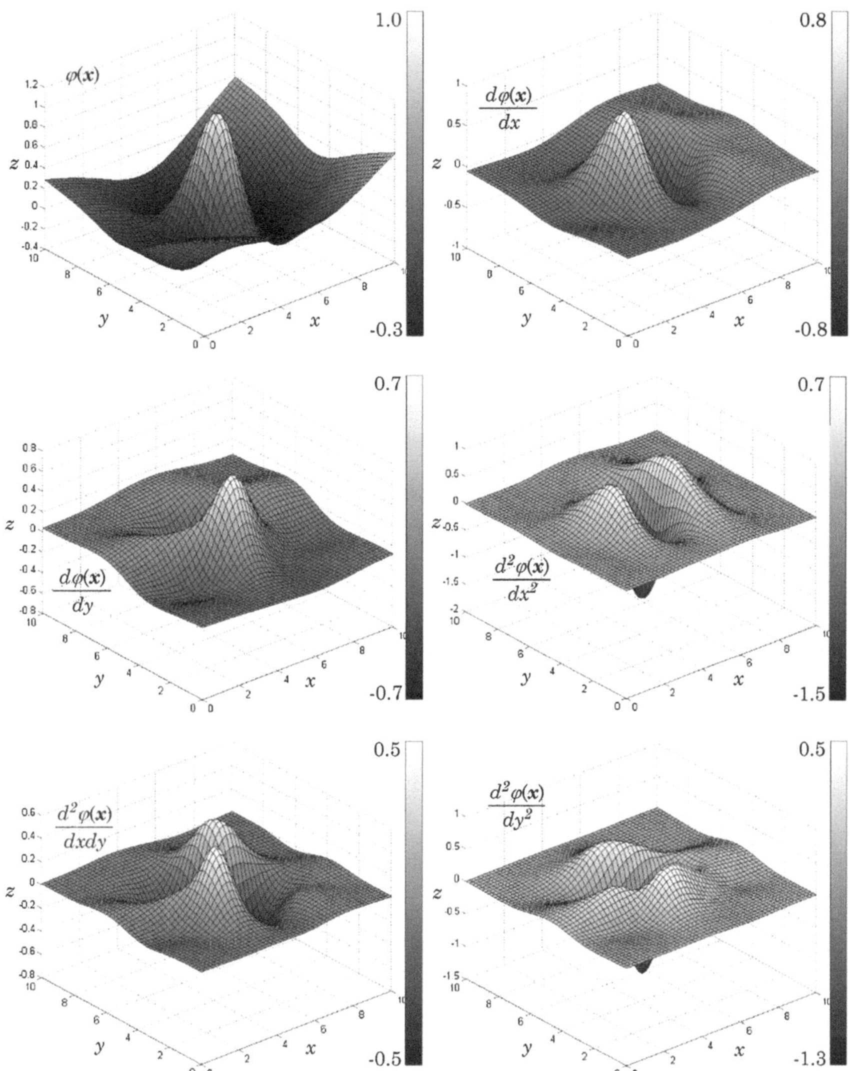

Fig. 4.27 Node $\boldsymbol{x}_1$ RPI shape function obtained for the nodal discretization X_2

For the nodal discretization X_2 the constructed RPI shape function of $\boldsymbol{x}_1$, and the respective first order and second order partial derivatives, are presented in Fig. 4.27. It is visible that the RPI are capable to produce smooth and continuous shape functions regardless the irregularity of the nodal discretization or the size of the support-domain.

References

1. Zienkiewicz OC, Taylor RL (1994) The finite element method, 4th edn. McGraw-Hill, London
2. Bathe KJ (1996) Finite element procedures. Prentice-Hall, Englewood Cliffs
3. Nguyen VP, Rabczuk T, Bordas S, Duflot M (2008) Meshless methods: a review and computer implementation aspects. Math Comput Simul 79(3):763–813
4. Liu GR (2002) Mesh free methods-moving beyond the finite element method. CRC Press, Boca Raton
5. Liu GR, Gu YT (2005) An introduction to meshfree methods and their programming. Springer, Netherlands
6. Lancaster P, Salkauskas K (1981) Surfaces generation by moving least squares methods. Math Comput 37:141–158
7. Nayroles B, Touzot G, Villon P (1992) Generalizing the finite element method: diffuse approximation and diffuse elements. Comput Mech 10:307–318
8. Belytschko T, Lu YY, Gu L (1994) Element-free Galerkin method. Int J Numer Meth Eng 37:229–256
9. Atluri SN, Zhu T (1998) A new meshless local Petrov-Galerkin (MLPG) approach in computational mechanics. Comput Mech 22(2):117–127
10. Oñate E, Perazzo F, Miquel J (2001) A finite point method for elasticity problems. Comput Struct 79(22–25):2151–2163
11. Fleming M, Chu YA, Moran B, Belytschko T (1997) Enriched element free Galerkin methods for crack tip fields. Int J Numer Meth Eng 40:1483–1504
12. Belytschko T, Krongauz Y, Organ D, Fleming M, Krysl P (1996) Meshless methods: an overview and recent developments. Comput Methods Appl Mech Eng 139(1):3–47
13. Dolbow J, Belytschko T (1998) An introduction to programming the meshless element free Galerkin method. Arch Comput Mech 5(3):207–241
14. Belinha J, Dinis LMJS (2006) Elasto-plastic analysis of plates using the element free Galerkin method. Eng Comput 23(3):525–551
15. Belinha J, Dinis LMJS (2006) Analysis of plates and laminates using the element free Galerkin method. Comput Struct 84(22–23):1547–1559
16. Belinha J, Dinis LMJS (2007) Non linear analysis of plates and laminates using the element free Galerkin method. Compos Struct 78(3):337–350
17. Atluri SN, Kim HG, Cho JY (1999) A critical assessment of the truly meshless local Petrov-Galerkin (MLPG), and local boundary integral equation (LBIE) methods. Comput Mech 24(5):348–372
18. Krongauz Y, Belytschko T (1996) Enforcement of essential boundary conditions in meshless approximations using finite elements. Comput Methods Appl Mech Eng 131(1–2):133–145
19. Belytschko T, Gu L, Lu YY (1994) Fracture and crack growth by element free Galerkin methods. Model Simul Mater Sci Eng 2(3A):519–534
20. Belytschko T, Lu YY, Gu L (1995) Crack propagation by element-free Galerkin methods. Eng Fract Mech 51(2):295–315
21. Liu GR, Gu YT (2001) A point interpolation method for two-dimensional solids. Int J Numer Meth Eng 50:937–951
22. Wang JG, Liu GR (2002) A point interpolation meshless method based on radial basis functions. Int J Numer Meth Eng 54:1623–1648
23. Wang JG, Liu GR (2002) On the optimal shape parameters of radial basis functions used for 2-D meshless methods. Comput Methods Appl Mech Eng 191:2611–2630
24. Kansa EJ (1990) Multiquadrics—A scattered data approximation scheme with applications to computational fluid-dynamics—I surface approximations and partial derivative estimates. Comput Math Appl 19(8–9):127–145

25. Kansa EJ (1990) Multiquadrics—A scattered data approximation scheme with applications to computational fluid-dynamics—II solutions to parabolic, hyperbolic and elliptic partial differential equations. Comput Math Appl 19(8–9):147–161
26. Duan Y (2008) A note on the meshless method using radial basis functions. Comput Math Appl 55(19):66–75
27. Wu Z (1995) Compactly supported positive definite radial functions. Adv Comput Math 4:283–292
28. Wendland H (1998) Error estimates for interpolation by compactly supported radial basis functions of minimal degree. J Approximation Theor 93:258–272
29. Belinha J, Jorge RMN, Dinis LMJS (2013) The natural radial element method. Int J Numer Meth Eng 93(12):1286–1313
30. Belinha J, Jorge RMN, Dinis LMJS (2013) Composite laminated plate analysis using the natural radial element method. Compos Struct 103(1):50–67
31. Belinha J, Jorge RMN, Dinis LMJS (2013) Analysis of thick plates by the natural radial element method. Int J Mech Sci 76(1):33–48
32. Hardy RL (1990) Theory and applications of the multiquadrics—Biharmonic method (20 years of discovery 1968–1988). Comput Math Appl 19(8–9):163–208
33. Dinis LMJS, Jorge RMN, Belinha J (2007) Analysis of 3D solids using the natural neighbour radial point interpolation method. Comput Methods Appl Mech Eng 196(13–16):2009–2028
34. Dinis LMJS, Jorge RMN, Belinha J (2008) Analysis of plates and laminates using the natural neighbour radial point interpolation method. Eng Anal Boundary Elem 32(3):267–279
35. Moreira S, Belinha J, Dinis LMJS, Jorge RMN (2014) Analysis of laminated beams using the natural neighbour radial point interpolation method. Revista Internacional de Métodos Numéricos para Cálculo y Diseño en Ingeniería. http://dx.doi.org/10.1016/j.rimni.2013.02.002
36. Golberg MA, Chen CS, Bowman H (1999) Some recent results and proposals for the use of radial basis functions in the BEM. Eng Anal Boundary Elem 23:285–296
37. Duchon J (1976) Splines minimizing rotation invariant seminorms in Sobolev spaces. In: Schemmp W, Zeller K (eds) Constructive theory of functions of several variables. Lecture notes in Mathematics. Springer, Berlin
38. Wendland H (1995) Piecewise polynomial, positive definite and compactly supported radial functions of minimal degree. Adv Comput Math 4(1):389–396
39. Gu YT, Wang W, Zhang LC, Feng XQ (2011) An enriched radial point interpolation method (e-RPIM) for analysis of crack tip fields. Eng Fract Mech 78:175–190
40. Liu GR, Gu YT, Dai KY (2004) Assessment and applications of interpolation methods for computational mechanics. Int J Numer Meth Eng 59:1373–1379
41. Liu GR (2002) A point assembly method for stress analysis for two-dimensional solids. Int J Solid Struct 39:261–276
42. Gu YT (2005) Meshfree methods and their comparisons. Int J Comput Methods 2(4):477–515

Chapter 5
Solid Mechanics Problems

Abstract In this chapter basic solid mechanics benchmark examples are presented. Firstly, in order to determine the optimal shape parameters of the multiquadrics radial basis function (MQ-RBF), it is present an optimization test using a standard linear patch test. Additionally, a simple study regarding the numerical integration is performed, allowing to determine the most efficient integration scheme for the nodal based integration. These optimization tests are extended to the two-dimensional and three-dimensional analysis. Then, elastostatic and elastodynamic benchmark examples as presented. The obtained results permit to confirm the efficiency and accuracy of the natural neighbour radial point interpolation method (NNRPIM).

5.1 Solid Mechanics NNRPIM Flow Chart

In Table 5.1 it is presented a flowchart to analyse a solid mechanics problem using a meshless method, such as the element free Galerkin method (EFGM) or the radial point interpolation method (RPIM).

The NNRPIM numerical implementation for a generic elastostatic solid mechanics problem can be summarized with the steps suggested in Table 5.2.

The elastodynamic analysis can be numerically implemented using the same steps presented in Tables 5.1 and 5.2. The numerical dynamic analysis can be performed by substituting steps 8 and 9 by an eigenproblem solver, such as the Jacobi solver. In addition to the free vibration elastodynamic analysis, it is possible to modify Tables 5.1 and 5.2 to solve a transient elastodynamic analysis using the same numerical tools used by the finite element method (FEM) [1–5].

Any type of nonlinear analysis can be implemented by assuming a recurring process based in the basic flow charts presented in Tables 5.1 and 5.2 [6, 7].

J. Belinha, *Meshless Methods in Biomechanics*, Lecture Notes in Computational Vision and Biomechanics 16, DOI: 10.1007/978-3-319-06400-0_5,

Table 5.1 RPIM or EFGM linear elastostatic flow chart

Step	Action
1.	Discretize the problem domain with an adequate nodal distribution
2.	Construct an adequate background integration mesh using a Gauss-Legendre quadrature scheme
3.	Establish the influence-domain of each interest point
4.	Loop over the integration points set, in order to numerically integrate the terms of the Galerkin weak form expression
	4.1 Using the previously defined influence-domains, determine the nodes that directly influence each integration point
	4.2 Construct the RPI or MLS shape functions, and the respective partial derivatives, for each integration point
	4.3 Evaluate the stiffness, the mass and the body load at each integration point
	4.4 Assemble the contribution of each integration point in order to construct the complete system of equations
5.	Determine the external force vector on the natural boundaries
6.	Impose the displacement constrains directly on the stiffness matrix for the RPIM or enforce the displacement constrains using the Lagrange-Multipliers for the EFGM
7.	Solve the algebraic system of equation and obtain the nodal displacement field
8.	Evaluate the strain field and the stress field at each integration point

Table 5.2 NNRPIM linear elastostatic flow chart

Step	Action
1.	Discretize the problem domain with an adequate nodal distribution
2.	Determine the natural neighbours of each field node and construct the Voronoï diagram of the nodal distribution discretizing the problem domain
3.	Determine the integration points using the Voronoï cells
4.	Establish the influence-cells (first or second degree)
5.	Loop over the integration points set, in order to numerically integrate the terms of the Galerkin weak form expression
	5.1 Using the previously defined influence-cells, determine the nodes that directly influence each integration point
	5.2 Construct the RPI shape functions, and the respective partial derivatives, for each integration point
	5.3 Evaluate the stiffness, the mass and the body load at each integration point
	5.4 Assemble the contribution of each integration point in order to construct the complete system of equations
6.	Determine the external force vector on the natural boundaries
7.	Enforce the displacement constrains directly on the stiffness matrix
8.	Solve the algebraic system of equation and obtain the nodal displacement field
9.	Evaluate the strain field and the stress field at each integration point

5.2 RPI Shape Function Patch Test

The patch test [8] was originally designed to prove the convergence in non-conforming finite element formulation. Generally, the test consists on the imposition of a known displacement field in the boundary of the patch. In this book linear patch tests are used. If the prescribed field is reproduced in the interior of the patch then the test is verified. Although being a benchmark for the evaluation and validation of non-conforming elements, in the context of the meshless methods the relevance of the patch test, from the convergence point of view, is still an open issue.

To perform the patch test first the problem domain, $\Omega \subset \mathbb{R}^d$, is discretized with a nodal distribution, $\boldsymbol{X} = \{\boldsymbol{x}_1, \boldsymbol{x}_2, \ldots, \boldsymbol{x}_N\} \in \mathbb{R}^d$. Then the procedure described in Table 5.2 is performed and in step 7 a known displacement field is imposed in the essential boundary of the patch. Afterwards, in step 8, the displacement field $\boldsymbol{U} = \{\boldsymbol{u}_1, \boldsymbol{u}_2, \ldots, \boldsymbol{u}_N\}$ is obtained. Since in this book only the two-dimensional deformation theories (plane strain and plane stress) and the classical three-dimensional deformation theory are considered, the number of degrees of freedom in each node is equal to the domain dimensional space, being $\boldsymbol{u}_i = \{u_1, u_2, \ldots, u_d\}^{\mathrm{T}} \in \mathbb{R}^d$. The patch test exact solution is compared with the meshless solution using the following three-dimensional medium error expression,

$$E_{med} = \frac{1}{N}\sum_{i=1}^{N} \frac{\sqrt{\left((u_i)_{meshless} - (u_i)_{exact}\right)^2 + \left((v_i)_{meshless} - (v_i)_{exact}\right)^2 + \left((w_i)_{meshless} - (w_i)_{exact}\right)^2}}{\sqrt{(u_i)^2_{exact} + (v_i)^2_{exact} + (w_i)^2_{exact}}} \tag{5.1}$$

Being N the total number of nodes discretizing the problem domain. In this book only the MQ-RBF is considered to construct the RPI shape functions. The MQ-RBF expression, Eq. (4.108), require two shape parameters: γ and p. In order to obtain both shape parameters an optimization test will be performed using the linear patch test.

In the following subsection it is reproduced the patch test of the radial point interpolation method (RPIM) [9] in order to clearly explain the major differences between the RPIM and the NNRPIM and to introduce the NNRPIM procedure for the elastostatic and elastodynamic analysis.

5.2.1 RPIM Patch Test

In this book, the considered RPIM formulation is analogous with the one suggested in the literature [9, 10]. Consider a two-dimensional domain $\boldsymbol{x} \in \mathbb{R}^2 : x \in [0, 1], y \in [0, 1]$ discretized with two distinct nodal distributions: an irregular distribution,

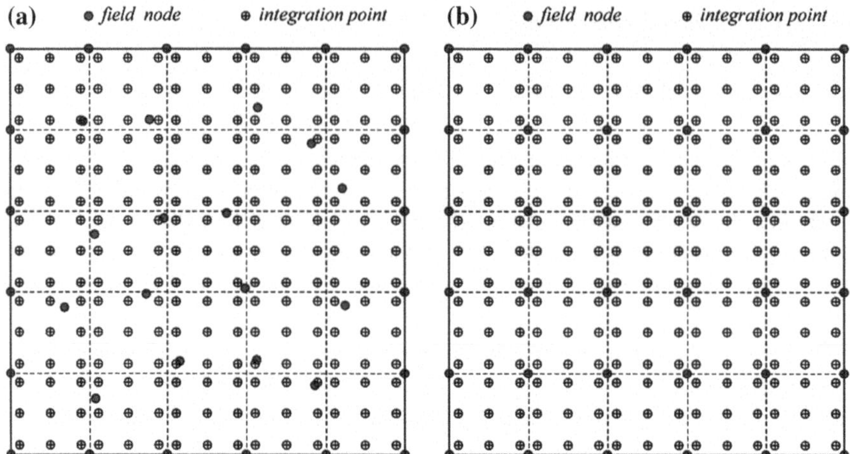

Fig. 5.1 **a** Irregular nodal distribution. **b** Regular nodal distribution

Table 5.3 Coordinates of the nodes on the patch domain

Node	Irregular		Regular		Node	Irregular		Regular	
	x	y	x	y		x	y	x	y
1	0.000	0.000	0.000	0.000	19	0.600	0.000	0.600	0.000
2	0.000	0.200	0.000	0.200	20	0.627	0.230	0.600	0.200
3	0.000	0.400	0.000	0.400	21	0.597	0.407	0.600	0.400
4	0.000	0.600	0.000	0.600	22	0.549	0.593	0.600	0.600
5	0.000	0.800	0.000	0.800	23	0.629	0.852	0.600	0.800
6	0.000	1.000	0.000	1.000	24	0.600	1.000	0.600	1.000
7	0.200	0.000	0.200	0.000	25	0.800	0.000	0.800	0.000
8	0.214	0.135	0.200	0.200	26	0.770	0.167	0.800	0.200
9	0.136	0.359	0.200	0.400	27	0.849	0.364	0.800	0.400
10	0.212	0.541	0.200	0.600	28	0.841	0.654	0.800	0.600
11	0.182	0.818	0.200	0.800	29	0.764	0.765	0.800	0.800
12	0.200	1.000	0.200	1.000	30	0.800	1.000	0.800	1.000
13	0.400	0.000	0.400	0.000	31	1.000	0.000	1.000	0.000
14	0.429	0.226	0.400	0.200	32	1.000	0.200	1.000	0.200
15	0.345	0.394	0.400	0.400	33	1.000	0.400	1.000	0.400
16	0.392	0.580	0.400	0.600	34	1.000	0.600	1.000	0.600
17	0.354	0.823	0.400	0.800	35	1.000	0.800	1.000	0.800
18	0.400	1.000	0.400	1.000	36	1.000	1.000	1.000	1.000

Fig. 5.1a, and a regular distribution, Fig. 5.1b. The coordinates of the field nodes are presented in Table 5.3.

The background integration mesh is independent from the nodal distribution and it is constructed using integration cells. In each integration cell, 3×3 integration points are distributed respecting the Gauss-Legendre quadrature scheme. In

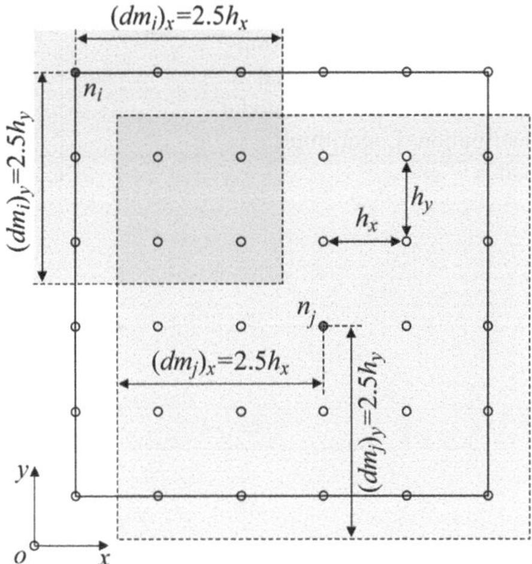

Fig. 5.2 Influence-domain examples

the present example, the same background integration mesh is considered for both irregular and regular nodal distributions, Fig. 5.1a, b.

It is assumed a constant square support-domain, $d_s = dm = 2.5\,h$, being h the mesh density parameter, which in this case is defined by the quadratic norm of the maximum distance between neighbour nodes, Fig. 5.2.

In order to obtain the optimized parameters γ and p, firstly the parameter p is fixed: $p = 1.0001$, and the parameter $\gamma \in \mathbb{R}$ is varied between $\gamma = [10^{-4}, 10^{1}]$. Afterwards, once the optimal γ_{opt} is achieved, the process is repeated but now only using the optimal shape parameter $\gamma = \gamma_{opt}$, and varying the shape parameter $p \in \mathbb{R}$ between $p = [10^{-4}, 5]$, until an optimal p_{opt} is obtained. The RPI shape functions are constructed using the linear polynomial basis, $\boldsymbol{p}(\boldsymbol{x}) = \{1\; x\; y\}$, and the MQ-RBF. To perform the patch test, the displacement field imposed on the boundary is defined by,

$$\boldsymbol{u}(\boldsymbol{x}) = \begin{cases} u(\boldsymbol{x}) = 0.1 + 0.1x \\ v(\boldsymbol{x}) = 0.1 + 0.1y \end{cases} \tag{5.2}$$

The square patch $1 \times 1\ \text{m}^2$ is analysed considering the plane stress deformation theory and unit thickness. The material properties of the square patch are: $E = 1$ Pa and $\upsilon = 0.3$. The problem is solved using the flow-chart of Table 5.1 and the medium displacement error, E_{med}, defined in Eq. (5.1) is calculated. The sum of the interior equivalent forces is given by the expression,

$$f_{tot} = \sum_{i=1}^{n} f_i^{eq} \tag{5.3}$$

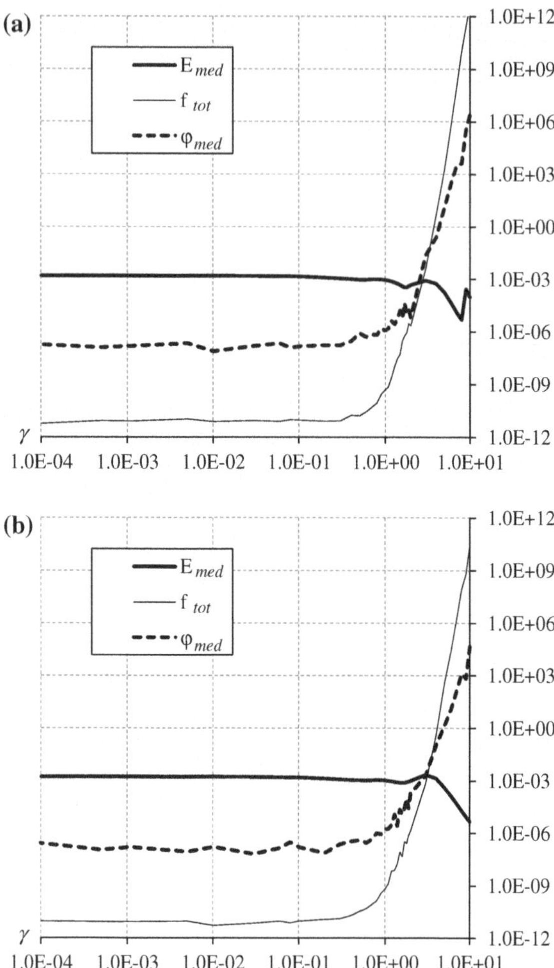

Fig. 5.3 Shape parameter γ effect on the RPIM solution accuracy. **a** Regular nodal distribution. **b** Irregular nodal distribution. Logarithmic scales

where n_t is the number of nodes that do not belong to the essential boundary and $\boldsymbol{f}^{eq}$ is the equivalent force obtained with,

$$\boldsymbol{f}^{eq} = \int_{\Omega} \boldsymbol{B}^{\mathrm{T}} \boldsymbol{\sigma} \, \mathrm{d}\Omega \tag{5.4}$$

In Fig. 5.3 the medium displacement error E_{med} is presented, together with f_{tot}, as a function of the shape parameter γ. It is visible that E_{med} stabilizes for values $\gamma \leq 1$. The stabilization and the minimization of f_{tot} is achieved for lower values of $\gamma : \gamma \leq 0.1$.

The shape parameters values suggested in the RPIM early works [9, 10] are $c = 1.43$ and $p = 1.03$. Notice that the shape parameter c is obtained with: $c = \gamma h$, being h in the present example a constant value: $h = h_x = h_y = 0.2\,m$, Fig. 5.2. Consequently the shape parameter γ suggested in the literature is $\gamma = 7.15$. It is visible in Fig. 5.3 that such values for the parameters conducts to a significantly low medium displacement error: $E_{med} \cong 10^{-5}$. However the same $\gamma = 7.15$ leads to a very high value of the sum of the interior equivalent forces: $f_{tot} \cong 10^9$, which is unacceptable, it should be zero.

This anomaly is explained with the effective lack of the Kronecker delta property on the RPIM shape functions for high values of γ. In Sect. 4.4.4 it was possible to observe in Fig. 4.23 that the value of the shape parameter γ regulates the silhouette of the shape function. Considering small values for γ the constructed shape function becomes cone-shaped. If γ is increased then the peak of the shape function becomes flat. As a consequence, the value of the shape parameter γ affects the accuracy of the solution. However the shape parameter γ cannot be null, because $\gamma = 0$ leads ill-conditioned or singular moment matrices [11, 12].

It is now understandable that for high values of γ the shape function does not pass exactly on the nodes. To prove the statement, during the previous analyses the shape function vector constructed for node n_{15}, $\boldsymbol{\varphi}(\boldsymbol{x}_{15})$, was saved. The $\boldsymbol{\varphi}(\boldsymbol{x}_{15})$ values on nodes n_9, n_{14}, n_{16} and n_{21} (all inside the support-domain of n_{15}) were used to calculate the following medium value,

$$\varphi_{med} = \frac{1}{4}(\varphi_9(\boldsymbol{x}_{15}) + \varphi_{14}(\boldsymbol{x}_{15}) + \varphi_{16}(\boldsymbol{x}_{15}) + \varphi_{21}(\boldsymbol{x}_{15})) \tag{5.5}$$

Note that these four nodes are the closest nodes to node n_{15} and if $\boldsymbol{\varphi}(\boldsymbol{x}_{15})$ possesses the Kronecker delta property, $\varphi_i(\boldsymbol{x}_{15})$ should be zero for all $i \neq 15$, and consequently $\varphi_{med} = 0$. However, it is visible in Fig. 5.3a, b that φ_{med} only stabilizes for $\gamma \leq 0.1$ and even for those values of γ, φ_{med} is different of zero, $\varphi_{med} \cong 10^{-7}$, indicating an effective lack of the Kronecker delta property.

As in the RPIM early works [9, 10], in this example the essential boundary conditions were directly imposed, as in FEM, since it was initially assumed that the RPI shape functions possess the Kronecker delta property. This is the reason why f_{tot} is different from zero. However, it is acceptable to consider, for $\gamma \leq 0.1$, that the RPIM shape functions possesses the Kronecker delta property.

Next, the optimal value of the shape parameter p is pursue. The procedure to find p_{opt} is similar with the previous shown example. Considering $\gamma_{opt} = 0.0001$, the shape parameter p is changed between $p = [10^{-4}, 5]$ until an optimal p is achieved. All other considerations regarding the material properties, the geometric and the boundary conditions remain the same as in previous analyses. Also the two nodal distributions previously considered and the respective background integration mesh Fig. 5.1, are once more assumed.

The obtained results are shown in Fig. 5.4. As it is possible to visualize, $p \cong 1$ is an optimal value since E_{med} show a minimum for this value for both the regular and the irregular nodal distribution. Additionally, when $p \cong 1$ is considered the sum of the interior equivalent forces f_{tot} present a very low value, as it should.

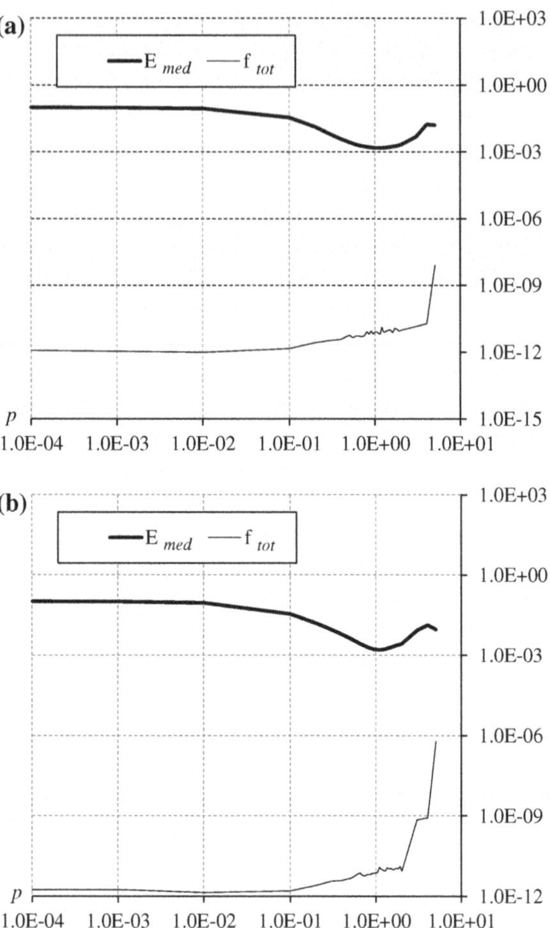

Fig. 5.4 Shape parameter p effect on the RPIM solution accuracy. **a** Regular nodal distribution. **b** Irregular nodal distribution. Logarithmic scales

Nevertheless, the obtained results also permitted to conclude that p cannot be equal to 1, or any other integer value, because when p assumes an integer value the moment matrix is singular. Others authors [9, 10] suggested a similar value for p, $p = 1.03$.

5.2.2 NNRPIM Patch Test

The process to obtain the optimal shape parameters γ and p using the NNRPIM is similar with the previously presented RPIM procedure. All previous example assumptions regarding: the problem discretization; the material properties; the patch geometry; and boundary conditions, remain unchanged.

Table 5.4 NNRPIM acronyms

Basis functions	Influence-cells	
	First degree	Second degree
MQ-RBF	V1P0	V2P0
MQ-RBF + constant unit basis	V1P1	V2P1
MQ-RBF + linear polynomial basis	V1P3	V2P3
MQ-RBF + quadratic polynomial basis	V1P6	V2P6

First it is presented the optimization study of the MQ-RBF shape parameters considering the NNRPIM formulation using first degree influence-cells. Afterwards the shape parameters optimization study is repeated for a NNRPIM formulation assuming the second degree influence-cells. In the present subsection only the nodal based basic integration scheme is considered, as shown in Section "Basic Integration Scheme". To obtain the RPI shape functions four combinations of basis functions are considered: (1) only the MQ-RBF; (2) MQ-RBF and an unit constant basis; (3) MQ-RBF and a linear polynomial basis; (4) MQ-RBF and a quadratic polynomial basis.

In order to identify the applied NNRPIM formulation, acronyms are used for a better understanding. In each acronym the first two characters identifies the degree of the influence-cell, first degree = "V1" and second degree = "V2". The polynomial basis is identified by "Pm", where m is the number of monomials in the polynomial basis. For example, the formulation using a second degree influence-cell and an unit constant basis is called: "V2P1". The complete list of acronyms is presented in Table 5.4.

In order to obtain the optimized MQ-RBF shape parameter γ, the MQ-RBF shape parameter p is fixed: $p = 1.0001$, and the parameter $\gamma \in \mathbb{R}$ is varied between $\gamma = [10^{-4}, 10^{1}]$. The solid domain of the unit square patch $1 \times 1\ \text{m}^2$ is discretized in the two distinct nodal distributions presented in Fig. 5.1. On the patch boundary it is imposed the displacement field defined in Eq. (5.2). It is considered a unit thickness and the material properties of the square patch are: $E = 1$ Pa and $\upsilon = 0.3$. The problem is solved using the flow-chart of Table 5.2 and considering the plane stress deformation theory.

The medium displacement error, E_{med}, defined in Eq. (5.1), is determined and the sum of the interior equivalent forces f_{tot} is obtained with Eq. (5.3).

In Fig. 5.5 the medium displacement error E_{med} is presented, together with f_{tot}, as function of the shape parameter γ, for the V1P0 NNRPIM formulation. The results regarding the V1P1 NNRPIM formulation and the V1P3 NNRPIM formulation are presented respectively in Figs. 5.6 and 5.7.

It was observed that using the quadratic polynomial basis, V1P6 NNRPIM formulation, generates a singular moment matrix, i.e., non invertible. Thus the study of this polynomial basis applied to this formulation was abandoned. As Fig. 5.5 indicates, for the V1P0 NNRPIM formulation the optimal values for the shape parameter γ are $\gamma \geq 2$, however the stress field produced with this

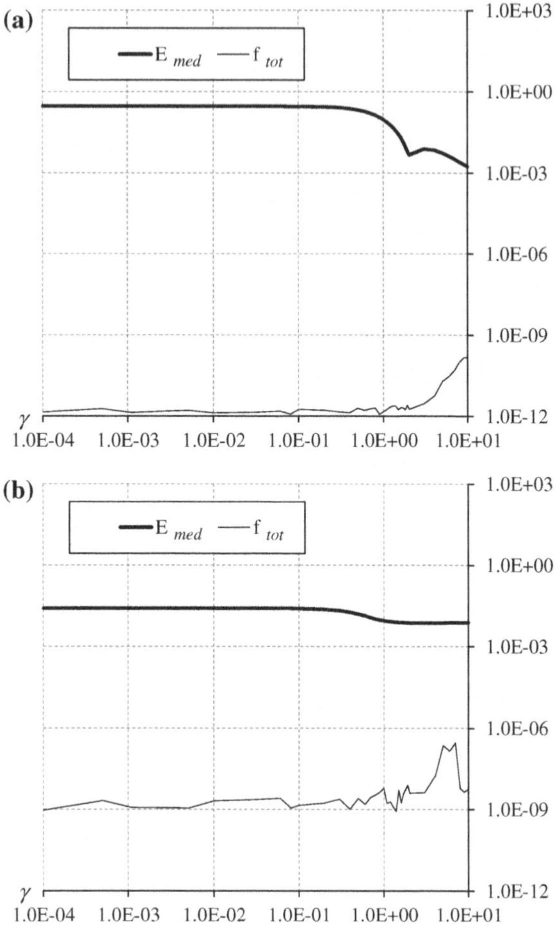

Fig. 5.5 Shape parameter γ effect on the V1P0 NNRPIM solution accuracy. **a** Regular nodal distribution. **b** Irregular nodal distribution. Logarithmic scales

formulation is highly irregular (it should be constant) and the observed field irregularity grows with the increasing of γ. Thus the optimal value suggested for the V1P0 NNRPIM formulation is $\gamma = 2$. The V1P1 NNRPIM formulation and the V1P3 NNRPIM formulation are quite similar, both show that the optimal value for the shape parameter γ is $\gamma \cong 0$. In further examples, for these two formulations, the shape parameter γ used is $\gamma = 0.0001$.

Using the obtained optimal MQ-RBF γ shape parameters for each respective NNRPIM formulation, the MQ-RBF shape parameter p is varied between $p = [10^{-4}, 5]$ until an optimal p is achieved for each NNRPIM formulation analysed. The results regarding the V1P0 NNRPIM formulation are shown in Fig. 5.8 and the results obtained with the V1P1 NNRPIM formulation and the V1P3 NNRPIM formulation are respectively presented in Figs. 5.9 and 5.10.

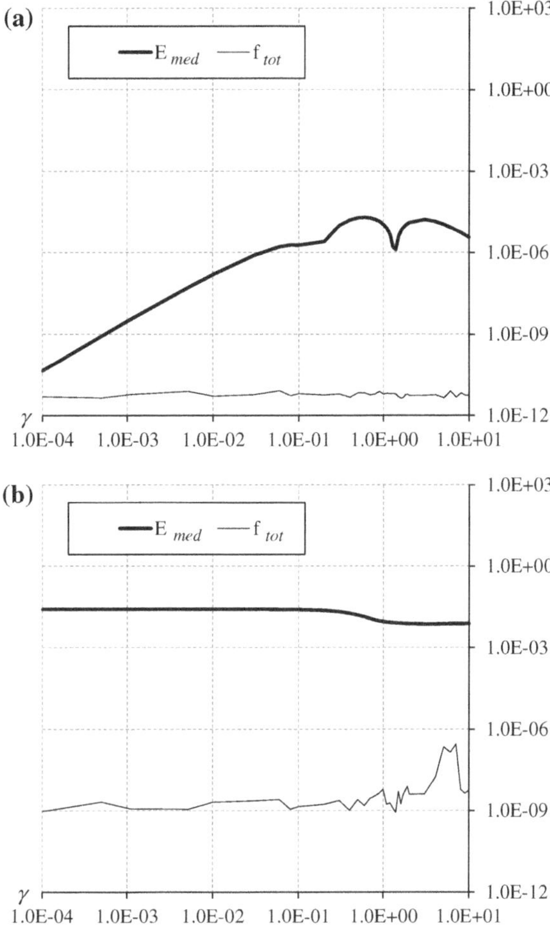

Fig. 5.6 Shape parameter γ effect on the V1P1 NNRPIM solution accuracy. **a** Regular nodal distribution. **b** Irregular nodal distribution. Logarithmic scales

As Fig. 5.8 shows, for the V1P0 NNRPIM formulation the optimization curves only stabilize for $p > 0.01$. Therefore, the optimal value suggested for the V1P0 NNRPIM formulation is $p_{opt} = 0.0001$. For the V1P1 NNRPIM formulation, Fig. 5.9, it is perceptible that the optimization curves only stabilize for $p > 0.1$, however an optimal value for the shape parameter p is achieved near the value 1.0. Thus, for the V1P1 NNRPIM formulation it is assumed $p_{opt} = 1.0001$. Regarding the V1P3 NNRPIM formulation, Fig. 5.10, the optimization curves become flat for $p > 0.1$ and it is possible to identify an optimal shape parameter near the value 3.0. Therefore, the optimal value suggested for the V1P3 NNRPIM formulation is $p_{opt} = 3.0001$.

It is important to refer that the V1P0 NNRPIM formulation it is not capable to produce an acceptable smooth stress field, for this reason the study of this formulation is abandoned.

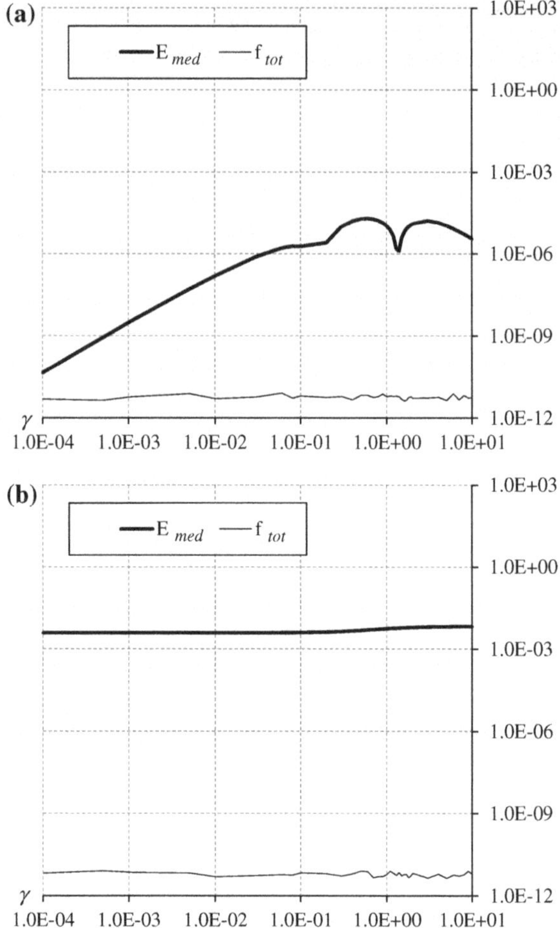

Fig. 5.7 Shape parameter γ effect on the V1P3 NNRPIM solution accuracy. **a** Regular nodal distribution. **b** Irregular nodal distribution. Logarithmic scales

Next, NNRPIM formulations considering second degree influence-cells are studied. The procedure is similar to the previous NNRPIM formulations optimization studies. For V2P0 NNRPIM formulation, the results of the patch test concerning the optimization of the MQ-RBF shape parameter γ are presented in Fig. 5.11. The MQ-RBF shape parameter γ optimization results regarding the V2P1, the V2P3 and the V2P6 NNRPIM formulations are shown respectively in Figs. 5.12, 5.13 and 5.14.

It is possible to observe in Figs. 5.11 and 5.12 that the optimization curves of the V2P0 and V2P1 NNRPIM formulations stabilize for shape parameters values $\gamma \leq 0.1$. Regarding the V2P3 and V2P6 NNRPIM formulations, Figs. 5.13 and 5.14 show that these NNRPIM formulations are not capable to construct RPI shape functions possessing the Kronecker delta property when $\gamma > 0.1$. Therefore, for all

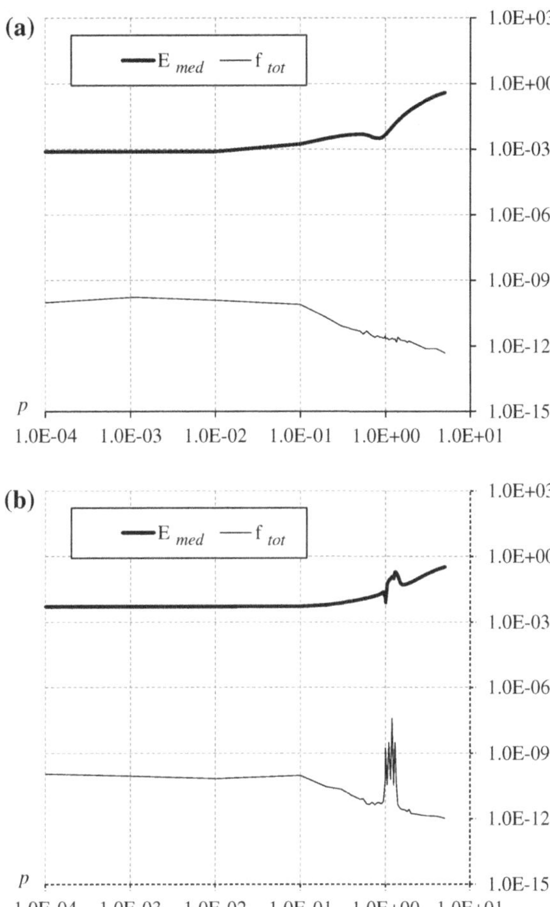

Fig. 5.8 Shape parameter p effect on the V1P0 NNRPIM solution accuracy. **a** Regular nodal distribution. **b** Irregular nodal distribution. Logarithmic scales

the studied NNRPIM formulations using second degree influence-cells the chosen optimal MQ-RBF γ shape parameter is $\gamma_{opt} = 0.0001$.

Afterwards, using the obtained optimal MQ-RBF γ shape parameters for each respective NNRPIM formulation, the MQ-RBF shape parameter p is varied between $p = [10^{-4}, 5]$ until an optimal p is achieved for each NNRPIM formulation analysed. The results for the V2P0 NNRPIM formulation are shown in Fig. 5.15 and the results obtained with the V2P1, V2P3 and V2P6 NNRPIM formulations are respectively presented in Figs. 5.16, 5.17 and 5.18. Once again it is desirable to avoid integer values to the shape parameter p, since those values lead to a singular moment matrix.

As Fig. 5.15 shows, for the V2P0 NNRPIM formulation the optimization curves shows an optimal value for the shape parameter p near the value 1.0. For the V2P1 NNRPIM formulation, Fig. 5.16, it is possible to identify the same optimal value. Therefore, for both V2P0 and V2P1 NNRPIM formulations the

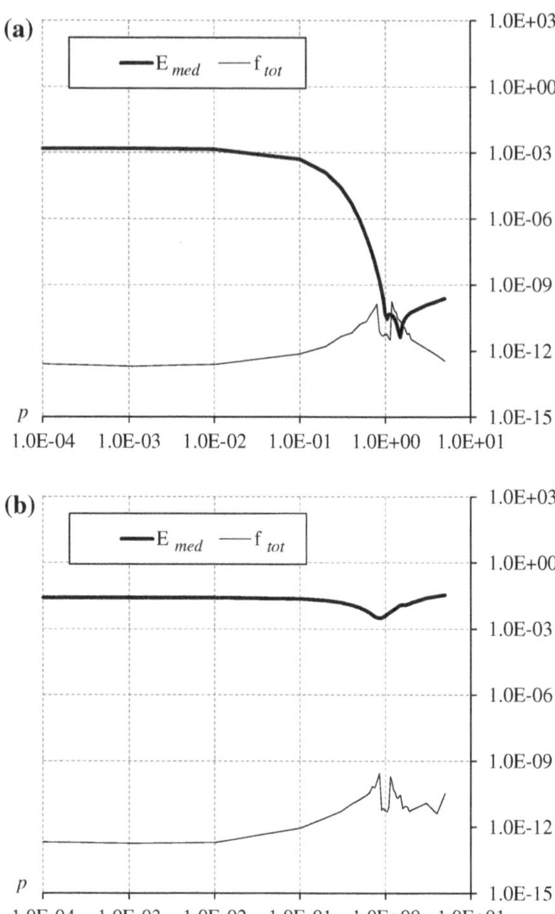

Fig. 5.9 Shape parameter p effect on the V1P1 NNRPIM solution accuracy. **a** Regular nodal distribution. **b** Irregular nodal distribution. Logarithmic scales

chosen optimal value is $p_{opt} = 1.0001$. Regarding the V2P3 NNRPIM formulation, Fig. 5.17, the optimization curves only stabilize for $p > 0.1$, however it is possible to identify an optimal value for the shape parameter p near the value 1.0. Although for $p \cong 1$ the f_{tot} value as not reached the minimum, it is already sufficiently low: $f_{tot} \cong 10^{-11}$.

For the V2P6 NNRPIM formulation it is perceptible in Fig. 5.18 that the optimization curves become flat for $p \geq 0.1$. It is not possible to identify a clear optimal value, therefore for the V2P6 NNRPIM formulation $p_{opt} = 0.1$.

It was observed that the stress field obtained with the NNRPIM, using the second degree influence-cells, is constant, as it should be, unlike the stress field obtained with the NNRPIM considering first degree influence-cells. This is an important feature, which influences the accuracy and smoothness of the stress field.

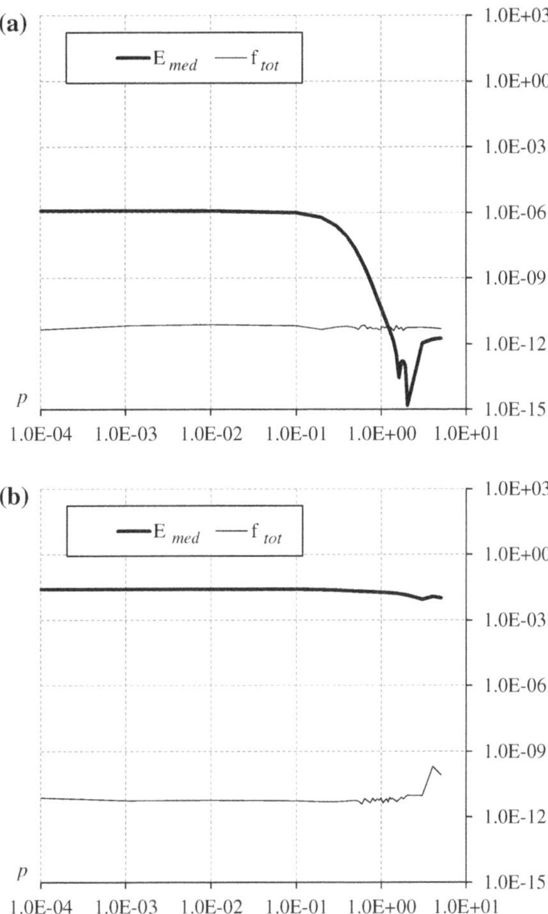

Fig. 5.10 Shape parameter p effect on the V1P3 NNRPIM solution accuracy. **a** Regular nodal distribution. **b** Irregular nodal distribution. Logarithmic scales

In Table 5.5 the obtained E_{med} and f_{tot} using the optimal MQ-RBF shape parameters for the distinct NNRPIM formulations are summarized.

As Table 5.5 shows, for irregular nodal distributions, the lowest medium displacement error E_{med} is obtained with the V2P1 NNRPIM formulation, is very close with the error obtained with the RPIM. Notice as well that generally the error obtained for regular meshes with NNRPIM is lower than the error obtained with RPIM.

In this section the patch test study will continue considering only the V1P0, V1P1, V2P0 and V2P1 NNRPIM formulations, since the lowest medium displacement errors are obtained with these four NNRPIM formulations. It is important to refer that the RPIM analysis is an optimized analysis because adequate MQ-RBF shape parameters and proper integration schemes were applied [9, 10]. However in the case of the NNRPIM it was applied only the basic nodal based integration

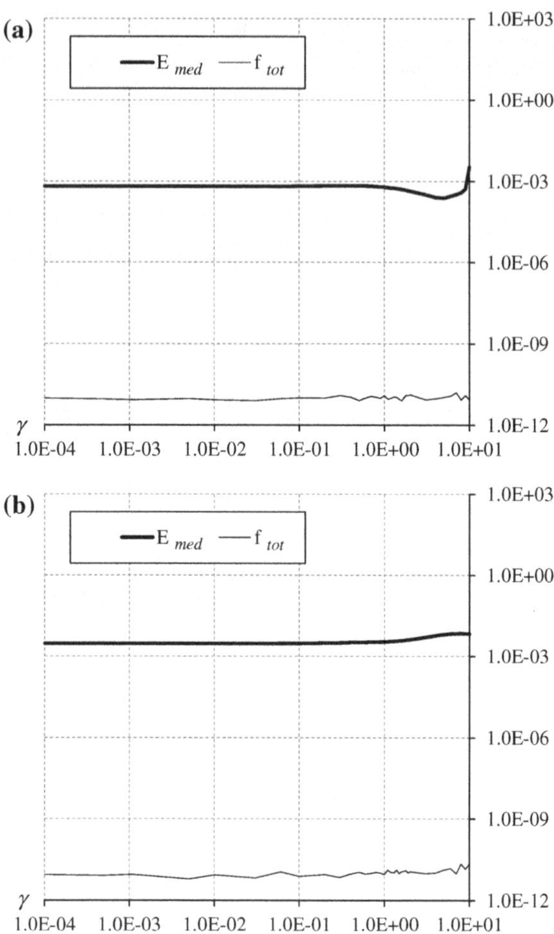

Fig. 5.11 Shape parameter γ effect on the V2P0 NNRPIM solution accuracy. **a** Regular nodal distribution. **b** Irregular nodal distribution. Logarithmic scales

scheme, Section "Basic Integration Scheme". Thus, using the obtained optimal MQ-RBF shape parameters a study to find the optimal integration scheme is carried out.

The irregular nodal distribution presented in Fig. 5.1a is used and the optimal shape parameters presented in Table 5.5 are applied to the RPIM and NNRPIM formulations. The problem is studied for various integration schemes considering each RPIM and NNRPIM formulation previously presented.

For the RPIM analysis several background integration meshes are constructed. Considering the quadrilateral integrations cells presented in Fig. 5.1a, the number of integration points of the Gauss-Legendre quadrature scheme is varied from 1×1 integration points per integration cell to 5×5 integration points per integration cell. The medium displacement errors, E_{med}, for each analysis are presented in Fig. 5.19.

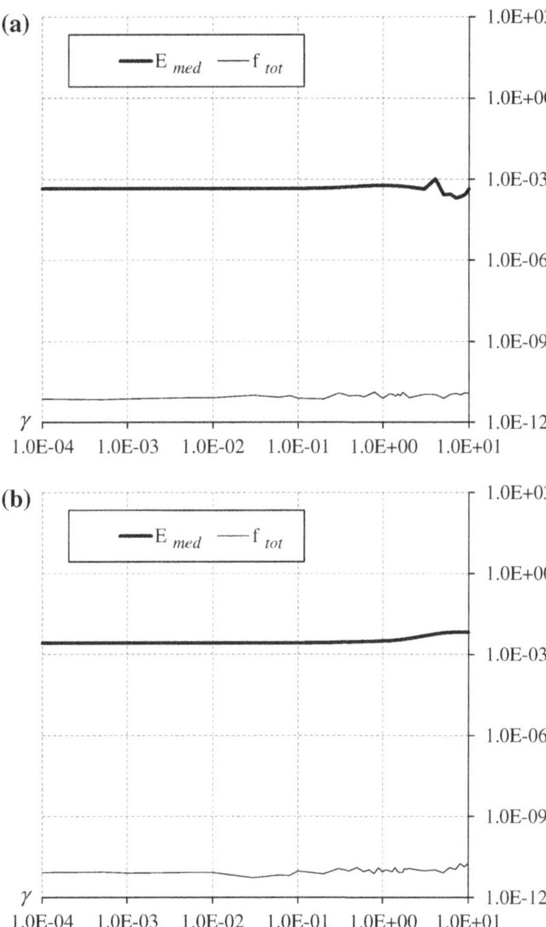

Fig. 5.12 Shape parameter γ effect on the V2P1 NNRPIM solution accuracy. **a** Regular nodal distribution. **b** Irregular nodal distribution. Logarithmic scales

It is possible to confirm that, for the RPIM formulation, 3×3 Gauss-Legendre integration points per integration cell is sufficient to integrate accurately the integro-differential equations ruling the present solid mechanics problem. Using more integration points per cell does not increase the precision of the solution.

Regarding the patch test analysis using the NNRPIM formulation, several numerical integration schemes were used. The medium displacement errors, E_{med}, for each analysis are presented in Fig. 5.20. The first integration scheme considered was the simple nodal based integration scheme described in Section "Basic Integration Scheme". In Fig. 5.20 this simple integration scheme is called "basic". The other $k \times k$ integrations schemes presented in Fig. 5.20 are obtained following the procedure indicated in Section "Gauss-Legendre Quadrature Integration Scheme".

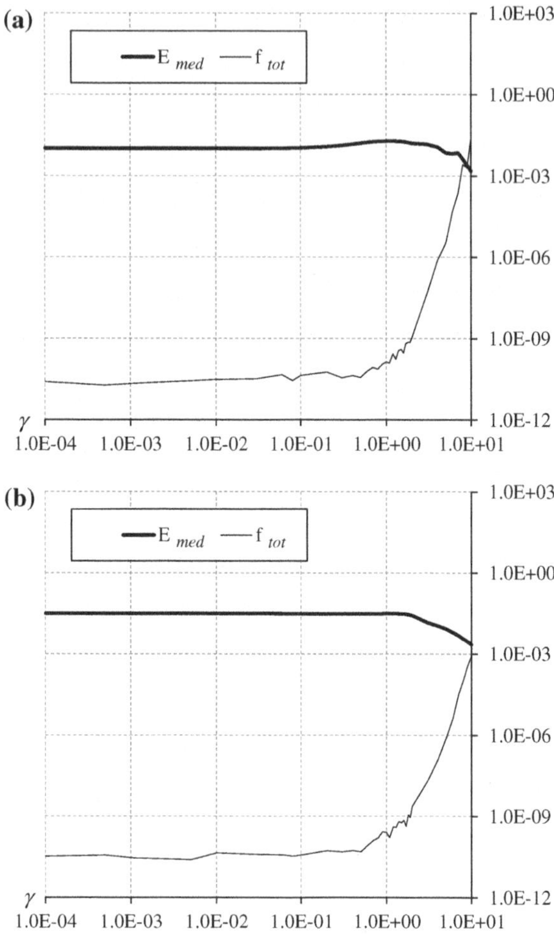

Fig. 5.13 Shape parameter γ effect on the V2P3 NNRPIM solution accuracy. **a** Regular nodal distribution. **b** Irregular nodal distribution. Logarithmic scales

Notice that the number of integration points used in the distinct NNRPIM integration schemes increases by a factor of 4, which means that the computational cost will increase in the same proportionality. It is visible that the V2P0 and V2P1 NNRPIM solutions only stabilize for integration schemes greater than "1×1", however it is reasonable to use the "basic" integration scheme as an acceptable numerical integration scheme, since it minimizes the computational cost without risking to much the method accuracy.

In furthers two-dimensional numerical analysis only the V2P1 NNRPIM formulation combined with the "basic" nodal based integration scheme, is considered.

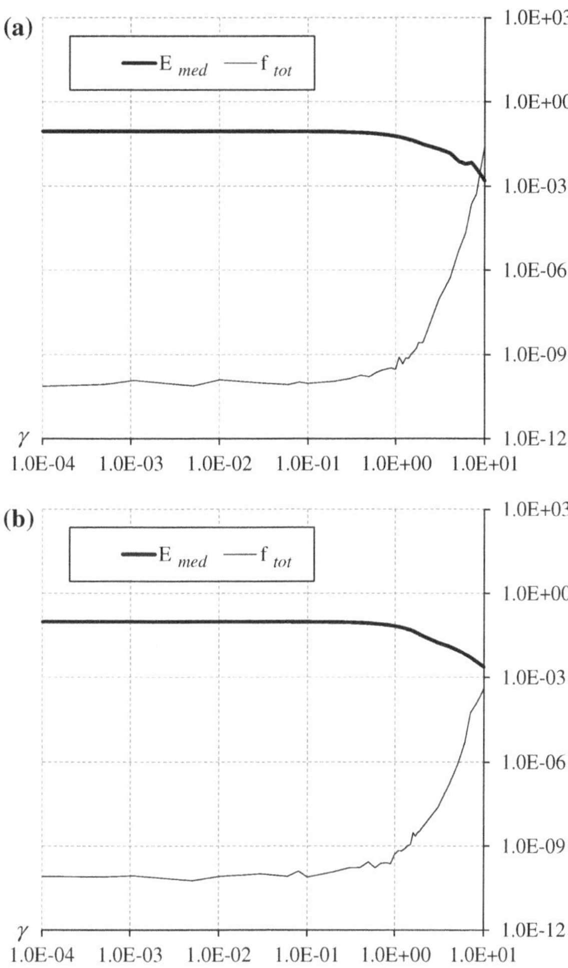

Fig. 5.14 Shape parameter γ effect on the V2P6 NNRPIM solution accuracy. **a** Regular nodal distribution. **b** Irregular nodal distribution. Logarithmic scales

5.2.3 3D NNRPIM Patch Test

The three-dimensional patch test was also studied, an unit cubic solid with a volume of $1.0 \times 1.0 \times 1.0\,\text{m}^3$, and with the same material properties as the previously two-dimensional example, was submitted to an imposed displacement in all six plane boundaries,

$$\begin{cases} u = 0.1 + 0.1x \\ v = 0.1 + 0.2y \\ w = 0.1 + 0.3z \end{cases} \tag{5.6}$$

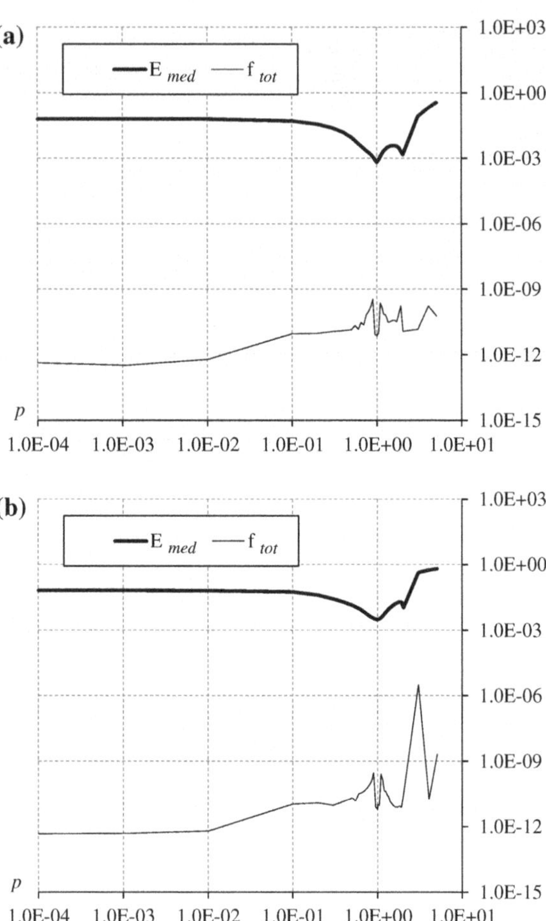

Fig. 5.15 Shape parameter p effect on the V2P0 NNRPIM solution accuracy. **a** Regular nodal distribution. **b** Irregular nodal distribution. Logarithmic scales

The three-dimensional domain $\boldsymbol{x} \in \mathbb{R}^3 : x \in [0, 1], y \in [0, 1], z \in [0, 1]$ was discretized with two distinct nodal distributions: a regular and an irregular nodal distribution with $6 \times 6 \times 6$ nodes. The three-dimensional nodal distributions follow the layout of the two-dimensional nodal distributions presented in Fig. 5.1.

Once more the procedure used to obtain the optimal MQ-RBF shape parameters for the two-dimensional space is considered. First, the shape parameter p is fixed: $p = 1.0001$, and the parameter $\gamma \in \mathbb{R}$ is varied between $\gamma = [10^{-4}, 10^1]$. Then, using the optimal shape parameter $\gamma = \gamma_{opt}$, the shape parameter $p \in \mathbb{R}$ is changed between $p = [10^{-4}, 5]$ until an optimal p_{opt} is reached. The material properties of the cubic patch $1 \times 1 \times 1\,\text{m}^3$ are: $E = 1$ Pa and $\upsilon = 0.3$. The problem is solved using the flow-chart of Table 5.2. The medium displacement error, E_{med}, and the sum of the interior equivalent forces f_{tot} are determined using respectively Eqs. (5.1) and (5.3). The optimal MQ-RBF shape parameters were obtained [13] and the final results are presented in Table 5.6.

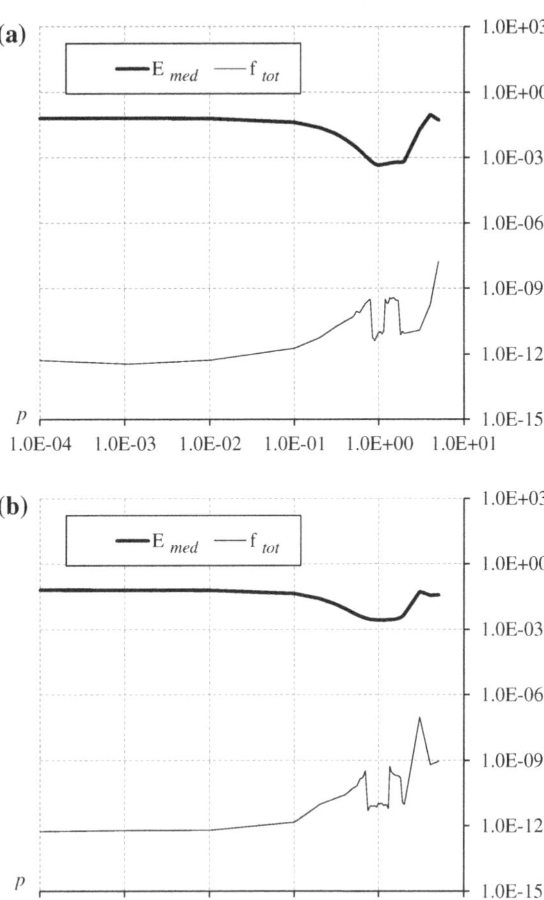

Fig. 5.16 Shape parameter p effect on the V2P1 NNRPIM solution accuracy. **a** Regular nodal distribution. **b** Irregular nodal distribution. Logarithmic scales

In the present three-dimensional study it was observed that the V1P0, V1P3, V2P3 and V2P6 NNRPIM formulations produce highly irregular stress fields. Since a linear displacement field is imposed in the cubic patch, constant stress fields should be obtained. In the Table 5.6 it is also perceptible that the referred V1P3, V2P3 and V2P6 NNRPIM formulations, when compared with the others NNRPIM formulations, present a higher medium displacement error. As so, in further three dimensional analyses only the V1P1, V2P0 or the V2P1 NNRPIM formulations are considered. Due to the significant computational cost, in this work, for the three-dimensional analysis the used integration scheme is the "basic" nodal based integration scheme.

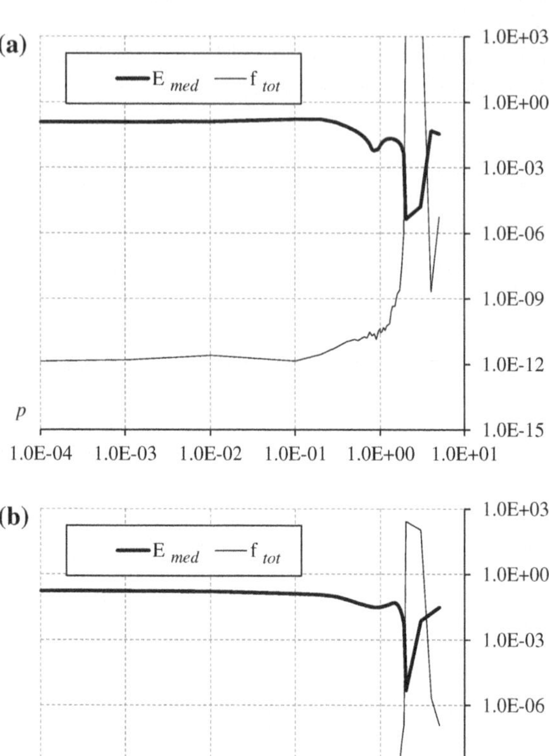

Fig. 5.17 Shape parameter p effect on the V2P3 NNRPIM solution accuracy. **a** Regular nodal distribution. **b** Irregular nodal distribution. Logarithmic scales

5.3 Elastostatic Numerical Examples

In this section three benchmark elastostatic numerical problems are analysed. All the solid mechanics examples studied in the present section are solved using the RPIM formulation suggested in [9, 10] and the NNRPIM formulation suggested in [13].

Therefore, for the RPIM formulation the RPI shape functions are constructed using the MQ-RBF and a linear polynomial basis. The MQ-RBF shape parameters are: $c = 1.43$ and $p = 1.03$. The nodal connectivity is imposed using square support-domains: $d_s = 2.5\,h$, as Fig. 5.2 shows. The background integration mesh respects the layout presented in Fig. 5.1 and for each integration cell are used 3×3 Gauss-Legendre quadrature integration points.

Regarding the NNRPIM formulation, the RPI shape functions are constructed using the MQ-RBF and a constant unit basis. The MQ-RBF shape parameters are:

Fig. 5.18 Shape parameter p effect on the V2P6 NNRPIM solution accuracy. **a** Regular nodal distribution. **b** Irregular nodal distribution. Logarithmic scales

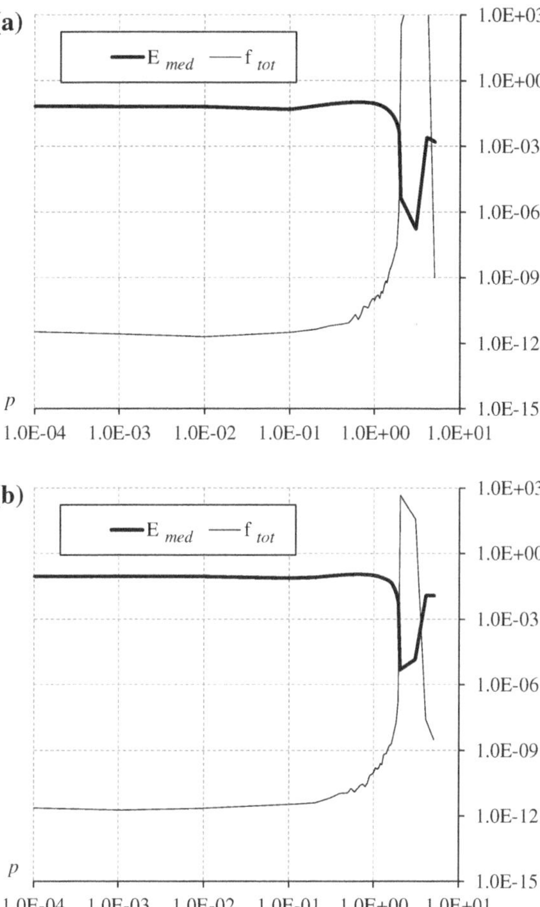

Table 5.5 Results obtained with the optimal MQ-RBF shape parameters for a 2D analysis

Meshless method	Shape parameters		Regular nodal mesh		Irregular nodal mesh	
	γ	p	E_{med}	f_{tot}	E_{med}	f_{tot}
RPIM	0.0001	1.0001	1.51E−03	8.08E−12	2.91E−03	7.27E−12
V1P0	2.0000	0.0001	8.00E−04	9.87E−11	5.10E−03	1.17E−10
V1P1	0.0001	1.0001	4.47E−11	5.07E−12	4.02E−03	5.44E−12
V1P3	0.0001	3.0001	1.09E−12	5.79E−12	8.89E−03	9.41E−12
V2P0	0.0001	1.0001	6.51E−04	9.63E−12	3.04E−03	8.77E−12
V2P1	0.0001	1.0001	4.59E−04	7.26E−12	2.67E−03	8.40E−12
V2P3	0.0001	1.0001	7.02E−03	3.12E−11	3.07E−02	2.99E−11
V2P6	0.0001	0.1000	4.74E−02	3.17E−12	7.43E−02	3.38E−12

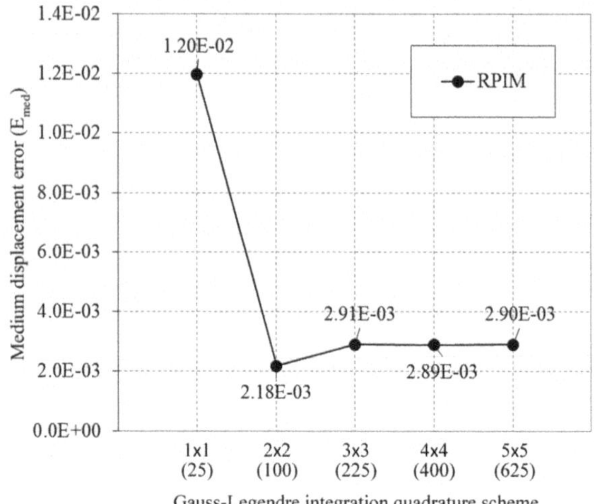

Fig. 5.19 Effect of the integration scheme on the RPIM solution accuracy

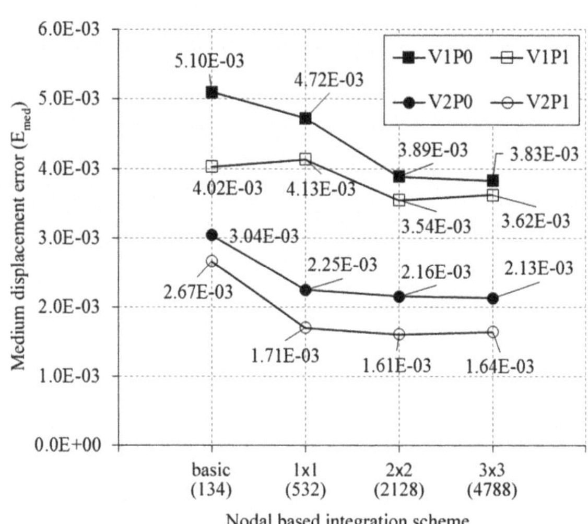

Fig. 5.20 Effect of the integration scheme on the NNRPIM solution accuracy

Table 5.6 Results obtained with the optimal MQ-RBF shape parameters for a 3D analysis

NNRPIM formulation	Shape parameters		Regular nodal mesh		Irregular nodal mesh	
	γ	p	E_{med}	f_{tot}	E_{med}	f_{tot}
V1P0	0.0001	1.0001	7.66E−04	2.26E−14	1.65E−02	1.36E−10
V1P1	0.0001	1.0001	1.75E−03	3.18E−14	1.37E−02	4.33E−10
V1p3	0.0001	1.0001	1.60E−02	2.60E−13	3.41E−02	4.02E−10
V2P0	0.0001	1.0001	3.32E−03	6.61E−14	8.00E−03	3.69E−10
V2P1	0.0001	1.0001	3.20E−03	7.16E−14	7.30E−03	2.89E−09
V2P3	0.0001	1.0001	3.31E−02	4.41E−13	4.61E−02	6.39E−08
V2P6	0.0001	1.0001	5.31E−02	9.20E−13	5.68E−02	4.06E−09

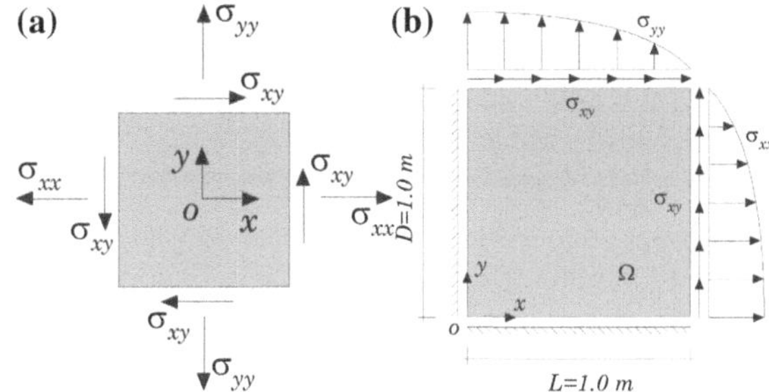

Fig. 5.21 **a** Generic solid mechanics problem. **b** Quarter of the square plate under parabolic stress

$\gamma = 0.0001$ and $p = 1.0001$. The nodal connectivity is enforced using the second degree influence-cells and the integration points are obtained using the "basic" nodal based integration scheme.

Besides the RPIM and the NNRPIM, the presented benchmark examples were analysed with the finite element method (FEM) [1, 14]. For the two-dimensional analyses, it were considered discretizations with quadrilateral elements (FEM-4n) and triangular elements (FEM-3n). For the three-dimensional analyses are considered hexahedral elements (FEM-8n) and tetrahedral elements (FEM-4n).

5.3.1 Square Plate Under Parabolic Stress

The proposed example combines several objectives: to analyse the behaviour of the NNRPIM when random irregular nodal distributions are used in the analysis; to compare the computational cost of the NNRPIM with other numerical methods; and to study the convergence and accuracy level of the NNRPIM.

In this example it is considered the solid domain $\Omega \subset \mathbb{R}^2$ represented in Fig. 5.21a, with the following material properties: $E = 1$ kPa and $\upsilon = 0.3$. In the natural boundary, $\Gamma_t \in \Omega$, of the solid domain the following stress field is applied,

$$\begin{aligned} \sigma_{xx}(\boldsymbol{x}) &= \sigma_0 \cdot \left(\frac{x^2}{L^2} - \frac{y^2}{D^2}\right) \\ \sigma_{yy}(\boldsymbol{x}) &= \sigma_0 \cdot \left(\frac{(L^2 - 2\,D^2)\cdot x^2}{L^2 \cdot D^2} + \frac{y^2}{L^2}\right) \\ \sigma_{xy}(\boldsymbol{x}) &= -\sigma_0 \cdot \left(\frac{2\,x \cdot y}{L^2}\right) \end{aligned} \tag{5.7}$$

being $\sigma_0 = 100$ Pa. Due the problem symmetry the study of the complete problem can be reduced to the analysis of only one quarter of the solid domain, Fig. 5.21b. Therefore, the displacement constrains on the essential boundary, $\Gamma_u \in \Omega$, must be considered as represented in Fig. 5.21b: $u = 0 : \forall y \in \mathbb{R} \wedge x = 0$ and

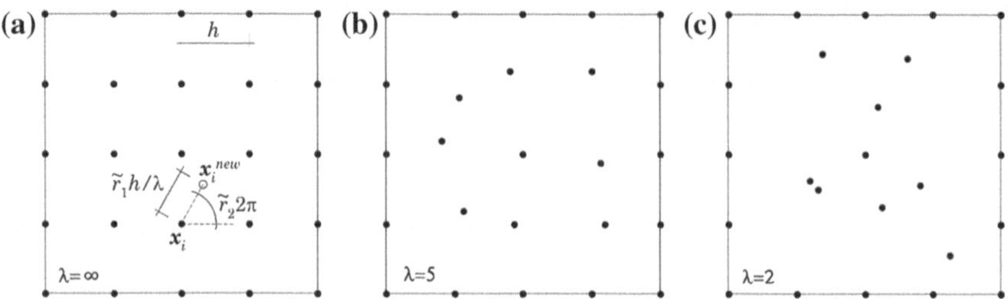

Fig. 5.22 **a** 5 × 5 regular nodal distribution. **b** Irregular nodal distribution with $\lambda = 5$. **c** Irregular nodal distribution with $\lambda = 2$

$v = 0 : \forall x \in \mathbb{R} \wedge y = 0$, being $\boldsymbol{u} = \{u, v\}$. The analytical displacement field [15], which can be obtained from the analytical stress field of Eq. (5.7), is described as,

$$u(\boldsymbol{x}) = \frac{\sigma_0}{E} \cdot \left(\frac{x^3}{3\,L^2} - \frac{x \cdot y^2}{D^2} - \upsilon \cdot \left(\frac{x^3 \cdot (L^2 - 2\,D^2)}{3\,L^2 \cdot D^2} + \frac{x \cdot y^2}{L^2} \right) \right)$$
$$v(\boldsymbol{x}) = \frac{\sigma_0}{E} \cdot \left(\frac{x^2 \cdot y \cdot (L^2 - 2\,D^2)}{L^2 \cdot D^2} + \frac{y^3}{3\,L^2} - \upsilon \cdot \left(\frac{x^2 \cdot y}{L^2} - \frac{y^3}{3\,D^2} \right) \right) \quad (5.8)$$

Firstly it is studied the behaviour of the NNRPIM when random irregular nodal distributions are used in the analysis. To create random irregular nodal distributions the following procedure is used. A uniform nodal distribution is constructed, with all nodes equally spaced and aligned, as in Fig. 5.22a, then all the nodes $\boldsymbol{x} \in \Omega \backslash \Gamma$ are affected with,

$$x_i^{new} = x_i + \frac{\tilde{r}_1 \cdot h}{\lambda} \cos(2\,\tilde{r}_2 \cdot \pi)$$
$$y_i^{new} = y_i + \frac{\tilde{r}_1 \cdot h}{\lambda} \sin(2\,\tilde{r}_2 \cdot \pi) \quad (5.9)$$

Being $\boldsymbol{x}_i$ the initial node n_i coordinates, $\boldsymbol{x}_i^{new}$ the new obtained node n_i coordinates and h is the distance shown in Fig. 5.22a. The random parameter is defined by $\tilde{r} \sim N(0, 1)$ and λ is a parameter that controls the irregularity level of the nodal distribution. The three nodal distributions presented in Fig. 5.22 show the effect of the irregularity parameter λ on the nodal distribution; notice that if $\lambda = \infty$ the nodal distribution is perfectly regular, Fig. 5.22a, and with the decrease of λ the nodal distribution becomes more and more irregular, Fig. 5.22b, c.

The proposed mechanical problem was analysed considering several irregular nodal distributions with $21 \times 21 = 441$ nodes, varying the irregularity parameter from $\lambda = 100$ (practically a regular mesh) to $\lambda = 2$ (extremely irregular mesh). Each irregular nodal distribution was used to analyse the problem considering the NNRPIM, the FEM-4n, the FEM-3n and the RPIM. The results of the medium displacement error, Eq. (5.1), are presented in Fig. 5.23a. The results regarding the

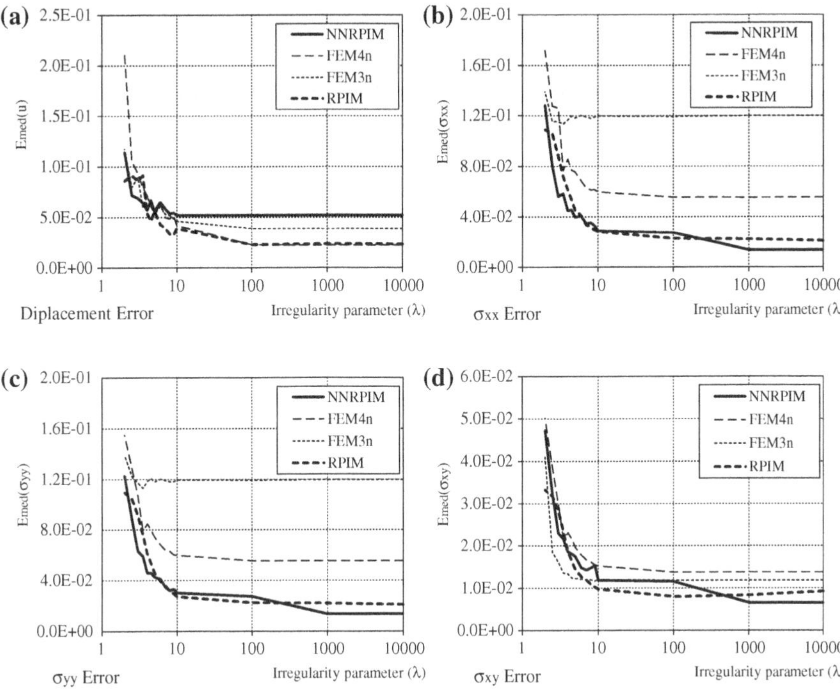

Fig. 5.23 Medium errors obtained for the following variables: **a** displacement; **b** normal stress σ_{XX}; **c** normal stress σ_{YY} and **d** shear stress σ_{XY}. Logarithmic scale

Table 5.7 Numerical method analysis phase and the respective description and acronym

Order	Phase	Phase description	Acronym
1.	Pre-process	Definition of the nodal distribution. Influence-domains are found (meshless methods) or elements are established (FEM). The integration mesh is determined	PP
2.	Shape function	The shape-functions are determined for all integration points	SF
3.	Stiffness matrix	Determination of the local stiffness matrix and assemblage of the global stiffness matrix	SM
4.	Natural boundary	The natural boundary is found and submitted to tension	NB
5.	Essential boundary	Constrained displacements are imposed on the essential boundary	EB
6.	Displacement field	Displacement field vector determination and medium error determination	DV
7.	Strain and stress field	Determination of the strain and stress field vector and the respective medium error	SV

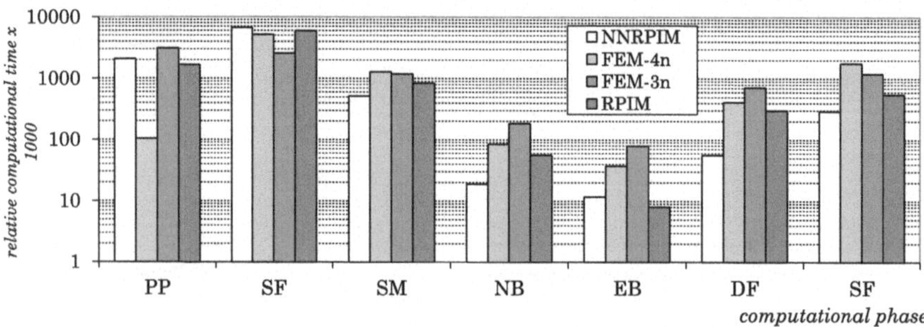

Fig. 5.24 Relative computational time for each analysis phase

medium stress error are presented in Fig. 5.23b–d, for the normal stresses σ_{xx} and σ_{yy} and the shear stress σ_{xy}. The medium stress errors is obtained with the expression,

$$E_{med}(\sigma) = \frac{1}{Q}\sum_{i=1}^{Q} \frac{\sqrt{\left((\sigma_i)_{meshless} - (\sigma_i)_{exact}\right)^2}}{\sqrt{(\sigma_i)_{exact}^{\;2}}} \tag{5.10}$$

Being Q the total number of interest points discretizing the problem domain. The results show that the NNRPIM presents a good behaviour when the problem domain is discretized with an irregular nodal distribution.

The next study regards the analysis of the computational effort. In order to compare the computational cost of each numerical method used in the analysis, the NNRPIM, the FEM and the RPIM flow-charts respect the phase sequence presented in Table 5.7.

In this study a regular nodal distribution with $21 \times 21 = 441$ nodes was used to analyse the previous example, Fig. 5.21b, considering the same material, geometric and boundary conditions. The time t_i^{NNRPIM} spent in each phase i is saved and in the end the total time of the analysis is obtained, $T^{NNRPIM} = \sum_{i=1}^{7} t_i^{NNRPIM}$. Then, the relative computational time is obtained, $\bar{t}_i^{NNRPIM} = t_i^{NNRPIM} / T^{NNRPIM}$. In Fig. 5.24 are presented the relative computational time $\bar{t}_i^{NNRPIM}$ (multiplied by a factor of 1,000) for each analysis phase. It is possible to observe that the NNRPIM presents a higher relative computational cost for the pre-processing phase (PP), in which the natural neighbours are determined and the integration mesh is constructed. For all the other phases the NNRPIM present a competitive relative computational cost. Considering that $T^{NNRPIM} \cong 3 \cdot T^{FEM}$, it is clear that if future works attend to reduce the computational cost of the NNRPIM pre-processing phase then the total computational cost of the NNRPIM should drop drastically.

In order to study the convergence of the RPIM and NNRPIM, the example illustrated in Fig. 5.21b is once again analysed, with the same material, geometric and boundary conditions considered previously. The solid domain is now

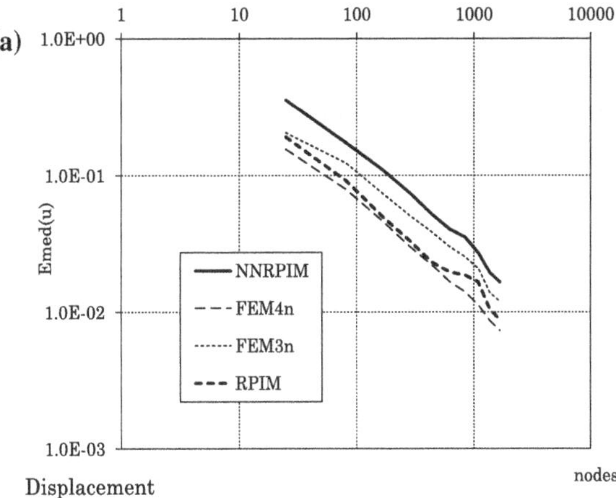

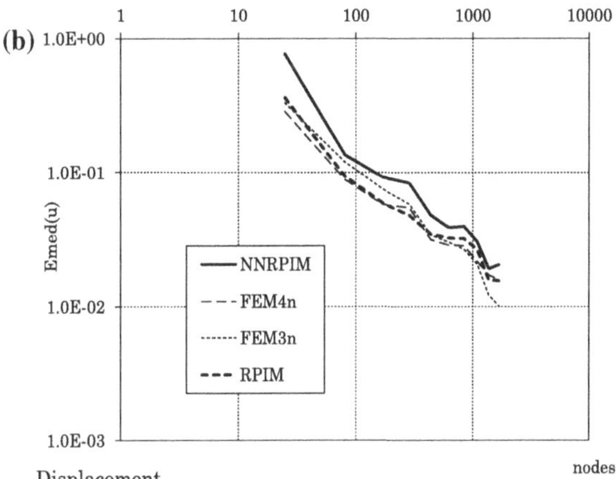

Fig. 5.25 Medium displacement errors obtained for the: **a** regular nodal distribution; **b** irregular nodal distribution. Logarithmic scales

discretized in increasingly denser nodal distributions. Two types of nodal distributions were considered, regular nodal distributions ($\lambda = \infty$) and random irregular nodal distributions with $\lambda = 5$. In order to obtain a reliable comparison, each irregular nodal distribution generated was used to analyse the problem with the NNRPIM, the FEM and the RPIM.

In Fig. 5.25 are presented the obtained results regarding the medium displacement errors: regular nodal distributions Fig. 5.25a and irregular nodal distributions Fig. 5.25b. It is visible the similitude between the RPIM and the NNRPIM convergence lines.

In Figs. 5.26, 5.27 and 5.28 are presented the obtained results regarding the medium stress errors, which are obtained with Eq. (5.10). Notice that regarding the stress field, when regular nodal distributions are used, the NNRPIM converges to a more accurate final solution and also that the results obtained with the NNRPIM are always more accurate than the solution of any other studied numerical method.

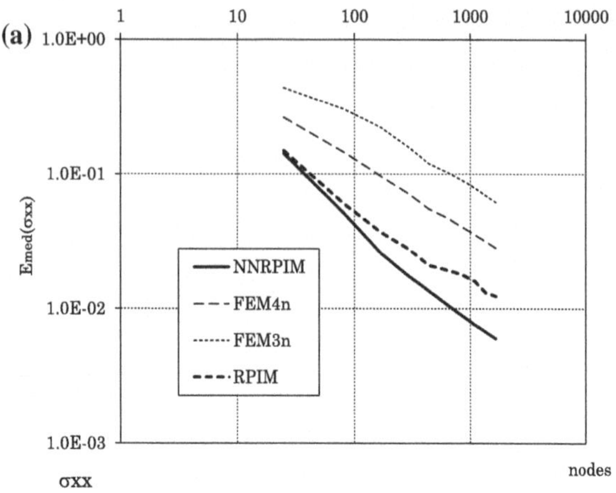

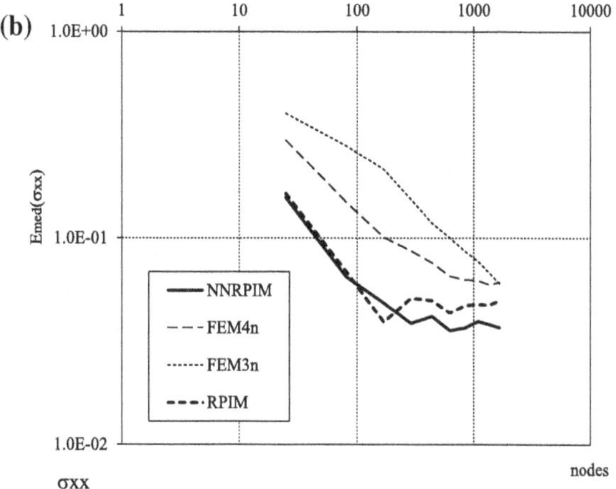

Fig. 5.26 Medium normal stress σ_{XX} errors obtained for the: **a** regular nodal distribution; **b** irregular nodal distribution. Logarithmic scales

In Fig. 5.29 it is presented the stress distribution along interest lines of the solid domain $\Omega \subset \mathbb{R}^2$ considering a regular nodal discretization of $41 \times 41 = 1{,}681$ nodes. In Fig. 5.29a it is possible to observe the development of the normal stress σ_{xx} along the line $x = 0 \wedge y \in [0, D]$. The distribution of the shear stress σ_{xy} is presented along the line $x = L/2 \wedge y \in [0, D]$ in Fig. 5.29b. Notice that the stress field obtained with the NNRPIM and the RPIM is very close with the analytical solution. In Fig. 5.30 the displacement field and the stress field distributions for the complete problem domain $\Omega \subset \mathbb{R}^2$ are displayed and it is possible to observe the extremely smooth variable field produced with the NNRPIM.

The same problem is now analysed considering $\Omega \subset \mathbb{R}^3$, as represented in Fig. 5.31a. The material properties are: $E = 1\,\text{kPa}$ and $\upsilon = 0.3$. In the natural boundary, $\Gamma_t \in \Omega$, of the solid domain the stress field defined in Eq. (5.7) is applied neglecting the stress in the oz direction. The following displacement imposition on

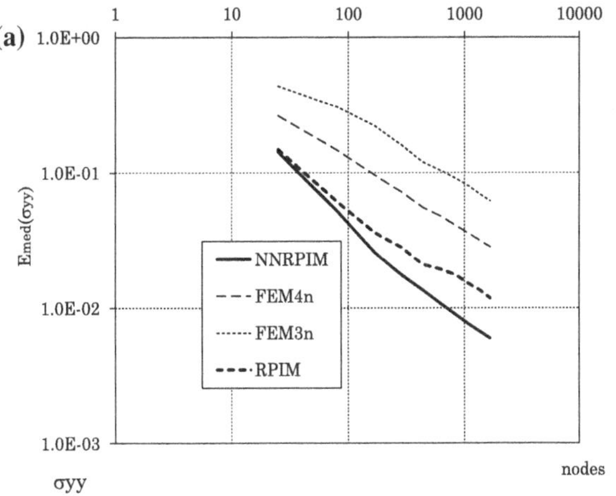

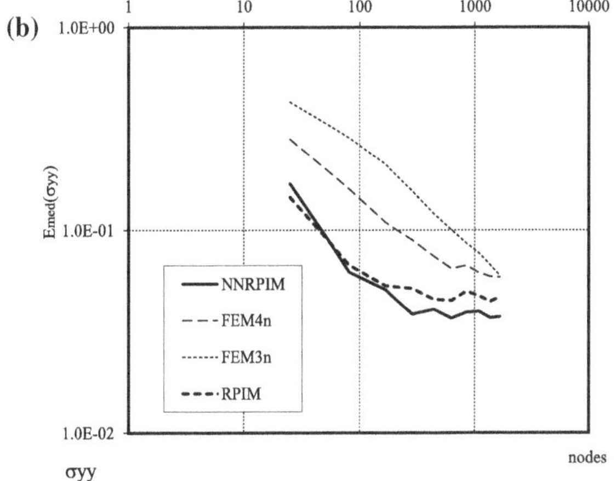

Fig. 5.27 Medium normal stress σ_{YY} errors obtained for the: **a** regular nodal distribution; **b** irregular nodal distribution. Logarithmic scales

the essential boundary, $\Gamma_u \in \Omega$, is considered: $u = 0 : \forall\{y, z\} \in \mathbb{R}^2 \wedge x = 0$, $v = 0 : \forall\{z, x\} \in \mathbb{R}^2 \wedge y = 0$ and $w = 0 : \forall\{x, y\} \in \mathbb{R}^2 \wedge z = 0$ being $\boldsymbol{u} = \{u, v, w\}$. The analytical displacement field [15] is obtained from the analytical stress field, Eq. (5.7), neglecting the displacement on the oz direction. The three-dimensional solid domain is discretized in increasingly denser nodal meshes, using the same procedure as in previous study. Examples of meshes are presented in Fig. 5.31b. In order to maintain a well-balance nodal mesh, although it can be irregular, the plate thickness H depends on the inter-nodal average distance h indicated in Fig. 5.31b.

As in the two-dimensional study, to obtain a reliable comparison, each irregular nodal distribution generated was used to analyse the problem with the NNRPIM, the FEM and the RPIM. The obtained medium displacement errors for the regular nodal distributions and for the irregular nodal distributions are presented in Fig. 5.32a, b.

The obtained results regarding the medium stress errors, determined with Eq. (5.10), are presented in Figs. 5.33, 5.34 and 5.35. It is perceptible that the

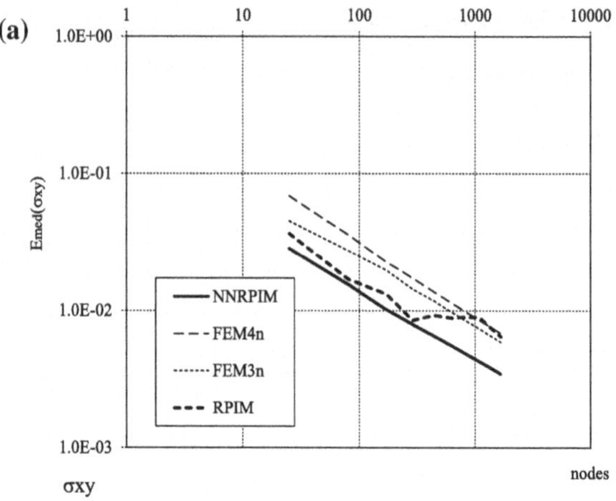

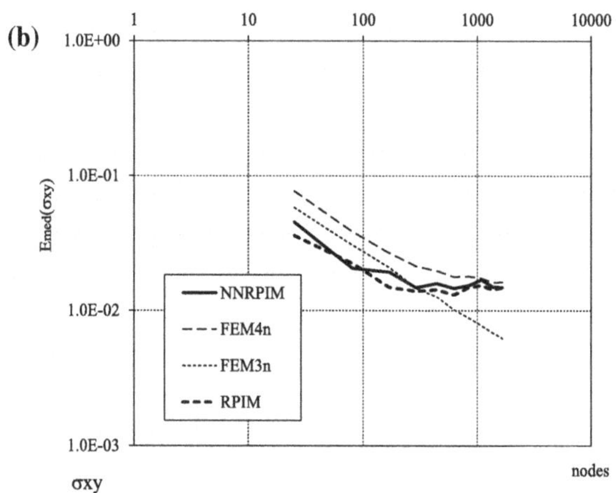

Fig. 5.28 Medium shear stress σ_{XY} errors obtained for the: **a** regular nodal distribution; **b** irregular nodal distribution. Logarithmic scales

NNRPIM shows a higher convergence rate when compared with all the remaining numerical methods. Additionally, the NNRPIM converges to a more accurate final solution and the results obtained with the NNRPIM are always more accurate than the results of the other studied numerical methods.

5.3.2 Cantilever Beam

Consider the cantilever beam defined by the domain $\Omega \subset \mathbb{R}^3$ represented in Fig. 5.36. The assumed material properties are: $E = 1$ kPa and $\upsilon = 0.3$. In the natural boundary, $\boldsymbol{x} \subset \Gamma_t \in \Omega : \forall \{y, z\} \in \mathbb{R}^2 \wedge \{x = 0 \vee x = L\}$, of the solid domain the following stress field is applied,

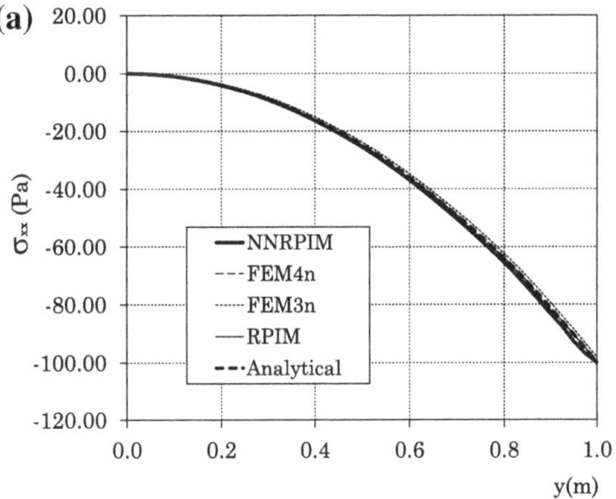

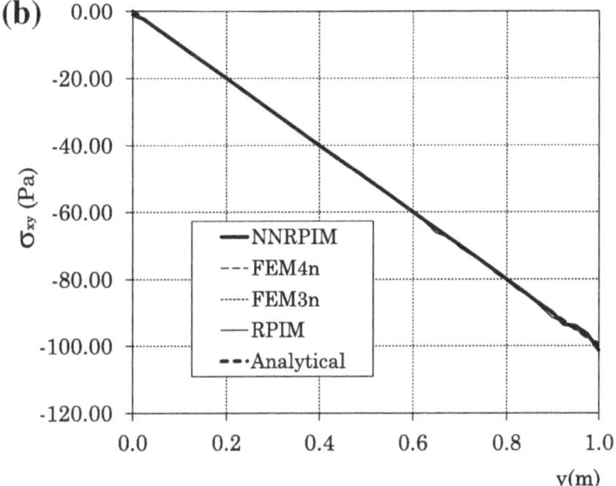

Fig. 5.29 **a** Normal stress σ_{XX} obtained along x = 0 and y = [0, D]. **b** Shear stress σ_{XY} obtained along x = L/2 and y = [0, D]

$$
\begin{aligned}
\sigma_{xx}(\boldsymbol{x}) &= -\frac{P \cdot (L - x) \cdot y}{I} \\
\sigma_{yy}(\boldsymbol{x}) &= 0 \\
\sigma_{xy}(\boldsymbol{x}) &= -\frac{P \cdot D^2}{8I} \cdot \left(1 - \frac{4y^2}{D^2}\right)
\end{aligned}
\tag{5.11}
$$

being $D = 1$ m the beam height, $L = 2$ m the beam length, $I = D^3/12$ and $P = 10\,N$. For the three-dimensional analysis, it is assumed for all domain boundaries $\sigma_{zx} = \sigma_{yz} = \sigma_{zz} = 0$. The displacement constrains on the essential boundary, $\Gamma_u \in \Omega$, must be considered as represented in Fig. 5.36: $u = 0 : \forall z \in \mathbb{R} \wedge x = 0 \wedge \{y = -D/2 \vee y = D/2\}$ and $\{u, v\} = \{0, 0\} : \forall z \in \mathbb{R} \wedge \{x, y\} = \{0, 0\}$, being $\boldsymbol{u} = \{u, v, w\}$.

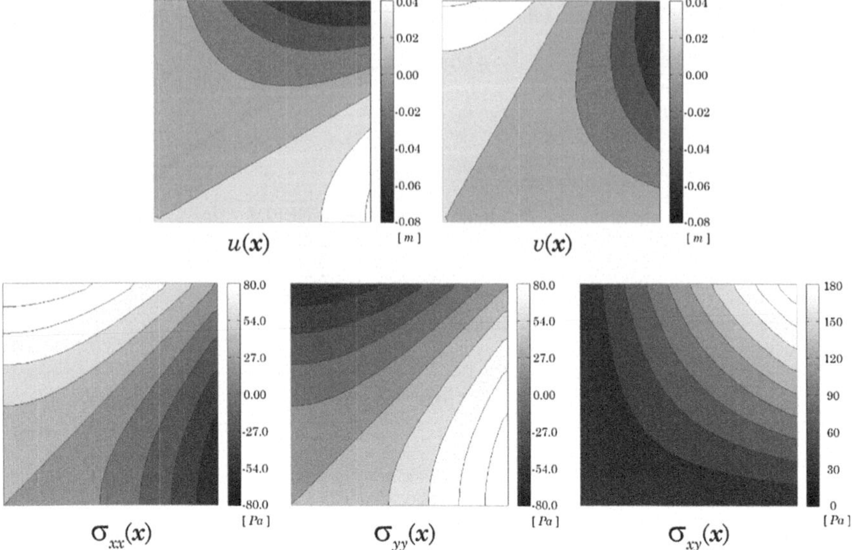

Fig. 5.30 Displacement field and stress field distribution obtained with the NNRPIM

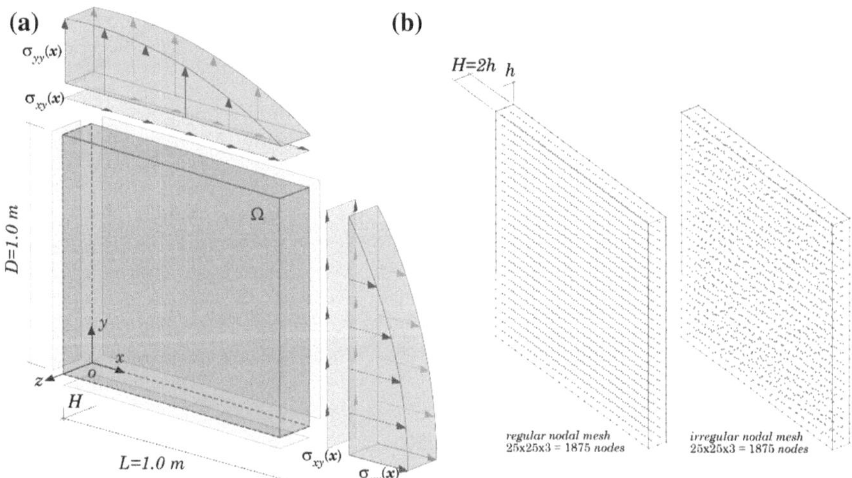

Fig. 5.31 Three-dimensional model of the square plate under parabolic stress and examples of regular and irregular nodal distributions

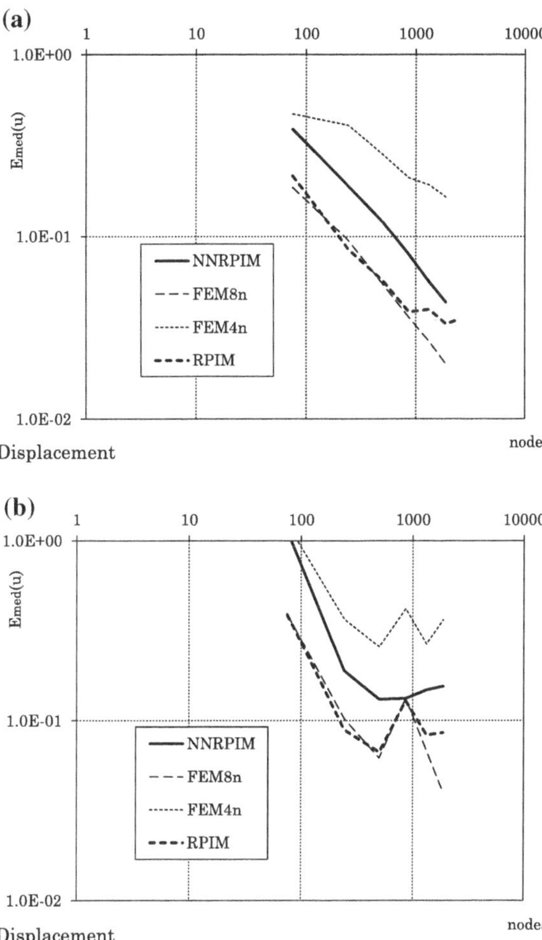

Fig. 5.32 Medium displacement errors obtained for the: **a** regular nodal distribution; **b** irregular nodal distribution. Logarithmic scales

The analytical displacement field [15] is obtained from the analytical stress field on Eq. (5.11).

$$
\begin{aligned}
u(\boldsymbol{x}) &= -\frac{2P}{E \cdot D^3}\left[3x \cdot (2L - x) \cdot y + (2 + \upsilon) \cdot \left(y^2 - \frac{D^2}{4}\right) \cdot y\right] \\
v(\boldsymbol{x}) &= \frac{2P}{E \cdot D^3}\left[x^2 \cdot (3L - x) + 3\upsilon \cdot (L - x) \cdot y^2 + x \cdot (4 + 5\upsilon) \cdot \frac{D^2}{4}\right]
\end{aligned} \tag{5.12}
$$

The problem was studied considering the two-dimensional analysis and also the three-dimensional analysis. Several regular and irregular nodal meshes, increasingly denser, were used to discretize the solid domain. In Fig. 5.37 two-dimensional and three-dimensional nodal discretizations examples are presented. As in previous example, in order to preserve a well-balance three-dimensional nodal

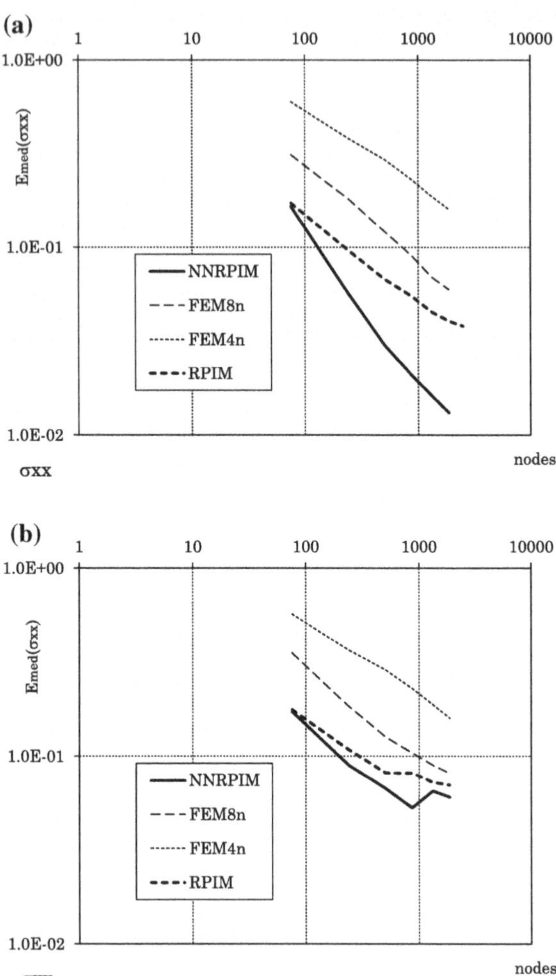

Fig. 5.33 Medium normal stress σ_{XX} errors obtained for the: **a** regular nodal distribution; **b** irregular nodal distribution. Logarithmic scales

distribution the beam thickness H depends on the inter-nodal average distance h indicated in Fig. 5.37.

The problem was analysed considering the FEM, the NNRPIM and the RPIM. The results regarding the displacement medium error, Eq. (5.1), are presented in Fig. 5.38 for the two-dimensional analysis. Notice that the NNRPIM converges with a higher rate when compared with the RPIM, regardless the use of regular or irregular nodal distributions. It is also visible in Fig. 5.38 that compared with the RPIM, the NNRPIM converges to a solution with a much lower medium displacement error.

In Figs. 5.39 and 5.40 are presented the stress field medium error obtained with Eq. (5.10). Comparing the NNRPIM stress results with the solutions obtained with

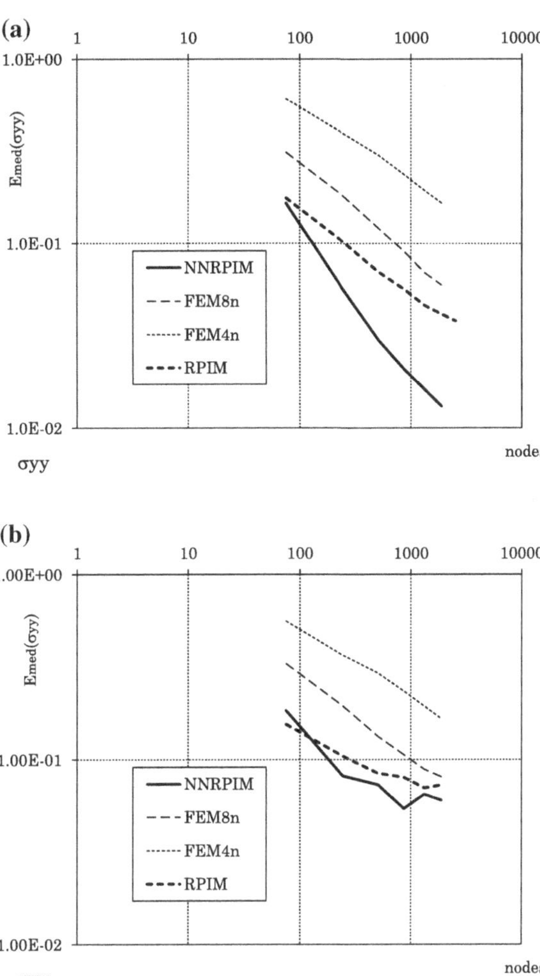

Fig. 5.34 Medium normal stress σ_{YY} errors obtained for the: **a** regular nodal distribution; **b** irregular nodal distribution. Logarithmic scales

other numerical approaches, it is possible to confirm that the NNRPIM is capable to produce more accurate stress fields.

In Figs. 5.41, 5.42 and 5.43 the results for the three-dimensional analysis are presented. The three-dimensional solution was compared with the analytical solution presented in Eqs. (5.11) and (5.12), neglecting the displacements and the stress components in the oz direction. It is possible to observe that for the present analysis the NNRPIM results generally are better than the RPIM results, consistently presenting a high convergence rate, indicating that higher accuracy can be achieved if denser meshes are used.

In Fig. 5.44a it is presented the obtained normal stress σ_{xx} distribution along the interest line $x = 0 \wedge y \in [0, D]$ of the solid domain $\Omega \subset \mathbb{R}^2$ considering a regular nodal discretization of $81 \times 41 = 3,321$ nodes. The distribution of the obtained

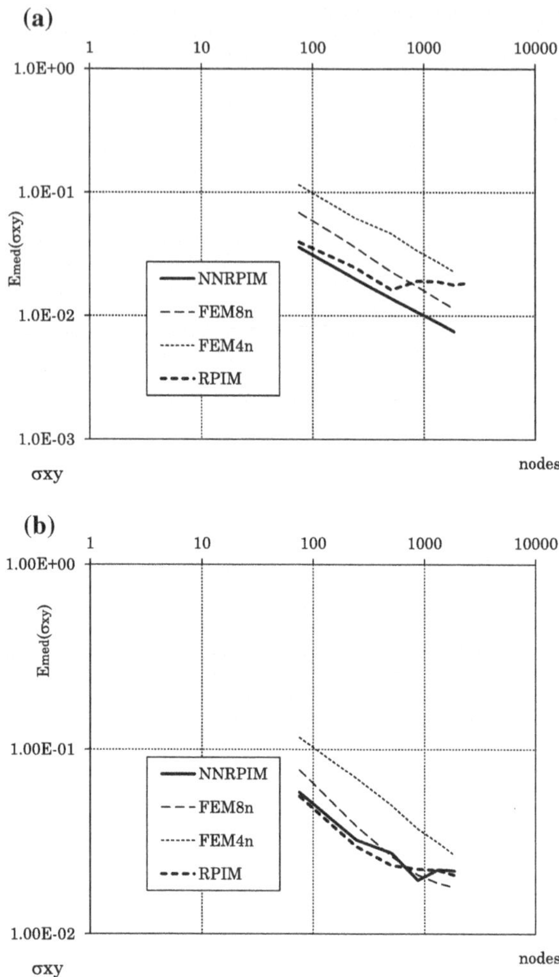

Fig. 5.35 Medium shear stress σ_{XY} errors obtained for the: **a** regular nodal distribution; **b** irregular nodal distribution. Logarithmic scales

shear stress σ_{xy} in the same interest line, $x = 0 \wedge y \in [0, D]$, is presented in Fig. 5.44b. The solution obtained with the NNRPIM and the RPIM is compared with the FEM solution and the analytical solution. In Fig. 5.44a, b it is possible to observe that the NNRPIM stress field is very close with the analytical solution. Notice that the NNRPIM stress field is always smoother and more accurate than the RPIM. In Fig. 5.45 it is presented the displacement field and the stress field distributions for the whole problem domain $\Omega \subset \mathbb{R}^2$. Once again it is possible to confirm that the NNRPIM produces smooth variable fields.

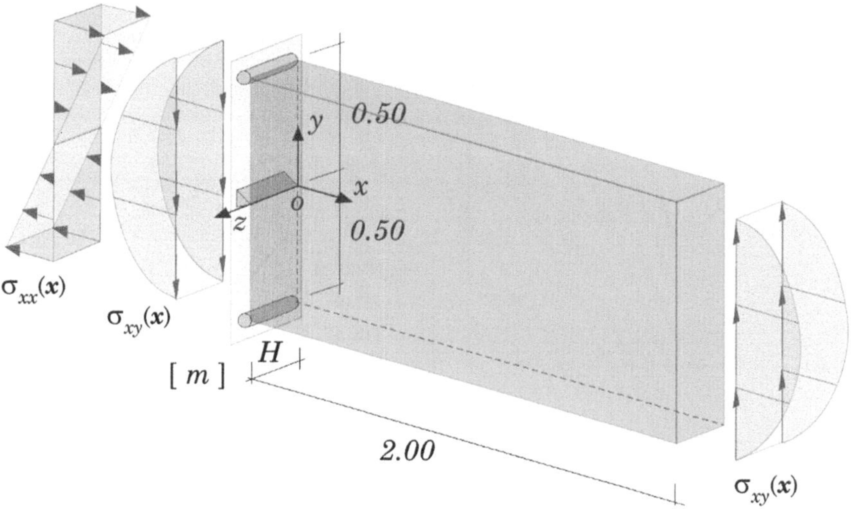

Fig. 5.36 Three-dimensional model of the analysed cantilever beam

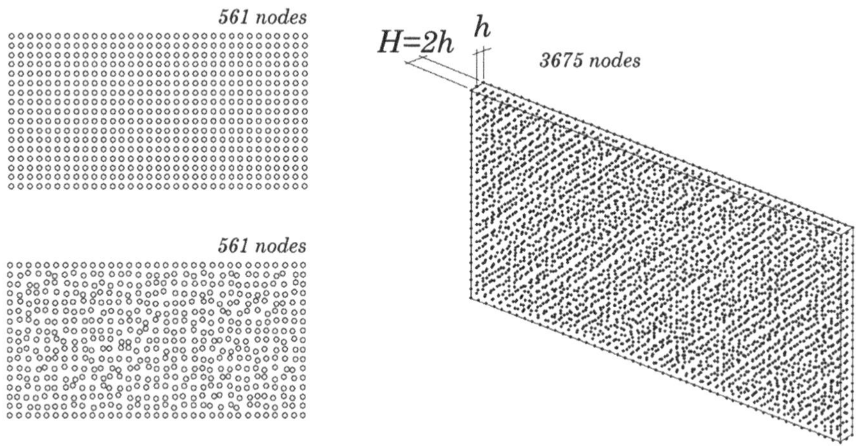

Fig. 5.37 Two-dimensional regular and irregular nodal distributions examples and a three-dimensional irregular nodal distribution example

5.3.3 Square Plate with Central Circular Hole

In Fig. 5.46 is represented a simplification of the complete square plate with a central circular hole. The following material properties are considered: $E = 1$ kPa and $\upsilon = 0.3$. Two natural boundaries, $\Gamma_t \in \Omega$, on the solid domain $\Omega \subset \mathbb{R}^3$ are considered: $\boldsymbol{x} \subset \Gamma_t \in \Omega : \forall \{y, z\} \in \mathbb{R}^2 \wedge x = L$ and $\boldsymbol{x} \subset \Gamma_t \in \Omega : \forall \{x, z\} \in \mathbb{R}^2 \wedge$

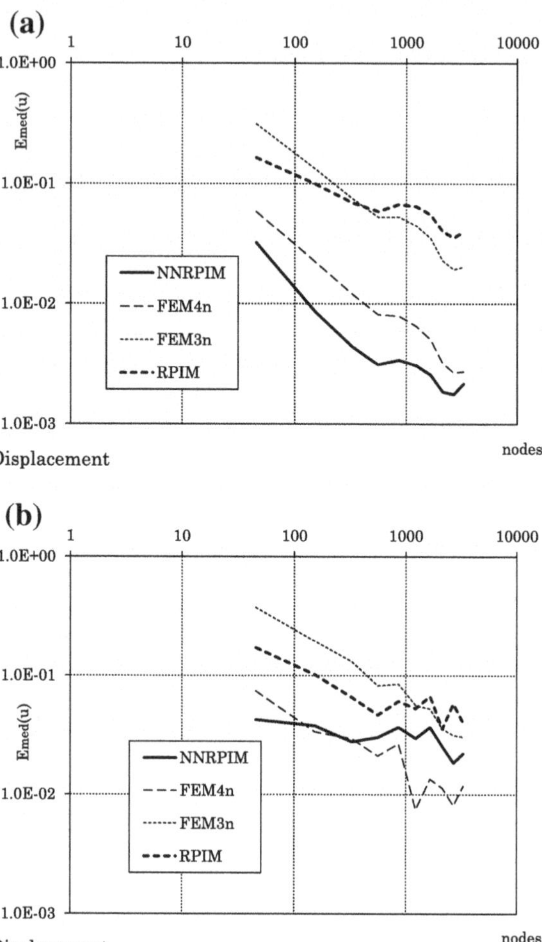

Fig. 5.38 Medium displacement errors obtained for the: **a** regular nodal distribution; **b** irregular nodal distribution. Logarithmic scales

$y = D$, being $L = 1$ m the plate length and $D = 1$ m the plate height. The following stress field is applied on $\Gamma_t \in \Omega$,

$$\begin{aligned}
\sigma_{xx}(\boldsymbol{x}) &= \sigma_0 \cdot \left[1 - \frac{a^2}{r^2} \cdot \left(\frac{3}{2}\cos(2\theta) + \cos(4\theta)\right) + \frac{3\,a^4}{2\,r^4}\cos(4\theta)\right] \\
\sigma_{yy}(\boldsymbol{x}) &= \sigma_0 \cdot \left[-\frac{a^2}{r^2} \cdot \left(\frac{1}{2}\cos(2\theta) - \cos(4\theta)\right) - \frac{3\,a^4}{2\,r^4}\cos(4\theta)\right] \\
\sigma_{xy}(\boldsymbol{x}) &= \sigma_0 \cdot \left[-\frac{a^2}{r^2} \cdot \left(\frac{1}{2}\sin(2\theta) + \sin(4\theta)\right) + \frac{3\,a^4}{2\,r^4}\sin(4\theta)\right]
\end{aligned} \tag{5.13}$$

Being $\sigma_0 = 100$ Pa the magnitude stress, $a = 0.25$ m the circular hole radius, r the Euclidean distance from the centre of the circular hole to the interest point

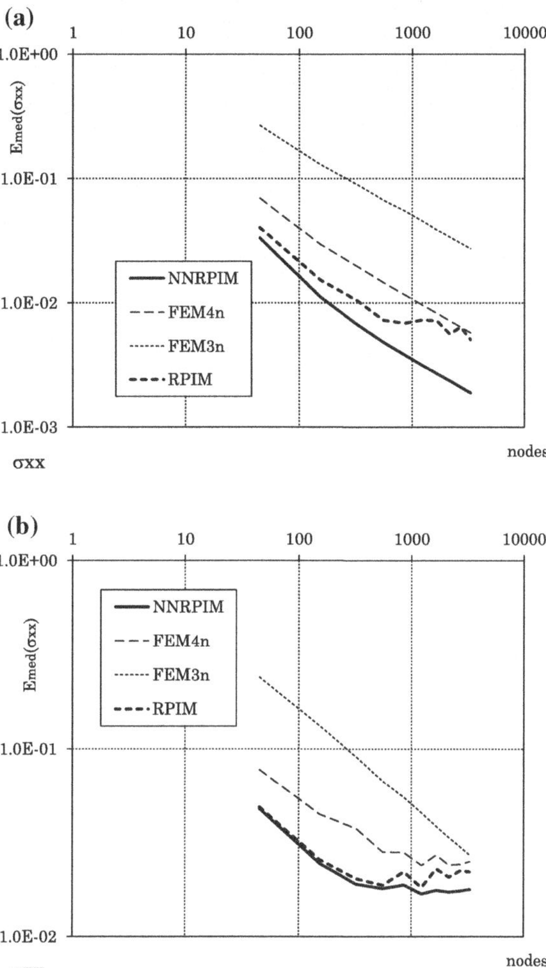

Fig. 5.39 Medium normal stress σ_{XX} errors obtained for the: **a** regular nodal distribution; **b** irregular nodal distribution. Logarithmic scales

P and θ the angle between the ox axis and the line oP, as Fig. 5.46 shows. As in previous example, for the three-dimensional analysis, it is assumed on all domain boundaries $\sigma_{zx} = \sigma_{yz} = \sigma_{zz} = 0$. The analytical displacement field is obtained from the analytical stress field on Eq. (5.13),

$$u(\boldsymbol{x}) = \frac{\sigma_0 \cdot (1+\bar{\upsilon})}{\bar{E}} \cdot \left((1-\bar{\upsilon}) \cdot \left(r + \frac{2a^2}{r} \right) \cos(\theta) + \left(\frac{a^2}{2r} - \frac{a^4}{2r^3} \right) \cos(3\theta) \right)$$
$$v(\boldsymbol{x}) = \frac{\sigma_0 \cdot (1+\bar{\upsilon})}{\bar{E}} \cdot \left(\left(-\bar{\upsilon} \cdot r - (1-2\bar{\upsilon}) \frac{a^2}{r} \right) \sin(\theta) + \left(\frac{a^2}{2r} - \frac{a^4}{2r^3} \right) \sin(3\theta) \right) \tag{5.14}$$

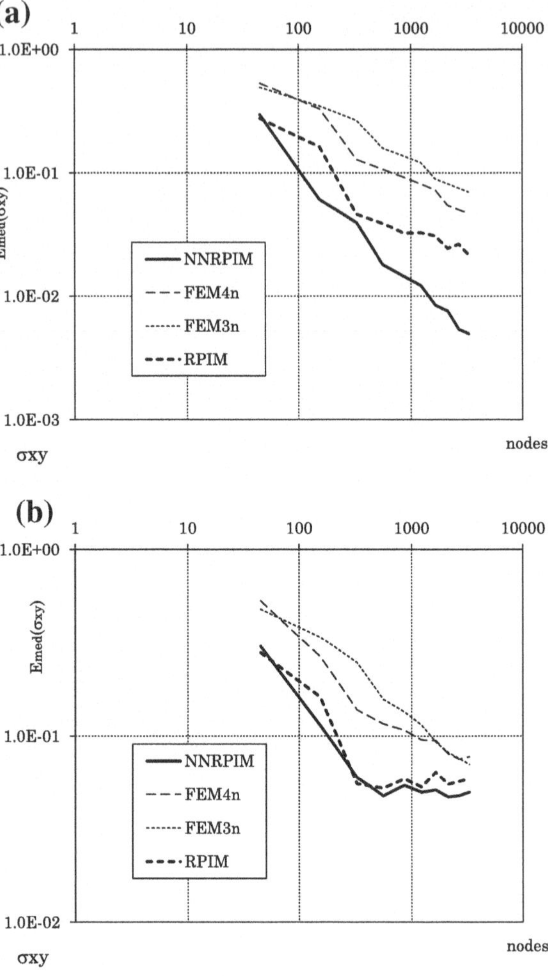

Fig. 5.40 Medium shear stress σ_{XY} errors obtained for the: **a** regular nodal distribution; **b** irregular nodal distribution. Logarithmic scales

Being $\bar{\upsilon} = \upsilon/(1+\upsilon)$ and $\bar{E} = E(1-\bar{\upsilon}^2)$. The displacement constrains on the essential boundary, $\Gamma_u \in \Omega$, must be considered as represented in Fig. 5.46: $u = 0 : \forall\{y,z\} \in \mathbb{R}^2 \wedge x = 0$ and $v = 0 : \forall\{z,x\} \in \mathbb{R}^2 \wedge y = 0$, being $\boldsymbol{u} = \{u, v, w\}$.

Once again the problem was studied considering the two-dimensional analysis and also the three-dimensional analysis, therefore several regular and irregular nodal distributions, increasingly denser, were used to discretize the solid domain. Nodal discretization examples are shown in Fig. 5.47. As in previous studies, in order to maintain a well-balance three-dimensional nodal distribution, the plate thickness H depends on the inter-nodal average distance h indicated in Fig. 5.47.

The problem was analysed considering the NNRPIM, the RPIM and the FEM. The results regarding the displacement medium error, Eq. (5.1), are presented in Fig. 5.48 for the two-dimensional analysis and the three-dimensional analysis.

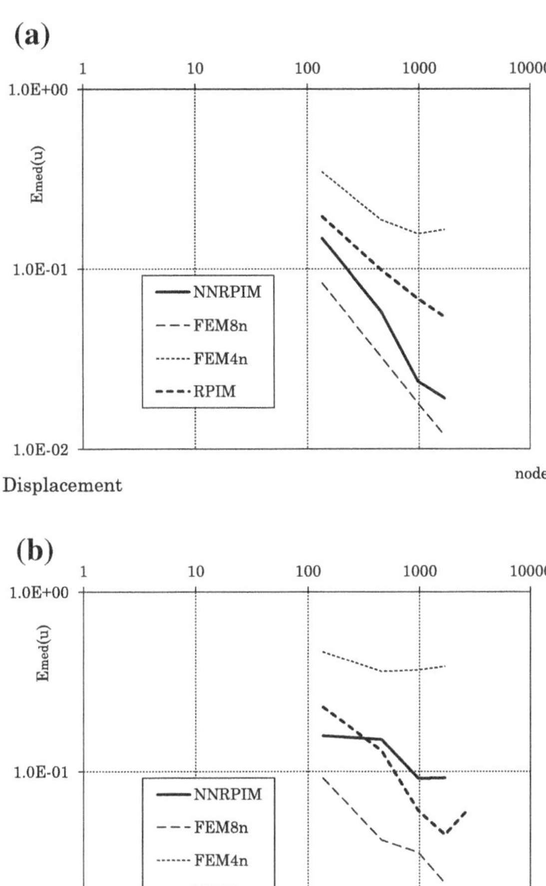

Fig. 5.41 Medium displacement errors obtained for the: **a** regular nodal distribution; **b** irregular nodal distribution. Logarithmic scales

Notice that the three-dimensional solution was compared with the analytical solution presented in Eq. (5.14) neglecting the displacements in the *oz* direction.

The results in Fig. 5.48 indicate that comparing the NNRPIM with the RPIM, the NNRPIM shows a better behaviour in the two-dimensional analysis. It is also visible that the NNRPIM presents a more predictable convergence path.

The obtained results regarding the medium stress errors, determined with Eq. (5.10), are presented in Figs. 5.49, 5.50 and 5.51. When compared with all the remaining numerical methods, it is perceptible that for the two-dimensional analysis the NNRPIM shows a higher convergence rate. Furthermore, for the two-dimensional analysis the NNRPIM converges to a more accurate final solution and the results obtained with the NNRPIM are always more accurate than the results of the other studied numerical methods.

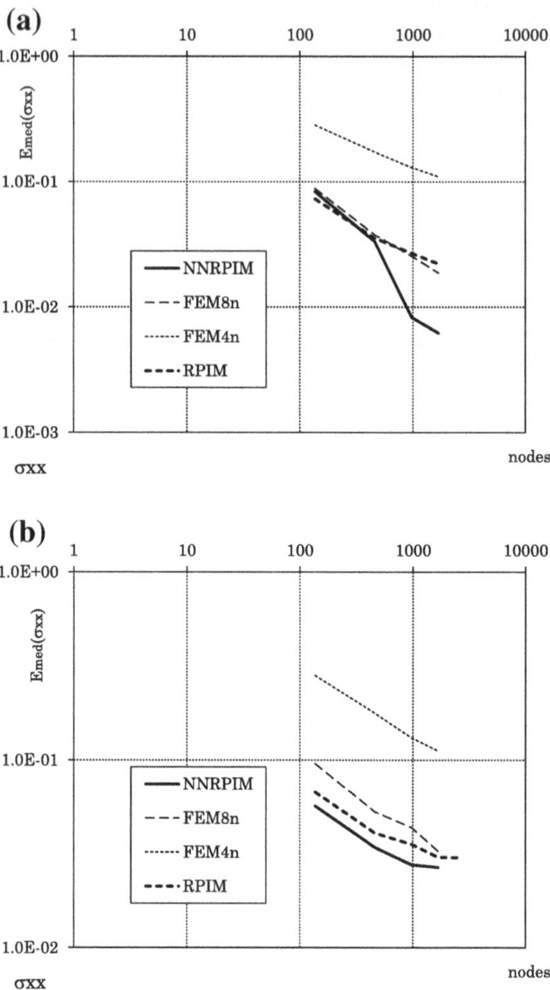

Fig. 5.42 Medium normal stress σ_{XX} errors obtained for the: **a** regular nodal distribution; **b** irregular nodal distribution. Logarithmic scales

The stress distribution along interest lines of the solid domain are displayed in Figs. 5.52 and 5.53. It was considered a two-dimensional analysis and a nodal discretization of 1,581 nodes. Along the interest line $x = 0 \wedge y \in [D/4, D]$, the normal stresses σ_{xx} and σ_{yy} distributions are presented in Fig. 5.52a, b, respectively. For the interest line $y = 0 \wedge x \in [L/4, L]$, are presented in Fig. 5.53a, b the normal stresses σ_{xx} and σ_{yy} distributions. The NNRPIM and the RPIM solutions are compared with the FEM solutions and the analytical solution. With Figs. 5.52 and 5.53 it is possible to observe that the NNRPIM stress field is smooth and always very close with the analytical solution.

In Fig. 5.54 are presented the displacement field and the stress field distributions for the whole problem domain $\Omega \subset \mathbb{R}^2$. It is clear in Fig. 5.54 that the variable fields obtained with the NNRPIM are smooth.

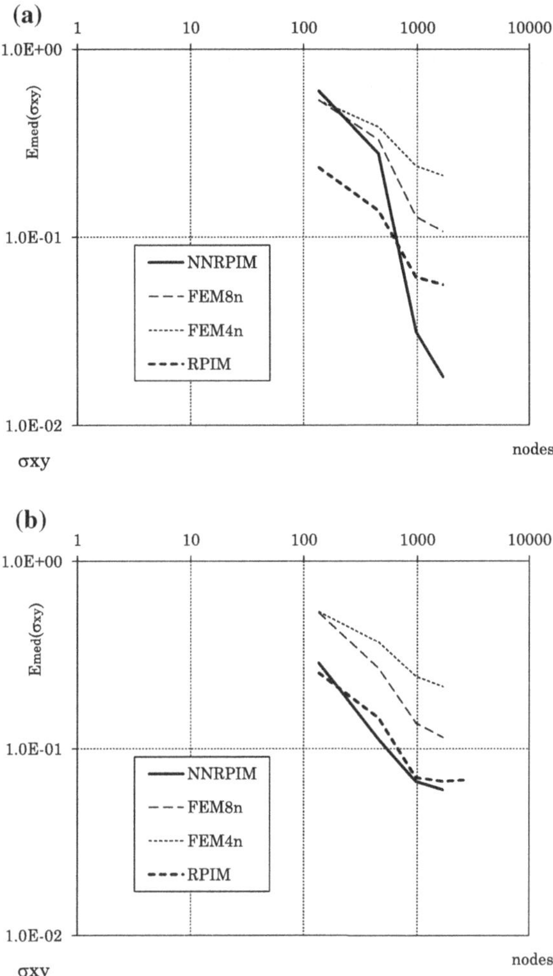

Fig. 5.43 Medium shear stress σ_{XY} errors obtained for the: **a** regular nodal distribution; **b** irregular nodal distribution. Logarithmic scales

5.4 Elastodynamic Numerical Examples

In this section in order to access the accuracy of the NNRPIM, in the context of vibration analysis, several numerical examples are presented and the numerical results are compared with analytical solutions and FEM solutions, which are available in the literature. Convergence studies of the first vibration frequencies errors are presented. All results are obtained using the consistent mass matrix. Firstly the free vibration 2D and 3D examples are presented and in the end a forced vibration example is shown.

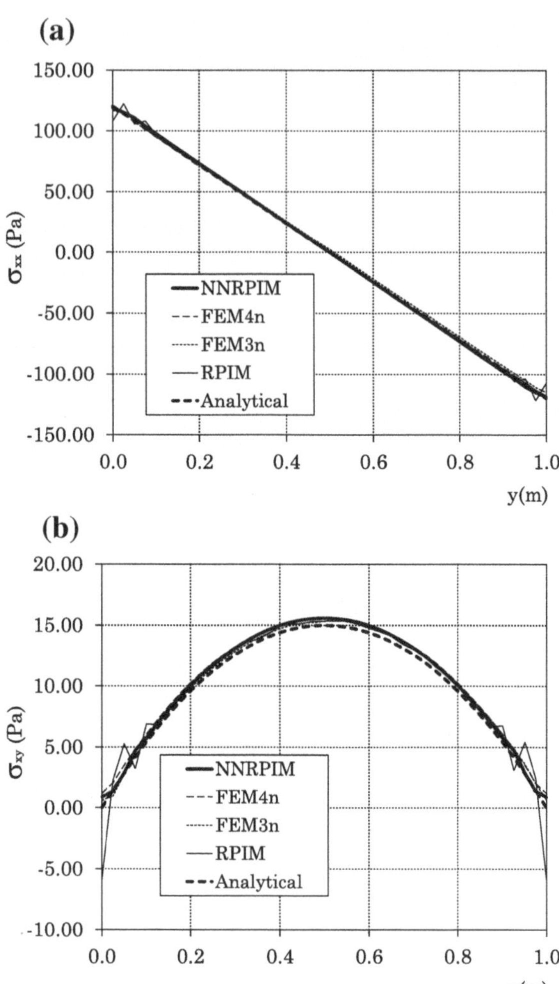

Fig. 5.44 **a** Normal stress σ_{XX} obtained along x = 0 and y = [0, D]. **b** Shear stress σ_{XY} obtained along x = L/2 and y = [0, D]

5.4.1 Free Vibration of a Cantilever Beam

The NNRPIM is applied to analyse the free vibration of a cantilever beam, Fig. 5.55a. The geometrical parameters of the beam are, $L = 100$ mm, $D = 10$ mm and thickness $h = 1$ mm. The material properties are Young's modulus $E = 2.1 \times 10^5$ N/mm^2, Poisson ratio $\upsilon = 0.3$ and mass density $\rho = 8.0 \times 10^{-10}$ kg s^2/mm^4.

Firstly some optimization studies are conducted considering the 2D plane stress deformation theory. In order to obtain the optimal integration scheme the problem is discretized in two distinct nodal discretizations, a regular nodal distribution and an irregular nodal distribution, both with 306 nodes, Fig. 5.55b, c.

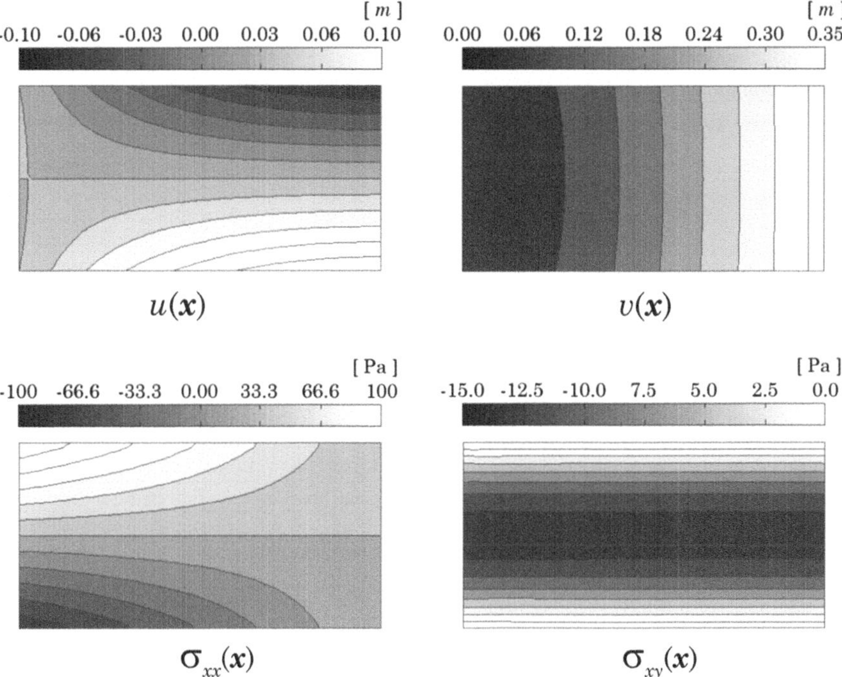

Fig. 5.45 Displacement field and stress field distribution obtained with the NNRPIM

Fig. 5.46 Three-dimensional model of square plate with a central hole

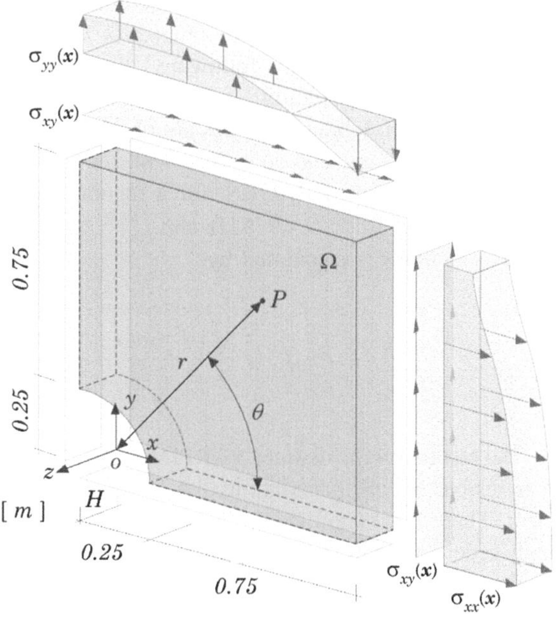

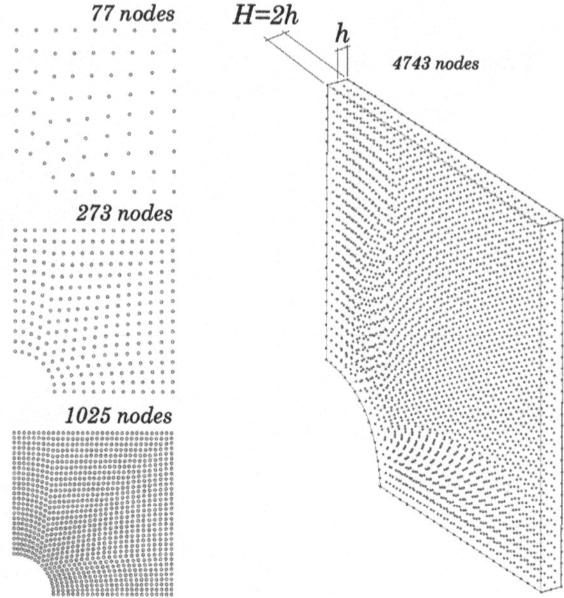

Fig. 5.47 Two-dimensional nodal distributions examples and a three-dimensional nodal distribution example

In Table 5.8 it is shown the number of integration points obtained for each integration scheme considered in the present two-dimensional analysis. The integration schemes respect the same nomenclature used in Sect. 5.2.2. The simple nodal based integration scheme described in Section "Basic Integration Scheme" is the first integration scheme considered. In Table 5.8 this simple integration scheme is called "basic". The other $k \times k$ integrations schemes presented in Table 5.8 are obtained following the procedure indicated in Section "Gauss-Legendre Quadrature Integration Scheme".

The first three vibration modes are obtained with the NNRPIM, for the distinct integration schemes, and compared with the first three vibration modes obtained with FEM software, ABAQUS, for a regular nodal distribution of 4,000 nodes, $f_1^{FEM} = 830\,\text{Hz}$, $f_2^{FEM} = 4{,}979\,\text{Hz}$ and $f_3^{FEM} = 12{,}826\,\text{Hz}$. The difference between the two methods is calculated by,

$$Error(f_i) = \frac{\sqrt{\left((f_i)_{meshless} - (f_i)_{FEM(ABAQUS)}\right)^2}}{\sqrt{\left((f_i)_{FEM(ABAQUS)}\right)^2}} \tag{5.15}$$

The results for the distinct NNRPIM formulations are presented in Fig. 5.56. As it is visible for all formulations the integration scheme 1×1 seems to be sufficient. There is no significant improvement in the accuracy of the solution when a high order integration scheme is used. Therefore, in further examples the integration scheme used in the three considered NNRPIM formulations is the 1×1 integration scheme.

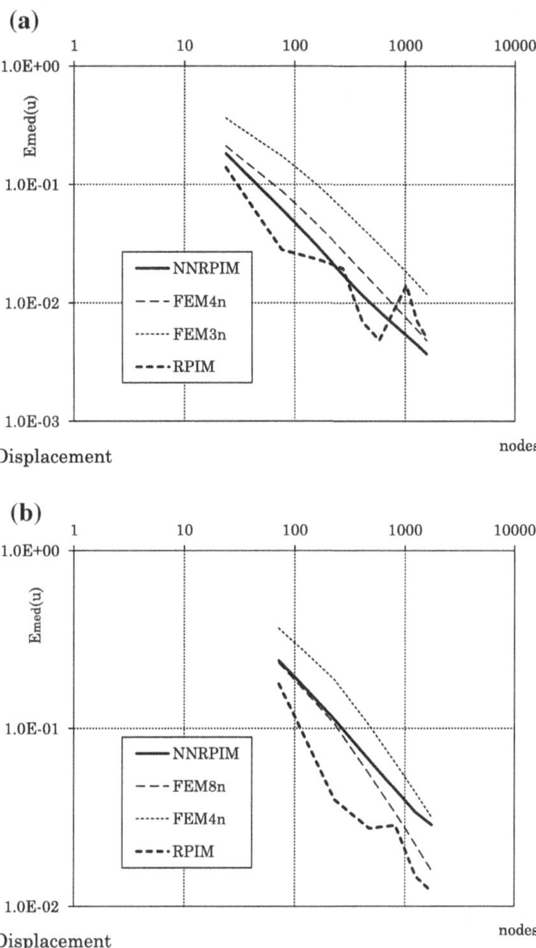

Fig. 5.48 Medium displacement errors obtained for the: **a** two-dimensional analysis; **b** three-dimensional analysis. Logarithmic scales

Using the same example, Fig. 5.55a, a convergence study was performed. The problem domain is discretized in several meshes and the results regarding the convergence of the NNRPIM in comparison with the FEM-ABAQUS solution for the first three vibration modes are presented in Fig. 5.57.

As Fig. 5.57 indicates, the V2P0 and the V2P1 NNRPIM formulations present a higher convergence rate when compared with the V1P1 NNRPIM formulation and there is no significant difference between the solution obtained with the regular nodal distribution or the irregular nodal distribution.

In the following study the ten first modes of vibration are obtained with the 3D NNRPIM analysis and with the 2D NNRPIM analysis. For the 2D approach the regular and irregular nodal distributions with 306 nodes presented in Fig. 5.55b, c are used and for the 3D case it was used the nodal discretization presented in

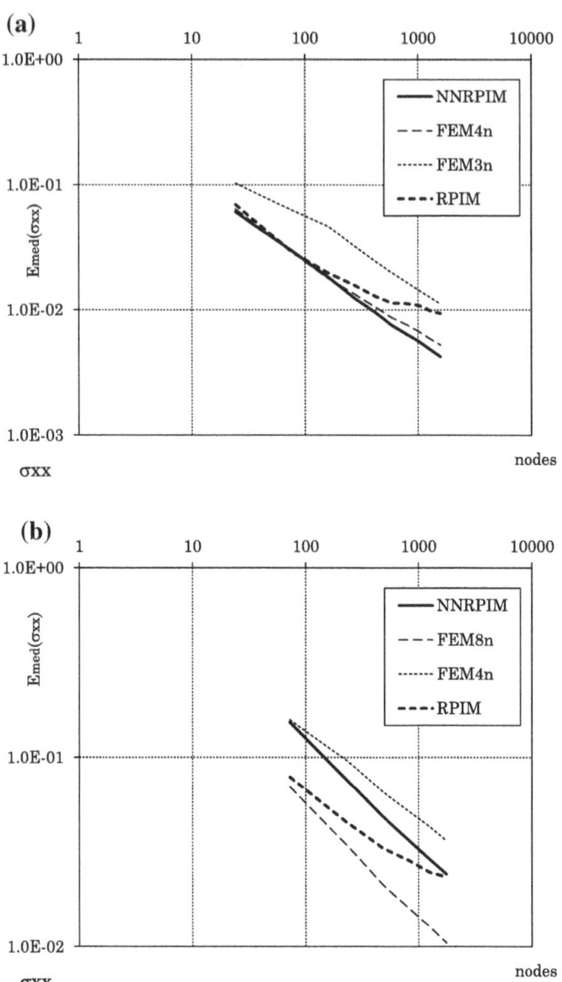

Fig. 5.49 Medium normal stress σ_{XX} errors obtained for the: **a** two-dimensional analysis; **b** three-dimensional analysis. Logarithmic scales

Fig. 5.55d. Only the V2P1 NNRPIM formulation is used, both for 2D analysis and in the 3D analysis.

Considering the 2D plane stress deformation theory, the problem was analysed, for the same regular discretization of 306 nodes, with the FEM-ABAQUS, with the MLPG [16], with the node-by-node meshless method of Nagashima [17] and with the LRPIM [18]. The obtained first ten natural frequencies are presented in Table 5.9. To clearly illustrate the differences between the obtained results, all the results are compared with the natural frequencies obtained for the 2D case with the FEM software, ABAQUS, for a regular discretization of 4,000 nodes, applying Eq. (5.15). The results are presented in Table 5.10.

In Tables 5.9 and 5.10 it is visible that for the 2D-NNRPIM there is no significant difference between the solutions using a regular or an irregular nodal

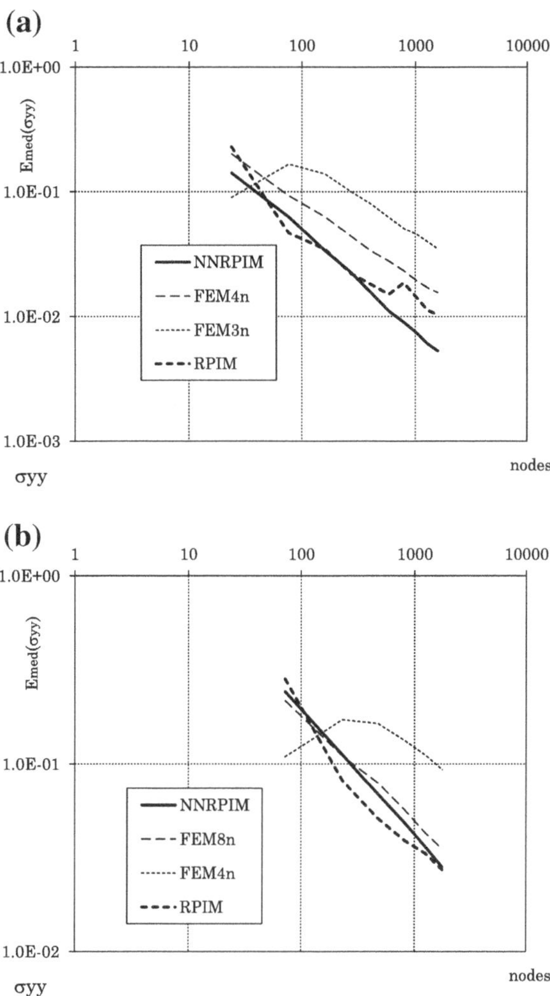

Fig. 5.50 Medium normal stress σ_{YY} errors obtained for the: **a** two-dimensional analysis; **b** three-dimensional analysis. Logarithmic scales

distribution. It is also visible that generally the 2D-NNPRIM formulation approaches more the FEM-4000n solution when compared with the MLPG, with the LRPIM and even with the FEM-306n solution for the same number of nodes. Also the results obtained with the 3D-NNRPIM when compared with the FEM-4000n solution are better than the 2D FEM-306n and the other meshless methods.

In Fig. 5.58 are represented the ten first modes of vibration of the considered cantilever beam obtained using the 2D-NNRPIM V2P1 formulation.

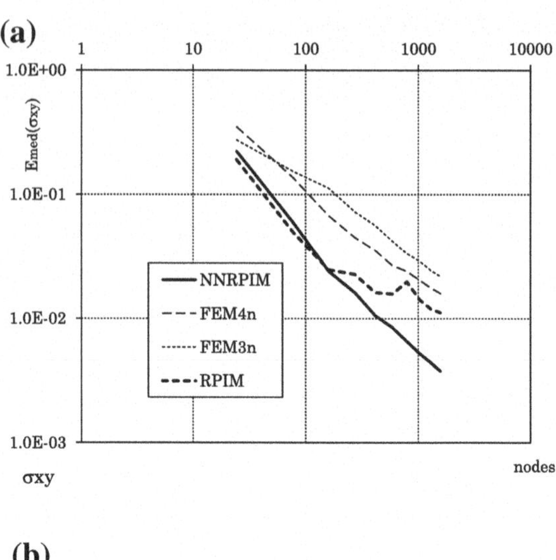

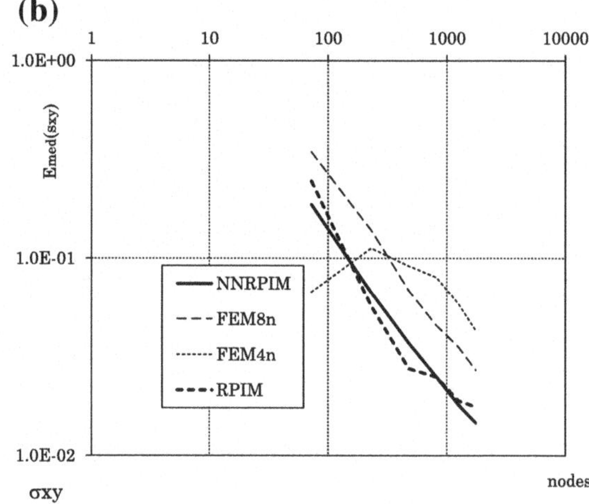

Fig. 5.51 Medium shear stress σ_{XY} errors obtained for the: **a** two-dimensional analysis; **b** three-dimensional analysis. Logarithmic scales

5.4.2 Free Vibration of Variable Cross Section Beams

In this example two distinct cantilever beams showing a variable cross section are studied, Fig. 5.59a, b. The geometrical parameters of beam A are, $L = 10$ m, $D_1 = 5$ m and $D_2 = 3$ m. For the beam B, $L_1 = 8$ m, $L_2 = 2$ m, $D_1 = 3$ m and $D_2 = 1$ m. For both beams the thickness is $h = 1$ m and the material properties are Young's modulus $E = 3.0 \times 10^7$ Pa, Poisson ratio $\upsilon = 0.3$ and mass density $\rho = 1.0$ kg/m^3. The problem is analysed considering a 2D and a 3D approach. The nodal arrangements of both beams are presented in Fig. 5.59. For the 2D analysis, beam A is discretized with 231 nodes and beam B with 287 nodes. In the 3D analysis 693 nodes are used for beam A and 861 nodes for beam B.

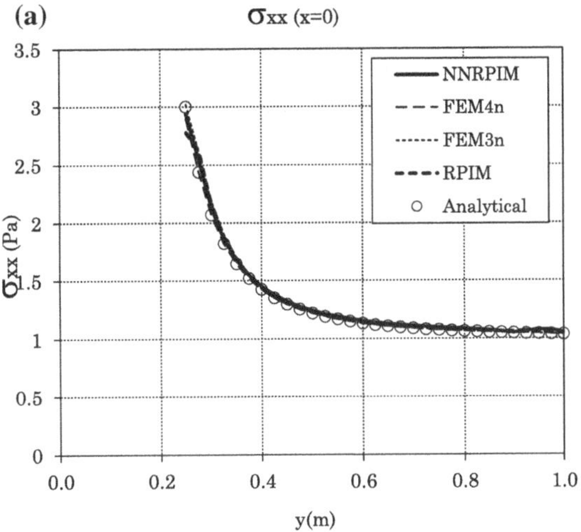

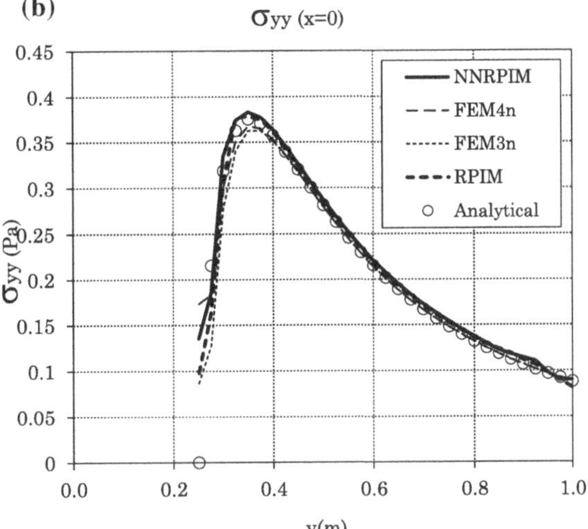

Fig. 5.52 **a** Normal stress σ_{XX} obtained along x = 0 and y = [0, D]. **b** Normal stress σ_{YY} obtained along x = 0 and y = [0, D]

The first five vibration frequencies (rad/s) are obtained for beam A with the three 2D-NNRPIM formulations, with the v2p1 3D-NNRPIM formulation, with the MLPG [16], with the LRPIM [18] and with the FEM-ABAQUS The beam A results are presented in Table 5.11. The results obtained for beam B are presented in Table 5.12. The results of the distinct analyses are obtained for the same nodal discretization.

It is visible that the results obtained with the NNRPIM are in a very good agreement with the other meshless method solutions (MPLG and LRPIM) and with the FEM solution. Notice that for the first vibration mode the 2D NNRPIM V2P1 formulation and the 3D NNRPIM are much more closer to the FEM results than all the others formulations.

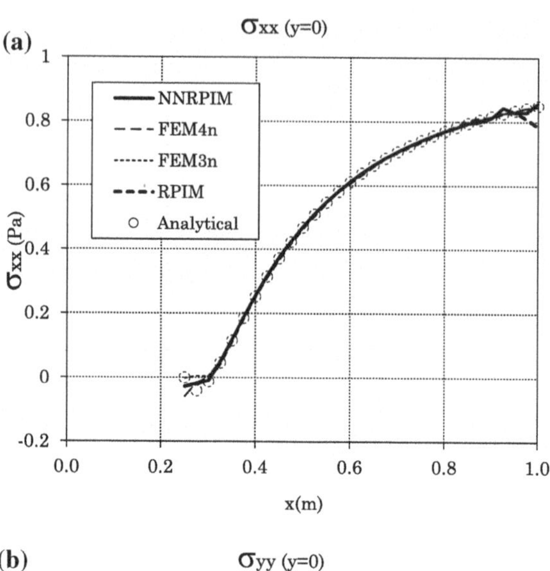

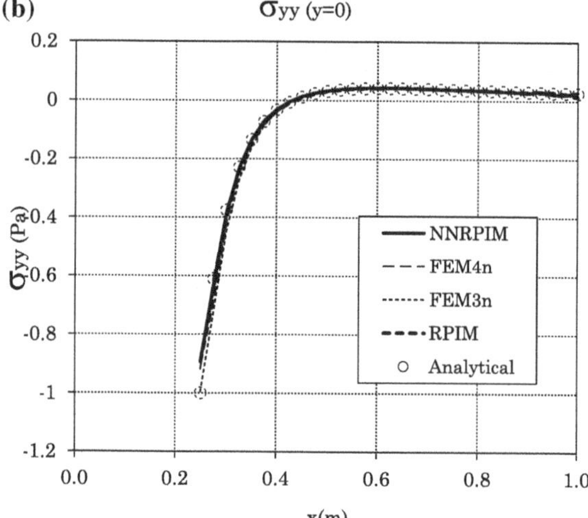

Fig. 5.53 **a** Normal stress σ_{XX} obtained along y = 0 and x = [0, L]. **b** Normal stress σ_{YY} obtained along y = 0 and x = [0, L]

In Fig. 5.60 the five first vibration modes of the two considered beams, with a variable cross section, obtained using the 2D-NNRPIM v2p1 formulation, are presented.

5.4.3 Free Vibration of a Shear Wall

A shear wall with four openings is studied in this example. The geometrical parameters of the shear wall are presented in Fig. 5.61 along with the considered nodal discretization (559 nodes). The material properties are Young's modulus $E = 1.0 \times 10^3$ Pa, Poisson ratio $\upsilon = 0.2$ and mass density $\rho = 1.0$ kg/m^3.

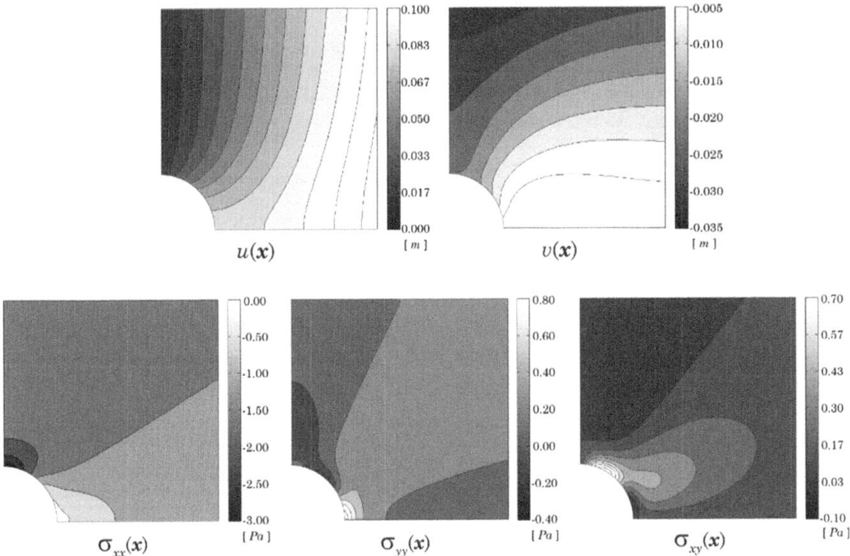

Fig. 5.54 Displacement field and stress field distribution obtained with the NNRPIM

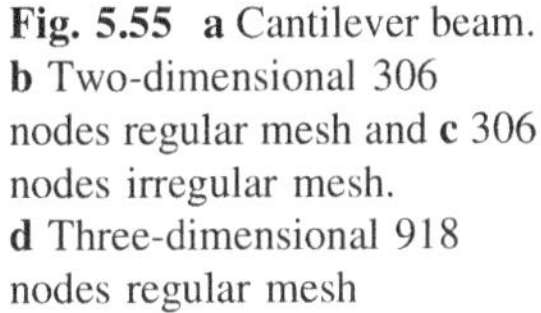

Fig. 5.55 **a** Cantilever beam. **b** Two-dimensional 306 nodes regular mesh and **c** 306 nodes irregular mesh. **d** Three-dimensional 918 nodes regular mesh

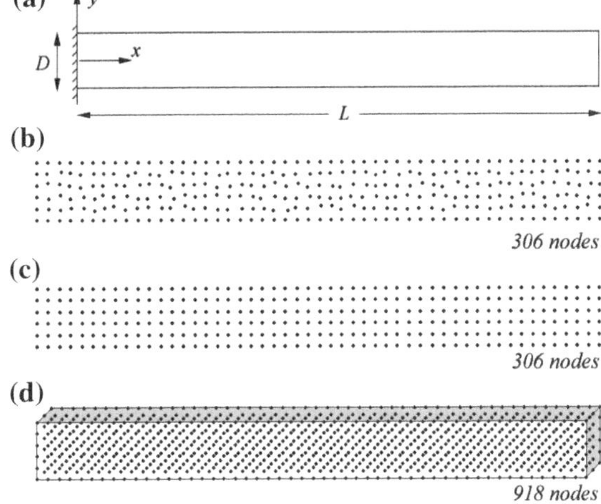

Table 5.8 Total number of integration points generated in the different integration schemes

Noda distribution	Integration scheme			
	"Basic"	1 × 1	2 × 2	3 × 3
Regular	2,000	6,000	24,000	54,000
Irregular	1,274	5,094	20,376	45,846

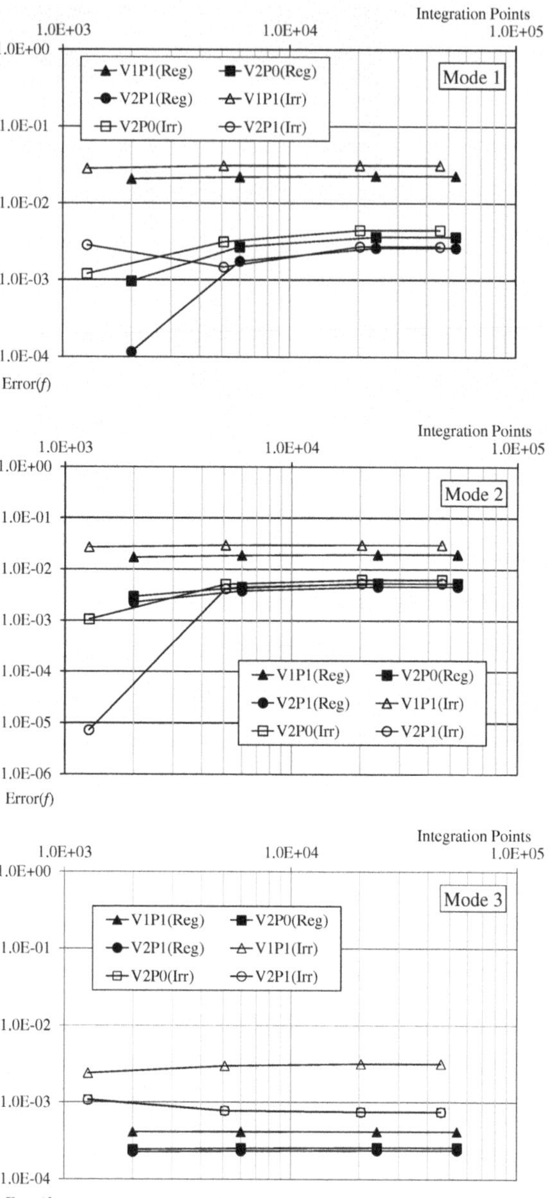

Fig. 5.56 First three vibration modes errors for distinct integration schemes for the regular and irregular nodal distributions. Logarithmic scales

The problem is analysed with the V1P1, V2P0 and V2P1 NNRPIM formulations and the obtained results are compared with the MLPG solution [16], with the LRPIM solution [18], with the Boundary Element Method solution (BEM) [19] and with the FEM-ABAQUS solution for the same nodal arrangement. The obtained results of the first eight vibration frequencies are presented in Table 5.13.

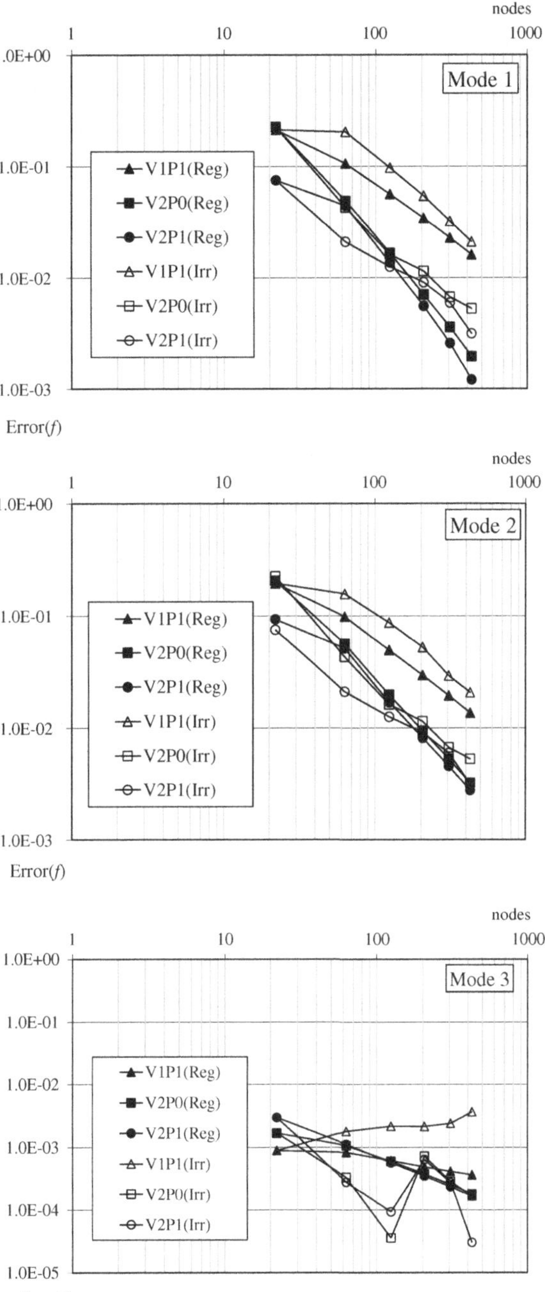

Fig. 5.57 Convergence of the first three vibration modes varying the number of nodes discretizing the problem domain. Logarithmic scales

Table 5.9 Obtained natural frequency

Mode	MLPG	Nagashima	LRPIM	Regular		Irregular		3D	FEM 2D	FEM 2D
		1999	p = 1.03; c = 1	RPIM p = 1.03; c = 0.1	NNRPIM V2P1	RPIM p = 1.03; c = 0.1	NNRPIM V2P1	NNRPIM V2P1	Abaqus 612dofs	Abaqus 8,000dofs
1	824.44	844.19	824.3	823.63	825.14	823.53	824.21	827.47	830	823
2	5,070.32	5,051.21	4,976.6	4,952.47	4,959.62	4,948.65	4,957.52	4,965.34	4,979	4,937
3	12,894.73	12,827.60	12,826.5	12,827.47	12,827.03	12,825.98	12,814.09	12,841.44	12,826	12,824
4	13,188.12	13,258.21	13,093.5	13,087.08	13,098.29	13,072.46	13,084.99	13,077.31	13,111	13,005
5	24,044.43	23,992.82	23,781.9	23,875.30	23,878.33	23,846.65	23,864.44	23,742.13	23,818	23,632
6	36,596.15	36,432.15	36,258.3	36,562.15	36,538.20	36,493.08	36,508.78	36,156.77	36,308	36,040
7	38,723.90	38,436.43	38,451.6	38,466.20	38,461.99	38,460.84	38,454.22	38,509.27	38,436	38,442
8	50,389.01	49,937.19	49,910.7	50,550.68	50,476.87	5,0425.99	50,419.92	49,681.99	49,958	49,616
9	64,413.89	63,901.16	63,987.8	64,049.22	64,032.88	64,036.17	64,018.23	64,013.33	63,917	63,955
10	64,937.83	64,085.90	64,334.8	65,442.82	65,294.11	65,252.03	65,188.17	64,128.20	64,348	63,967

Table 5.10 Errors on the obtained natural frequency between the meshless solutions and the FEM-ABAQUS (8,000 dofs) solution [2]

Mode	MLPG	Nagashima	LRPIM	Regular Mesh		Irregular Mesh		3D	FEM 2D
				RPIM	NNRPIM	RPIM	NNRPIM	NNRPIM	
		1999	p = 1.03; c = 1	p = 1.03; c = 0.1	V2P1	p = 1.03; c = 0.1	V2P1	V2P1	Abaqus 612dofs
1	1.75E−03	2.57E−02	1.58E−03	7.67E−04	2.60E−03	6.41E−04	1.47E−03	5.43E−03	8.51E−03
2	2.70E−02	2.31E−02	8.02E−03	3.13E−03	4.58E−03	2.36E−03	4.16E−03	5.74E−03	8.51E−03
3	5.52E−03	2.81E−04	1.95E−04	2.71E−04	2.36E−04	1.55E−04	7.73E−04	1.36E−03	1.56E−04
4	1.41E−02	1.95E−02	6.81E−03	6.31E−03	7.17E−03	5.19E−03	6.15E−03	5.56E−03	8.15E−03
5	1.75E−02	1.53E−02	6.34E−03	1.03E−02	1.04E−02	9.08E−03	9.84E−03	4.66E−03	7.87E−03
6	1.54E−02	1.09E−02	6.06E−03	1.45E−02	1.38E−02	1.26E−02	1.30E−02	3.24E−03	7.44E−03
7	7.33E−03	1.45E−04	2.50E−04	6.29E−04	5.20E−04	4.90E−04	3.18E−04	1.75E−03	1.56E−04
8	1.56E−02	6.47E−03	5.94E−03	1.88E−02	1.74E−02	1.63E−02	1.62E−02	1.33E−03	6.89E−03
9	7.18E−03	8.42E−04	5.13E−04	1.47E−03	1.22E−03	1.27E−03	9.89E−04	9.12E−04	5.94E−04
10	1.52E−02	1.86E−03	5.75E−03	2.31E−02	2.07E−02	2.01E−02	1.91E−02	2.52E−03	5.96E−03

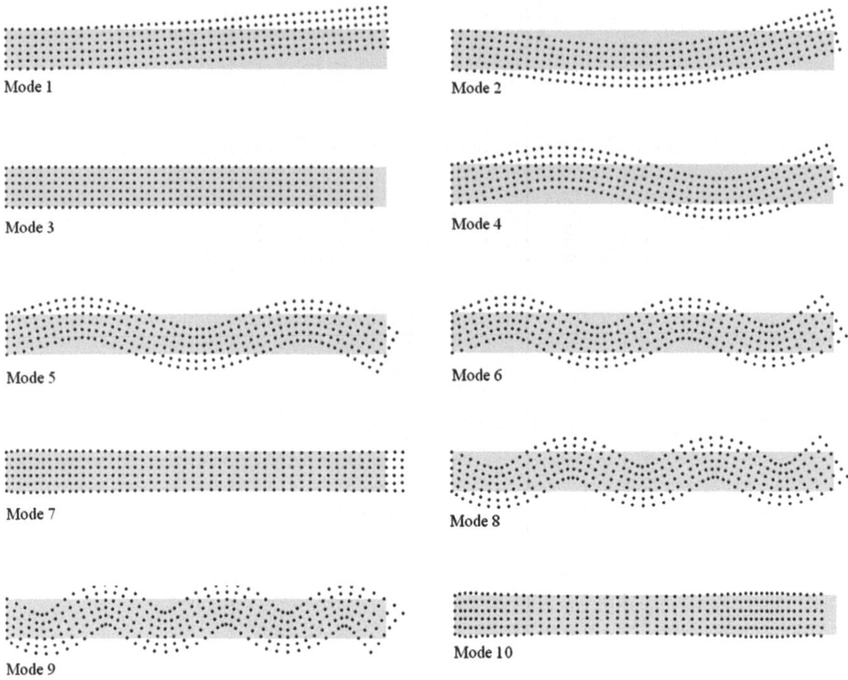

Fig. 5.58 Cantilever beam vibration modes [2]

Once again it is visible the good concordance between the distinct methods. In Fig. 5.62 the first eight vibration modes of the shear wall obtained using the V2P1 NNRPIM formulation are presented.

5.4.4 Forced Vibration of a Cantilever Beam

In this example a cantilever beam, Fig. 5.63, is subjected to a forced vibration. The geometrical parameters are, $L = 48$ m, $D = 12$ m and thickness $h = 1$ m. The material properties are Young's modulus $E = 3.0 \times 10^7$ Pa, Poisson ratio $\upsilon = 0.3$ and mass density $\rho = 1.0$ kg/m^3. The problem is analysed considering the 2D plane stress deformation theory and the 3D deformation theory. As Fig. 5.63 indicates, the beam is discretized with 297 nodes for the 2D analysis and with 891 nodes for the 3D study. The beam is subjected to a parabolic load at the free end, $f(t) = 1{,}000 \times g(t)N$, where $g(t)$ is a time functional.

The three distinct time-dependent loading conditions presented in Sect. 2.3.7 are considered. In this example, for the load case B it is considered: $t_i = 0.5$ s.

The dynamic response of the complete beam is obtained, i.e., the complete displacement field is obtained for each time step. However the most interesting

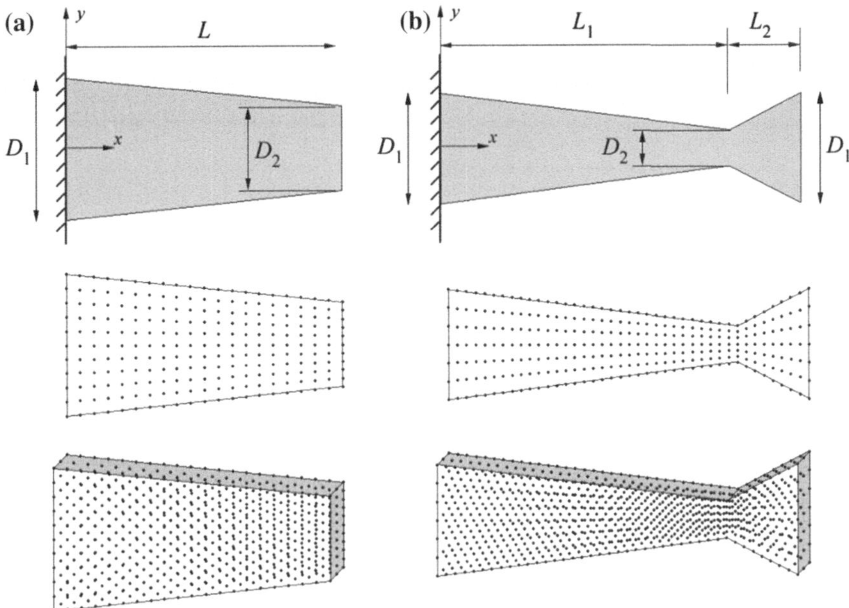

Fig. 5.59 Cross section of the variable beams and respective nodal arrangement. **a** Beam A and **b** Beam B

Table 5.11 Vibration frequencies (rad/s) of the cantilever beam with variable cross section—Beam A [2]

Mode	MLPG	LRPIM p = 1.03; c = 1	RPIM p = 1.03; c = 0.1	NNRPIM V1P1	NNRPIM V2P0	NNRPIM V2P1	NNRPIM V2P1 (3D)	FEM Abacus
1	263.21	262.13	261.54	263.42	263.10	262.58	262.18	262.09
2	923.03	920.81	922.48	918.93	925.95	925.21	920.78	981.93
3	953.45	952.06	952.15	952.30	952.40	952.43	953.9	951.86
4	1,855.14	1,854.32	1,871.07	1,844.59	1,875.62	1,874.89	1,862.35	1,850.92
5	2,589.78	2,589.87	2,589.61	2,578.56	2,590.02	2,589.76	2,593.4	2,578.63

Table 5.12 Vibration frequencies (rad/s) of the cantilever beam with variable cross section—Beam B [2]

Mode	NNRPIM V1P1	NNRPIM V2P0	NNRPIM V2P1	NNRPIM V2P1 (3D)	FEM 9n
1	133.14	132.39	132.21	134.01	131.59
2	484.23	485.39	484.63	476.9	469.88
3	878.34	878.43	878.63	891.45	878.31
4	1,105.35	1,119.01	1,117.20	1,075.85	1,062.42
5	1,908.63	1,945.77	1,942.27	1,883.92	1,863.21

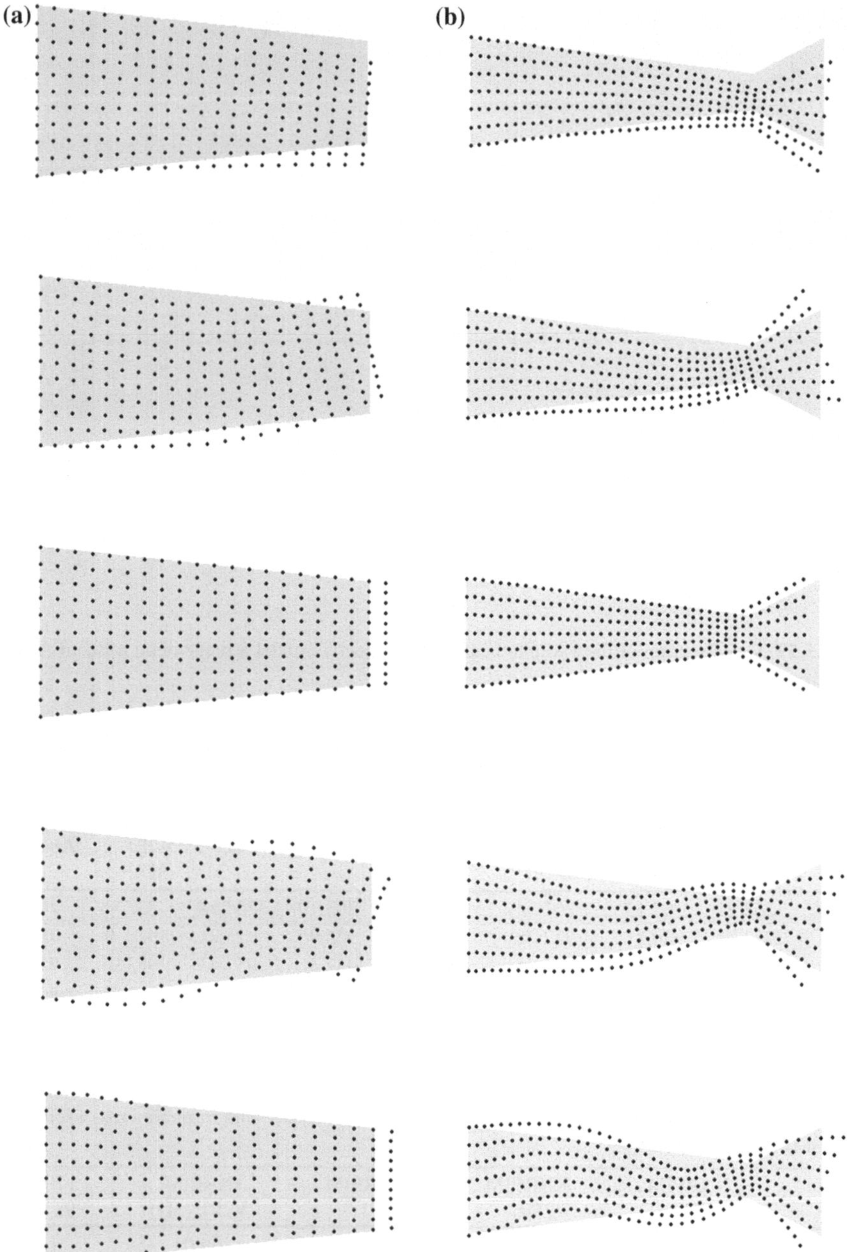

Fig. 5.60 Vibration modes of the Cantilever beam with a variable cross section. **a** Beam A. **b** Beam B [2]

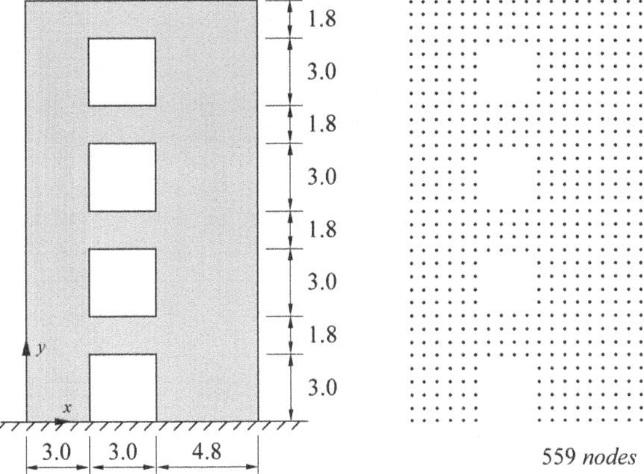

Fig. 5.61 Shear wall with four openings and respective discretization in 559 nodes

Table 5.13 Vibration frequencies (rad/s) of the shear wall with four openings [2]

Mode	MLPG	BEM	LRPIM p = 1.03; c = 1	NNRPIM V1P1	NNRPIM V2P0	NNRPIM V2P1	FEM Abaqus
1	2.069	2.079	2.086	2.070	2.097	2.098	2.073
2	7.154	7.181	7.152	7.033	7.109	7.110	7.096
3	7.742	7.644	7.647	7.645	7.647	7.647	7.625
4	12.163	11.833	12.019	12.030	12.346	12.353	11.938
5	15.587	15.947	15.628	15.121	15.417	15.418	15.341
6	18.731	18.644	18.548	18.141	18.391	18.385	18.345
7	20.573	20.268	20.085	19.621	19.944	19.937	19.876
8	23.081	22.765	22.564	21.712	22.103	22.097	22.210

point of the beam is point A, Fig. 5.63, localized on the middle node at the free end of the beam. The presented results show the variation of the vertical displacement of point A, u_y, with respect to the variable time, t. For comparison it is used the nine node finite element considering the same nodal arrangement.

The beam response to the dynamic load case A is presented in Fig. 5.64a. It should be noted that the static solution for this load case obtained with the V2P1 2D-NNRPIM in point A is $u_y^A = -8.91 \times 10^{-3}\,\mathrm{m}$ and the amplitude due the dynamic load is $u_y^A = -17.66 \times 10^{-3}\,\mathrm{m}$, which is double, as expected. The NNRPIM results are in a very good agreement with the FEM results.

The beam transient response when subjected to the load case B is presented in Fig. 5.64b. It is visible that in the period of time between $t = 0$ s and $t = 0.5$ s the beam response is identical to the load case A, as it should be, after $t = 0.5$ s the beam vibrates freely around the initial position. Once more the NNRPIM results are similar with the FEM results.

In load case C a simple harmonic dynamic load is applied to the beam. The first vibration frequency of the cantilever beam obtained with the FEM and with the NNRPIM are respectively $\omega_1^{FEM} = 27.26\,\mathrm{rad/s}$, $\omega_1^{v1p1} = 27.48\,\mathrm{rad/s}$,

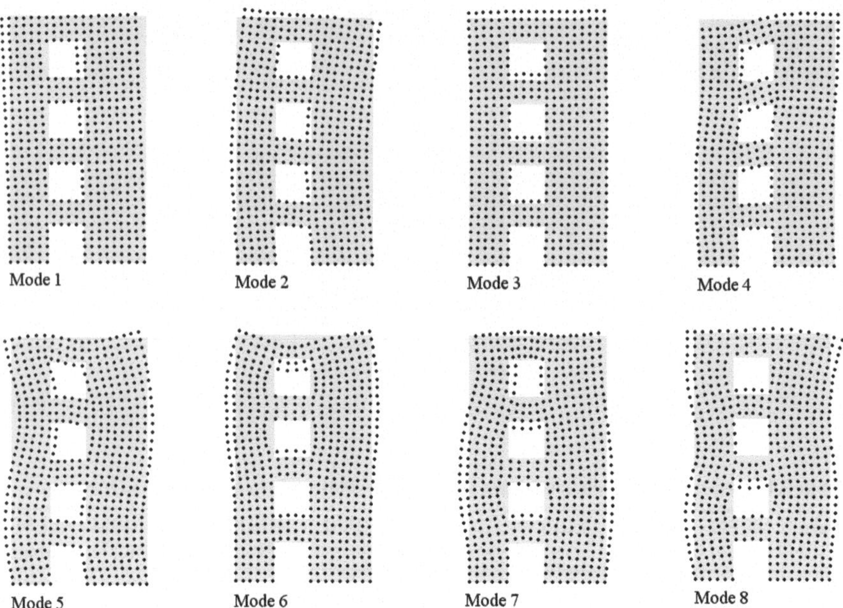

Fig. 5.62 Vibration modes of the shear wall with four openings [2]

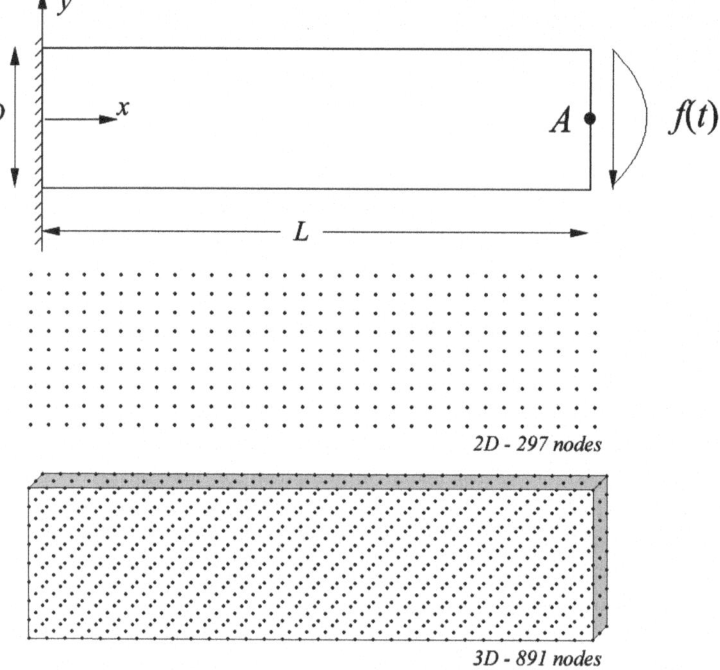

Fig. 5.63 Cantilever beam with an applied load at the free end and the used nodal distributions

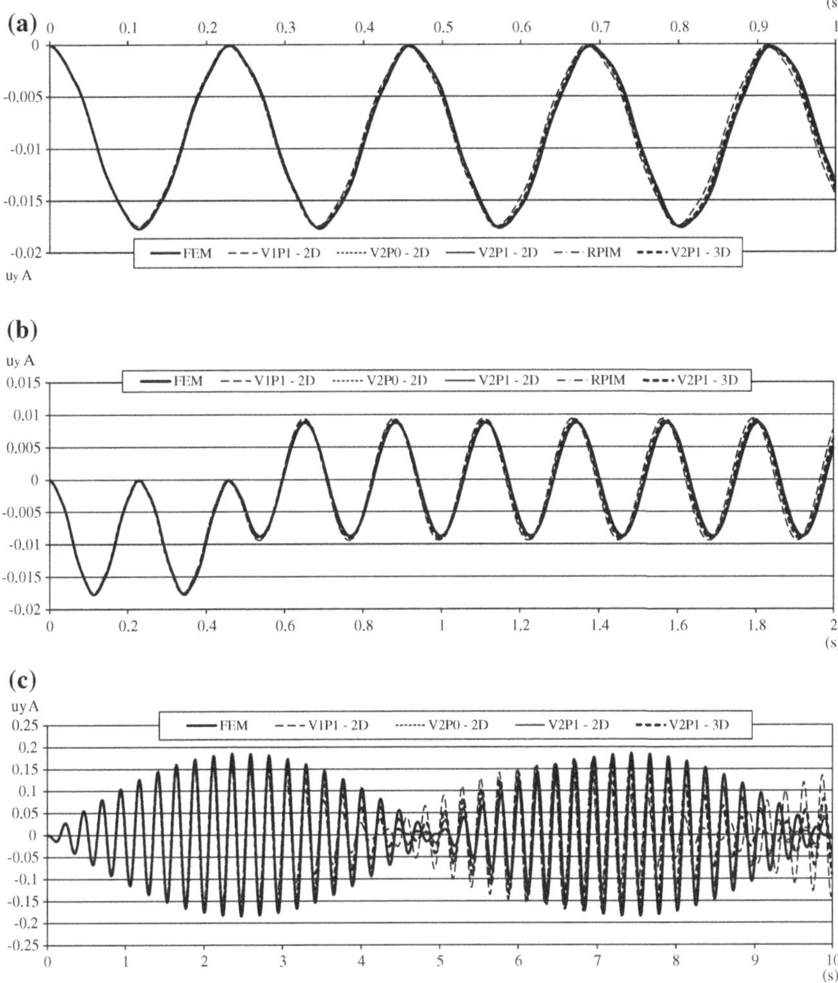

Fig. 5.64 Displacement at point A when it is applied **a** the load case A, **b** the load case B and **c** the load case C [2]

$\omega_1^{v2p0} = 27.32$ rad/s, $\omega_1^{v2p1} = 27.31$ rad/s and $\omega_1^{3D} = 27.37$ rad/s. The frequency of the dynamic load used in this example is $\gamma = 26$ rad/s. In Fig. 5.64c it is visible that the beam dynamic response obtained with the V1P1 NNRPIM formulation differs significantly from the other presented solutions. This variation can be explained with the considerably difference between the fundamental vibration frequency obtained from the V1P1 NNRPIM formulation and the frequency of the dynamic load used in this example.

Notice that the other numerical approaches present close fundamental frequencies between each other. Therefore, as expected, the beam dynamic responses

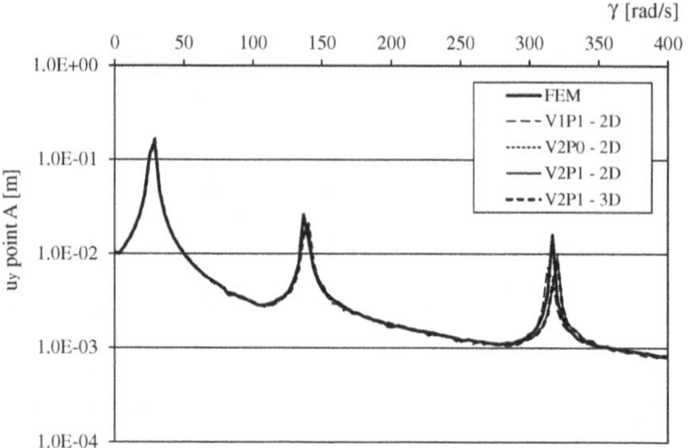

Fig. 5.65 Absolute maximum vertical displacement on point A versus frequency for the cantilever beam subjected to an uniform harmonic load [2]

obtained with the V2P0 and V2P1 NNRPIM (2D and 3D) formulations and the FEM are very similar.

In Fig. 5.65 the maximum value of the vertical displacement on point A versus the frequency of the dynamic load used, load case C, is presented. It is visible that the NNRPIM formulations present results almost coincident with the FEM.

References

1. Bathe KJ (1996) Finite element procedures. Prentice-Hall, Englewood Cliffs
2. Dinis LMJS, Jorge RMN, Belinha J (2009) The natural neighbour radial point interpolation method: dynamic applications. Eng Comput 26(8):911–949
3. Dinis LMJS, Jorge RMN and Belinha J (2011a) The dynamic analysis of thin structures using a radial interpolator meshless method. In: Vasques CMA, Dias Rodrigues J (eds) Vibration and structural acoustics analysis. Springer, Netherlands, pp 1–20
4. Dinis LMJS, Jorge RMN, Belinha J (2011) Static and dynamic analysis of laminated plates based on an unconstrained third order theory and using a radial point interpolator meshless method. Comput Struct 89(19–20):1771–1784
5. Dinis LMJS, Jorge RMN, Belinha J (2011) A natural neighbour meshless method with a 3D shell-like approach in the dynamic analysis of thin 3D structures. Thin-Walled Struct. 49(1):185–196
6. Dinis LMJS, Jorge RMN, Belinha J (2008) The radial natural neighbour interpolators extended to elastoplasticity. In: Ferreira AJM, Kansa EJ, Fasshauer GE, Leitao VMA (eds) Progress on meshless methods. Springer, Netherlands, pp 175–198
7. Dinis LMJS, Jorge RMN, Belinha J (2009) Large deformation applications with the radial natural neighbours interpolators. Comput Model Eng Sci 44(1):1–34
8. Irons BM, Razzaque A (1972) Experience with the patch test for convergence of finite elements. In: Aziz AK (ed) The Mathematical Foundations of the Finite Element Method with Applications to Partial Deferential Equations. Academic Press, New York
9. Wang JG, Liu GR (2002) A point interpolation meshless method based on radial basis functions. Int J Numer Meth Eng 54:1623–1648
10. Wang JG, Liu GR (2002) On the optimal shape parameters of radial basis functions used for 2-D meshless methods. Comput Methods Appl Mech Eng 191:2611–2630

11. Kansa EJ (1990) Multiquadrics—a scattered data approximation scheme with applications to computational fluid-dynamics—I surface approximations and partial derivative estimates. Comput Math Appl 19(8–9):127–145
12. Kansa EJ (1990) Multiquadrics—A scattered data approximation scheme with applications to computational fluid-dynamics—II solutions to parabolic, hyperbolic and elliptic partial differential equations. Comput Math Appl 19(8–9):147–161
13. Dinis LMJS, Jorge RMN, Belinha J (2007) Analysis of 3D solids using the natural neighbour radial point interpolation method. Comput Methods Appl Mech Eng 196(13–16):2009–2028
14. Zienkiewicz OC, Taylor RL (1994) The Finite Element Method, 4th edn. McGraw-Hill, London
15. Timoshenko S, Goodier JN (1970) Theory of elasticity, 3rd edn. McGraw Hill, Singapore
16. Gu YT, Liu GR (2001) A meshless local Petrov-Galerkin (MLPG) method for free and forced vibration analyses for solids. Comput Mech 27:188–198
17. Nagashima T (1999) Node-by-node meshless approach and its application to structural analyses. Int J Numer Meth Eng 46:341–385
18. Gu YT, Liu GR (2001) A local radial point interpolation method (LRPIM) for free vibration analyses of 2-D solids. J Sound Vib 246(1):29–46
19. Brebbia CA, Telles JC, Wrobel LC (1984) Boundary element techniques. Springer, Berlin

Chapter 6
Bone Tissue

Abstract In this chapter are introduced the basic concepts of the bone biology. Then, bone tissue phenomenological laws are presented, permitting to correlate the bone tissue local apparent density with the bone tissue local mechanical properties. The mathematical law proposed by the author is presented in detail. This chapter ends with an extensive presentation of the most relevant numerical approaches for the prediction of the bone tissue remodelling. Additionally, the bone tissue remodelling algorithm used in this book is presented explicitly.

6.1 Bone Biology: Basic Concepts

The bone, one of the hardest tissues on the human body, is a living-dynamic biomaterial tissue active during its life time. The bone tissue possesses many interesting structural properties, comparing with the steel the traction strength is similar, it is three times lighter and ten times more flexible. These properties are essentially due to its heterogeneous microstructure, composed of an organic part (mostly collagen, providing traction capacity) and a mineral part (providing stiffness and strength under compression). It possesses a remarkable remodelling capacity. Under permanent change in response to different signals, such as external loads or hormonal influence, it adapts, auto-repairs and changes the form and the internal biomaterial properties distribution.

The bone tissue is a connective tissue formed by cells, blood vessels, fibres and organic (collagen) and inorganic substances (carbonated hydroxyapatite). However, in opposition with other connective tissues the extracellular components suffer calcification, which confers hardness. The bone material properties are a consequence of the combination between the organic and inorganic substances. Due to collagen fibres the bone presents high values for the elastic modulus and the ultimate strength in tension. The high value for the ultimate strength in compression is given by the mineral components. In the end, the product is a light material with an unique microscopic and macroscopic layout, which permits to

J. Belinha, *Meshless Methods in Biomechanics*, Lecture Notes in Computational Vision and Biomechanics 16, DOI: 10.1007/978-3-319-06400-0_6,

maximize the resistance and minimize the weight. Bones perform several bio functions [1], such as:

Mechanical Functions:

Support—The skeleton provide a frame that keeps the body supported.
Protection—Bones serve to protect internal organs and soft tissues.
Movement—Bones, muscles, tendons, ligaments and joints function are combined together to generate and transfer forces so that located body parts, or the entire body, may be manipulated.

Metabolic Functions:

Mineral repository—Bones are an important mineral repository (calcium and phosphorus).
Blood cells producer—In the red marrow (hematopoietic) erythrocytes (red blood cells), leucocytes (with blood cells) and platelets are produced.
Energy repository—In the yellow marrow fat cells can be found, an energy depot.

Besides all this, the bone is a living tissue, in constant mutation, experimenting a continuous reconstruction and reformulation along its life span, due to mechanical, nutritional and endocrine stimulus [1]. This chapter describes the structure, the material composition and the mechanical properties of the bone tissue. Biologic material laws suitable to simulate the bone tissue are presented, as well as a new developed phenomenological law. Afterwards the bone tissue remodelling algorithm used in this book is presented.

6.1.1 Bone Morphology

The human skeleton structure has more than 200 bones classified, by shape, in four major types: long bone, short bone, flat bone and irregular bone [1]. These bone types are represented in Fig. 6.1.

In the diaphysis zone, the outer shell of the long bone is cortical bone (compact bone) and in the centre it can be found the medullar cavity, which contains in adulthood yellow marrow (mainly fat cells). In the epiphyses the outer shell of cortical bone is thin and in the core it can be found trabecular bone (spongy bone) and red marrow. The red marrow, unlike the yellow marrow which is just a soft tissue with energy depot functions, is a hematopoietic tissue and it is the site of production of red cells, platelets and most of the white blood cells. These features are represented in Fig. 6.2.

Short bones are characterized to be small and to have no preferential dimension. These bones consist of cancellous tissue (trabecular bone) enclosed within a thin layer of cortical bone. Examples are the carpals, the tarsals and the bones in the ankle and in the wrist.

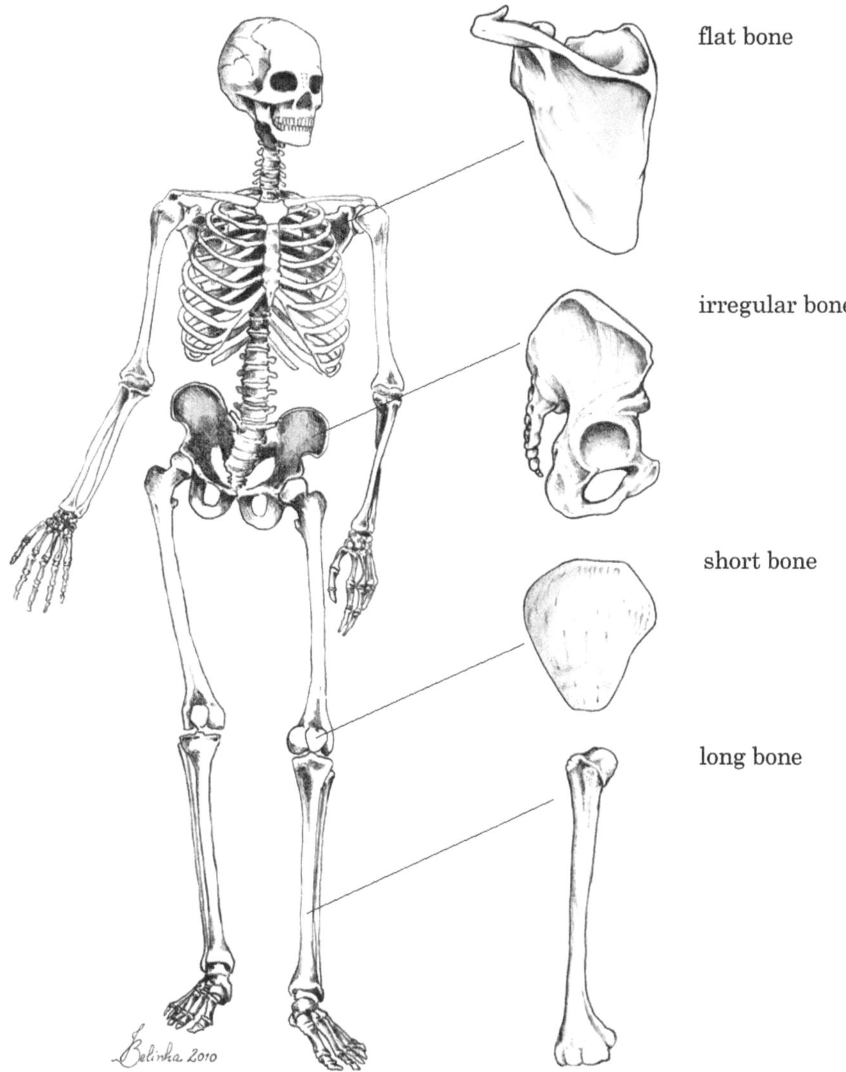

Fig. 6.1 Bone shape major groups

Flat bones do not have diaphysis or epiphyses. These bones are composed of two thin layers of cortical bone sandwiching trabecular bone tissue, which is the location of red bone marrow. In an adult, most red blood cells are formed in flat bones. These bones have generally a dimension much smaller than the other two. Examples are the cranial bones, sternum or the scapula.

Like the flat bones, irregular bones do not possess diaphysis or epiphyses and are assembled by two thin layers of cortical bone enclosing trabecular bone tissue. In these bones there is no preferential dimension or shape. Irregular bones serve

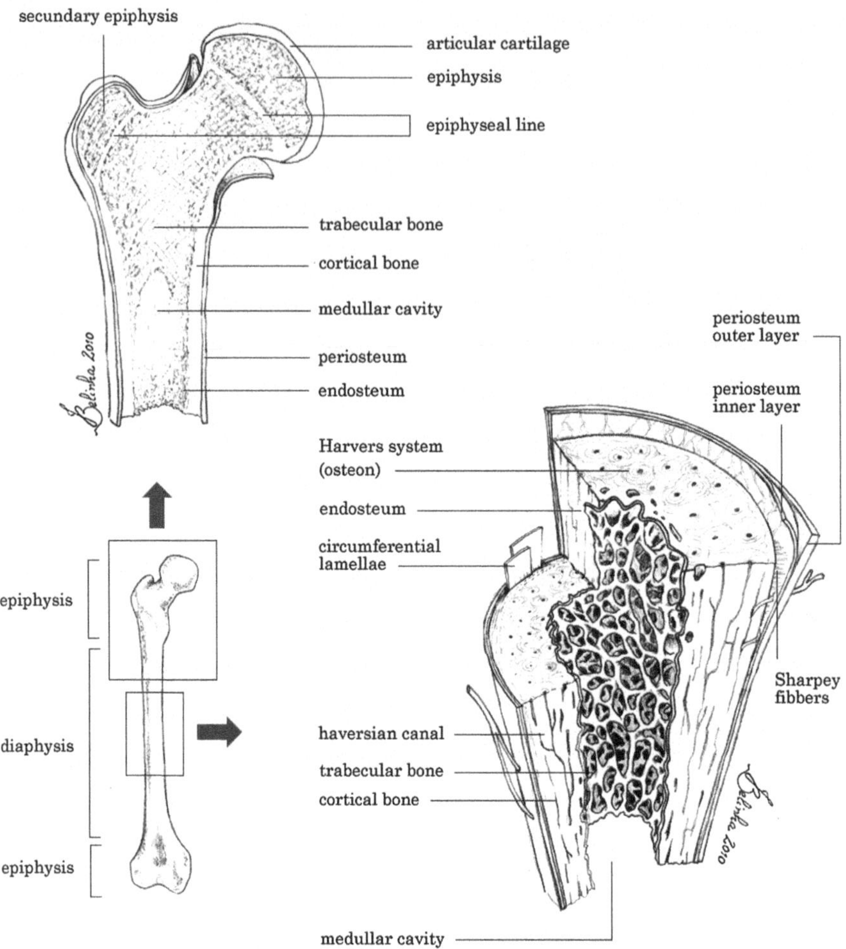

Fig. 6.2 Long bone. Human femur with the identification of the primary structures and inner features of a diaphysis section

various purposes in the body, such as protection of nervous tissue (the spinal cord is protected by the vertebrae), multiple muscle-skeletal anchor (the sacrum) or support of articulated structures (the hip bone or the zygomatic bone).

6.1.2 Composition and Structure of the Bone

6.1.2.1 Bone Matrix

In relation to the bone body dry weight, the bone matrix is proximally composed by 30 % of organic material and 70 % of inorganic material [1]. The bone organic part is composed majority by collagen (about 94 %) and small parts of noncollagenous proteins (about 4 %) and bone cells (about 2 %). The collagen present in the bone matrix is of type I, although small quantities of type V collagen can be found. Collagen fibres are arranged along lines of mechanical stress according to Wolff's Law [2]. It is the orientation and the quantity of these fibres that confers to the bone mechanical properties such as elasticity and high ultimate traction and shear stress. The collagen fibres are synthesized by the osteoblasts (bone cells). For instance, the cortical bone and the periosteum are bonded through collagen fibres, the Sharpey fibres.

Although it represents only 5 % of the organic bone matrix, the noncollagenous proteins play an important role in the bone metabolism and in the mineralization of the bone matrix. Only 2 % of the organic matrix are bone cells, but without such living elements the bone could not repair, rebuild, remodel, reshape itself or even grow.

The bone inorganic part consists mainly of a carbonate-rich hydroxyapatite, also called bone apatite, which is smaller and less perfect in crystal arrangement than pure hydroxyapatite, $Ca_{10}(PO_4)_6(OH)_2$. However it is due to this crystalline imperfections that bone apatite is biological more appropriate than pure hydroxyapatite, once it is more readily available for metabolic activity and for body fluid exchanges. Other minerals are also present in the bone mineral matrix, but in much lower percentage: sodium, magnesium, potassium, fluoride and chloride. The inorganic components confer to the bone structure the hardness, the stiffness and the high ultimate compression stress.

6.1.2.2 Bone Cells

The bone cells are divided, generally, in five cell-types: osteoprogenitors, osteoblasts, osteocytes, bone-lining cells and osteoclasts. The role of these cells in the bone reabsorption process is presented in Fig. 6.3.

The osteoprogenitors are immature non-specialized cells which differentiate to origin the osteoblasts [3]. These cells are in the genesis of all bone structures during the growth process, since the conception, and are generated in the bone marrow, or in other connective tissues. The osteoprogenitors can be found in the interior layer of the periosteum, in the endosteum and in the Haversian's and Volkmann's canals.

The osteoblasts are mononuclear cells responsible for the bone formation and for the mineralization of the osteoid matrix. These cells are in the origin of the osteocytes. The osteoblasts, produced by the osteoprogenitors in the periosteum, the endosteum and the bone marrow, can be found in the bone inner surface (endosteum) or outer surface (periosteum) in a laminar distribution. When active the osteoblasts are cuboidal or polygonal shaped [1]. The number of osteoblast cells tends to decrease as individuals become elder, therefore decreasing the natural renovation of the bone tissue [4].

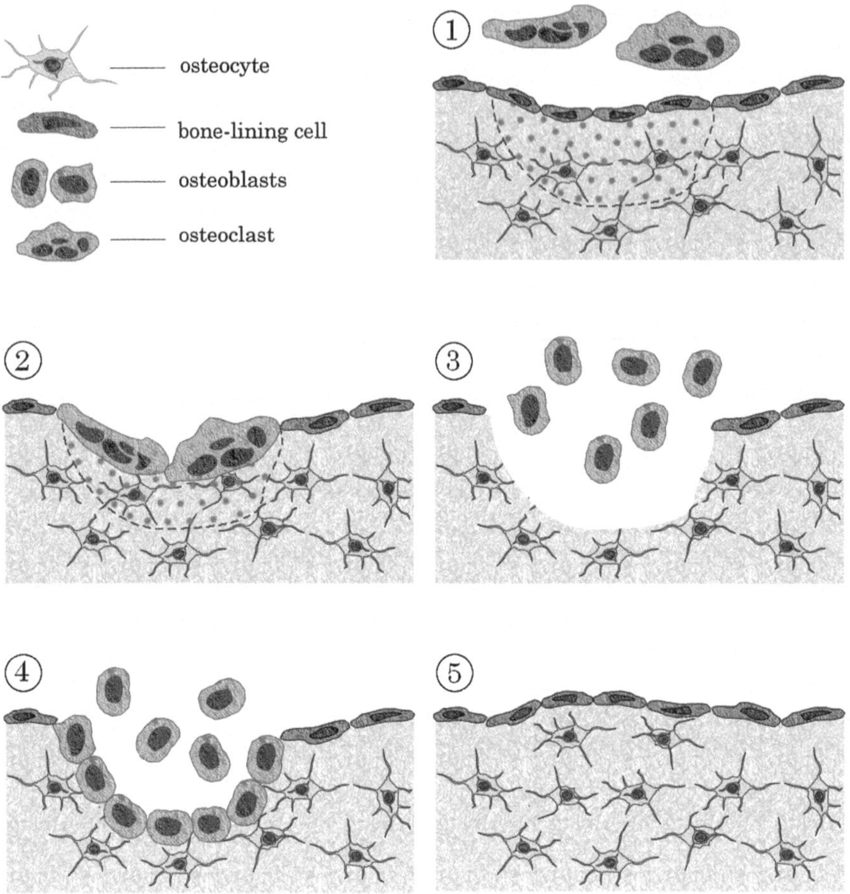

Fig. 6.3 Bone reabsorption process

The bone-lining cells are inactive elongated and thin osteoblasts. These cells remain inactive in the bone surface once the bone formation, or the bone remodelling, has finished. However the bone-lining cells can be reactivated in response to mechanical and/or chemical stimulation [5].

The osteocytes are mature cells derived from the osteoblasts. These cells can be found inside the mineralized bone matrix occupying small chambers, called lacunas. It are connected to each other by long cytoplasmic extensions, through thin canals called canaliculi, which are used in the exchange of nutrients, forming a complete and vast network. The osteocytes, through the canaliculi, conserve the connection with the outer osteoblasts and have as primary purpose the preservation of the surrounding mineralized bone matrix vitality. The osteocytes, using the connections with the outer cells, act as local sensors of the mechanical and chemical state of the bone and, in case of necessity, initiate the bone reabsorption process from the surface [6, 7]. These cells are affected directly by the damage in the mineralized bone matrix, once it destroys the cytoplasmic extensions, stopping the cellular metabolic process and the cellular fluid exchange. In the particular case of the bone micro damage, the destruction of the cytoplasmic extensions is the trigger button to activate the bone remodelling process. This process is also activated if the osteocytes are unused [8–10].

The osteoclasts are plurinuclear cells responsible for the reabsorption and/or elimination of the bone structure. These cells cause the demineralization of the mineralized bone matrix and destroy the bone organic matrix. The osteoclasts cells are formed by the fusion of precursor cells from the bone marrow or from other haematopoietic tissue. These cells are located in superficial depressions (Howship lacunas) which identify the bone reabsorption zones.

6.1.2.3 Reticular and Lamellar Bone

The bone tissue can be classified in reticular or lamellar bone tissue according with the collagen fibre organization inside the bone matrix. The reticular bone tissue is a neo-formed or immature bone, commonly referred as woven bone. In this bone tissue the collagen fibre arrangement shows an anisotropic layout, since all fibres are oriented in different directions (randomly), with no preferential orientation. In the fetal development or in a fracture repair, the reticular bone is the first to be formed. In this bone tissue active osteoblasts, excited by growing factors or stimulated by fracture damage, build more solid bone structures.

In a posterior phase, by the remodelling process, the reticular bone gives place to the lamellar bone. The lamellar bone is a mature bone tissue organized in thin layers glued by collagen fibres oriented in perfect alignment with each other. The osteocites in the interior of the lamellar bone are displayed in sandwich layers between the lamellas. The lamellar bone has the same chemical composition and material properties throughout the skeleton, regardless its mechanism of formation—intramembranous or endochondral—or its structural organization—cortical (compact) or trabecular (soft) bone [1], Fig. 6.4.

6.1.3 Cortical and Trabecular Bone

In adulthood the skeleton contains only two types of bone tissues, cortical bone (compact bone) and trabecular bone (soft or cancellous bone) [1].

6.1.3.1 Cortical Bone

The basic functional unit of cortical bone is the osteon (also known as the Havers system), which consists of concentric layers of bony lamellae surrounding a central Haversian canal. The Haversian canal contains the blood capillaries that supply the

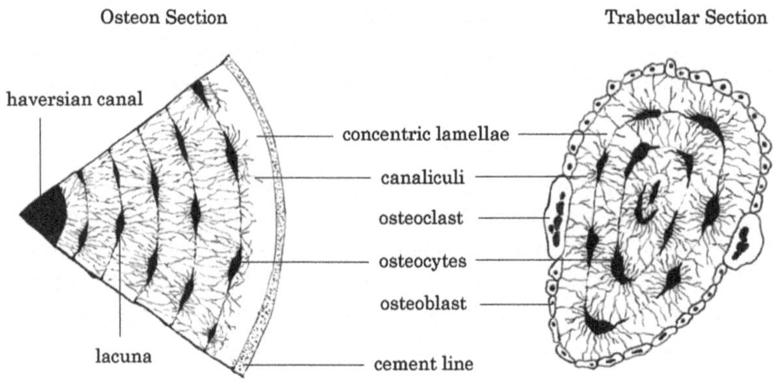

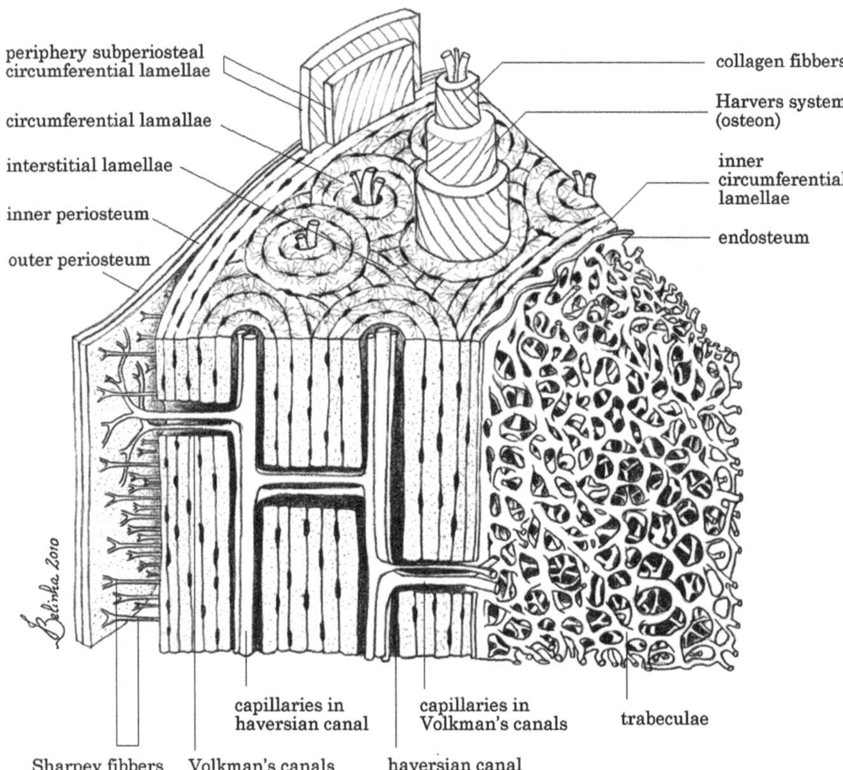

Fig. 6.4 Transversal cross-section pie-shaped of an osteon and transversal cross-section of a trabecular branch. Detailed view of a typical bone cortex

bone tissue. Cross branches, called Volkman's canals, connect the osteons with each other, conducting blood vessels. Each cylindrical lamella forming the osteon is shaped by osteocytes regularly distributed. The osteocytes communicate with each other by projecting throughout the lamellae very thin and long cytoplasmic extentions along the tiny canals called canaliculi. It is through the canaliculi that electro-chemical information flows and the nutrients and minerals reach all points of the bone matrix. Besides the osteons, long bones contain in the periphery subperiosteal circumferential lamellae, spawned by the inner layer of the periosteum.

6.1.3.2 Cortical Bone Remodelling

The remodelling process in cortical bone is activated by growing factors, or changing of load patterns or due to micro or macro damage of the bone tissue. The initial step is the removal of the inactive osteons, by osteoclast activity. Afterwards new lamellae are deposited in new concentric layers, from exterior to interior, until a new osteon is formed. In mature bone the recurring remodelling process causes the formation and destruction of many osteon generations. The new osteons are organized by complete circumferential layers, i.e., a complete Haversian system, this causes that some older osteons present portions of the outer border occupied by the outer border of a new osteon. In some old osteons, this process have occurred so many times that all it is left from the original lamellae are small portions of the circumferential lamellae. These small portions are called interstitial lamellae. Cortical bone is remodelled by bone cells lining the periosteal, the endosteal and the Haversian canal surfaces. The periosteal surface is responsible for the bone width growth. The endosteal activity determines the diameter of the medullary canal. The combined activity of the periosteal and endosteal structures determines the thickness of the bone cortex. The Haversian canal surface plays an important role in the remodelling process of the bone and it is responsible for the cortex density.

6.1.3.3 Trabecular Bone

The trabecular bone is composed of mature lamellar bone and, in contrast with the compact structure of cortical bone, it is a complex network of intersecting curved plates and tubes. The osteocytes within each trabecula are concentrically oriented and have a well-developed canalicular set of connections. In the case of a long bone, the trabecular bone is typically located at the proximal ends, here the arrangement of the trabeculae is relatively regular, reflecting the direction of the principal mechanical stresses to which this kind of bone is subjected. Trabecular bone constitutes 20 % of the skeletal bone mass, being the remaining 80 % occupied by cortical bone, nevertheless due to the trabeluae vast surface the surface-to-volume ratio is ten times higher than the cortical bone. It is because this high specific surface that the metabolic activity of trabecular bone is nearly eight times higher when compared with the cortical bone, which may help to explain why metabolic bone diseases have a greater effect on trabecular bone than on cortical bone.

Regardless of its relatively small volume and high apparent porosity, trabecular bone is well adapted to resist and conduct compressive loads, as can be seen by the vertebrae, handing over to the cortical bone other structural functions, as resisting bending and torsional stresses.

6.1.3.4 Trabecular Bone Remodelling

The border of the trabecula is filled with osteblasts and osteoclasts, ready and active. During the growth or remodelling of the trabecula the deposition of new bone by osteblast activity is counterbalanced by the removal of bone by osteoclast activity in the opposite surface of the trabecula, remodelling and reshaping the trabeculae configuration to a better stress path configuration or simply to repair micro-fractures. The diminishing of the bone mass with age is the result of the change in the balance between the rate of bone formation and the rate of bone reabsorption. Pathologic fractures occur when the bone density becomes so low that the skeleton can no longer withstand the mechanical stresses of everyday life. In a experimental study [11] it was showed that the compressive strength of bone is proportional to the square of its apparent density, e.g., if the bone density decreases by a factor of 3, its compressive strength decreases by a factor of 9. Both cortical and trabecular bone present analogous microscopic physical properties and biochemical composition. However, in order to suit local physical requirements, due to loads and to stresses distribution, the bone macroscopic structure is organized in order to maximize the strength and stiffness. For example, at the ends of long bones the thin cortical shell, inner supported by trabecular bone, is the optimal configuration to hold and distribute in the bone structure the concentrated loads applied. While the tubular cortical diaphysis is optimized to support the large torsional and bending loads. Despite being cortical or a trabecular structure, all normal adult bone is lamellar bone. In adults, only due to normal fracture healing, or in pathologic conditions, immature woven bone, or fibre bone, is seen.

6.2 Bone Tissue Mechanical Properties

As previously mentioned, bone tissue can be classified as cortical bone, highly densified bone tissue, and trabecular bone, which shows a considerably smaller apparent density. Although both types show the same molecular arrangement the mechanical behaviour is different. Many experimental studies show that the bone mechanical properties depend on the bone composition and on the bone porosity (directly related with the bone density) [12–17].

In the work of Carter and Hayes [12] it is stated that for trabecular and cortical bone the elastic modulus is closely related to the cube of the bone apparent density and that the strength is closely related to the square of the bone apparent density.

The apparent density, ρ_{app}, is defined as the wet mineralised mass of bone of the sample tissue, w_{sample}, over the volume occupied by the same sample tissue, V_{sample}.

$$\rho_{app} = w_{sample}/V_{sample} \tag{6.1}$$

The apparent density is dependent on the bone porosity, p, which can be obtained by the expression,

$$p = \frac{V_{holes}}{V_{sample}} \tag{6.2}$$

where the volume of bone holes, V_{holes}, is obtained by the Archimedes principle [18]. Thus, the apparent density can also be achieved using the porosity,

$$\rho_{app} = \rho_0 \cdot (1 - p) \tag{6.3}$$

Being ρ_0 the compact bone density, $\rho_0 = 2.1\,\mathrm{g/cm^3}$. The apparent density was the only variable accounted in these first models, therefore several authors started to study the influence of other variables in the mechanical properties of the cortical and trabecular bone. Thus, the apparent density combined with the bone mineral content, condensed in the bone ash density definition, was studied and correlation curves for the bone elasticity modulus and the bone ultimate compression stress were obtained [19– 23]. However these models have limitations, one of which is that none of these models take into account the influence of splitting the bone volume fraction from the ash fraction, therefore Hernandez [24] express the apparent density as a function of the bone volume fraction and the ash fraction, achieving a mathematical model that determine the elastic modulus and compressive strength with a 97 % correlation.

With these mathematical models one is capable to predict the bone main mechanical properties, however these correlations do not consider the influence of the bone micro-structure (cortical versus trabecular) or even the bone mechanical behaviour in distinct directions.

6.2.1 Lotz Material Law

The work of Lotz [25] was one of the firsts to consider the bone orthotropic behaviour. Lotz was able to determine the elasticity modulus and the ultimate compressive stress mathematical laws for both cortical and trabecular bone in the axial and transversal direction using as variable only the apparent density. The bone mechanical properties are approximated with the expressions,

Table 6.1 Coefficients of Lotz's law

Bone tissue	Direction	a_1	a_2	a_3	a_4
Cortical	Axial	2.065E+03	3.090E+00	7.240E+01	1.880E+00
	Transversal	2.314E+03	1.570E+00	3.700E+01	1.510E+00
Trabecular	Axial	1.904E+03	1.640E+00	4.080E+01	1.890E+00
	Transversal	1.157E+03	1.780E+00	2.140E+01	1.370E+00

$$E_i = a_1 \cdot \left(\rho_{app}\right)^{a_2} \tag{6.4}$$

$$\sigma_i^c = a_3 \cdot \left(\rho_{app}\right)^{a_4} \tag{6.5}$$

Being E_i the elasticity modulus and σ_i^c the ultimate compression stress in direction i, both are expressed in MPa and the apparent density ρ_{app} in g/cm^3. The coefficients a_j are presented in Table 6.1 for the axial and transversal direction for both cortical and trabecular bone.

Thus, besides considering the bone anisotropy, Lotz propose distinct material laws for the cortical and the trabecular bone.

6.2.2 Proposed Material Law

A recent experimental study by Zioupos [26], reinforces the idea that density is a salient property of bone and plays a crucial role in determining the mechanical properties of both its trabecular and cortical structural forms. The study, using the measured apparent density, was able to objectively isolate the bone in trabecular and compact forms. The results from the work of Zioupos show that the relation between E_i and ρ_{app} is not an increasing monotonic function, as it is Lotz law, but instead it is a 'boomerang'-like pattern. The experimental work of Zioupos also shows that the law governing the mechanical behaviour of the bone tissue is the same for cortical bone and trabecular bone. This experiment permits to differentiate the trabecular bone from the cortical bone based only in the apparent density measure,

$$\begin{aligned} \textit{trabecular bone} &\rightarrow \rho_{app} \leq 1.3\ \text{g/cm}^3 \\ \textit{cortical bone} &\rightarrow \rho_{app} > 1.3\ \text{g/cm}^3 \end{aligned} \tag{6.6}$$

The identified bone minimum and maximum apparent density in Zioupos work was: $\rho_{app}^{min} = 0.1\ \text{g/cm}^3$ and $\rho_{app}^{max} = 2.1\ \text{g/cm}^3$. In Fig. 6.5 it is presented the experimental data shown in Zioupos work [26]. Thus, following the conclusions of Zioupos [26], the author and co-workers [27, 28], proposed a new orthotropic mathematical law governing the mechanical behaviour of the bone tissue, which correlates the apparent density with the bone tissue mechanical properties, unifying

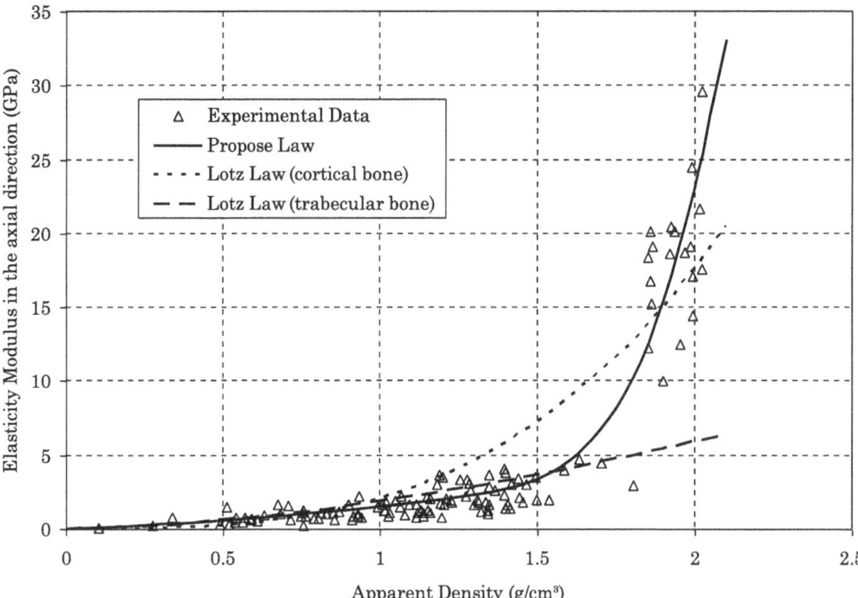

Fig. 6.5 Elasticity modulus in the axial direction. Experimental data obtain in Zioupos work compared with Lotz law for cortical and trabecular bone and with the mathematical model proposed in this book

Table 6.2 Coefficients of the propose bone tissue phenomenological model

Coefficient	$j = 0$	$j = 1$	$j = 2$	$j = 3$
a_j	0.000E+00	7.216E+02	8.059E+02	0.000E+00
b_j	−1.770E+05	3.861E+05	−2.798E+05	6.836E+04
c_j	0.000E+00	0.000E+00	2.004E+03	−1.442E+02
d_j	0.000E+00	0.000E+00	2.680E+01	2.035E+01
e_j	0.000E+00	0.000E+00	2.501E+01	1.247E+00

in the same mathematical law the cortical and trabecular bone tissue. The bone elasticity modulus for the axial direction is obtained using the approximation curve,

$$E_{axial} = \begin{cases} \sum_{j=0}^{3} a_j \cdot (\rho_{app})^j & \text{if} \quad \rho_{app} \leq 1.3\,\text{g/cm}^3 \\ \sum_{j=0}^{3} b_j \cdot (\rho_{app})^j & \text{if} \quad \rho_{app} > 1.3\,\text{g/cm}^3 \end{cases} \tag{6.7}$$

being the coefficients a_j and b_j presented in Table 6.2. This curve presents a 95 % correlation with the experimental data, Fig. 6.5. Since in Zioupos work only the

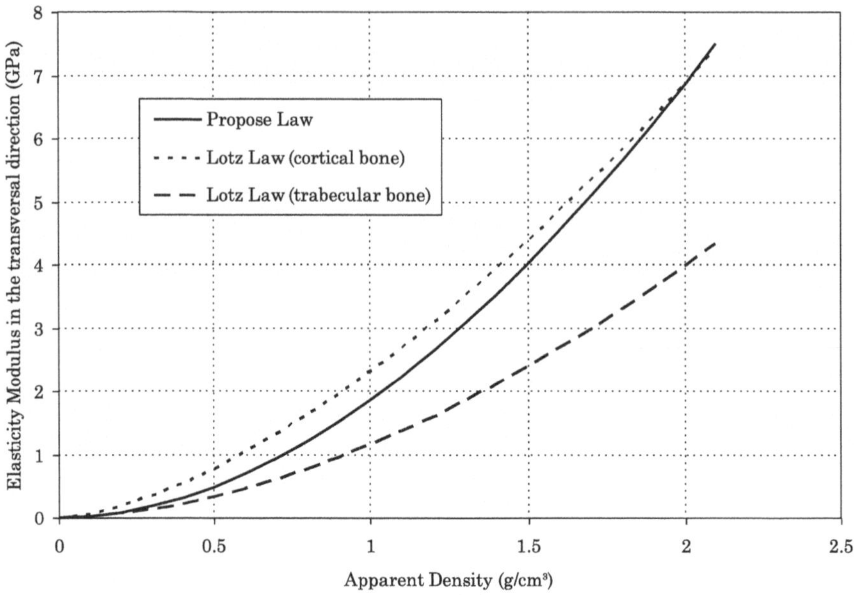

Fig. 6.6 Elasticity modulus in the transversal direction. Lotz law for cortical and trabecular bone compared with the mathematical model proposed in this book

bone elasticity modulus in the axial direction was analysed, the other curves, related with the elasticity modulus in the transversal direction and the ultimate compression stress in the axial and transversal direction, were obtained based in the values suggested in [25]. Similarly to the curve of the elasticity modulus in the axial direction, the other curves suggested in this book unify in just one curve the cortical and the trabecular bone curves from Lotz's law,

$$E_{trans} = \sum_{j=0}^{3} c_j \cdot (\rho_{app})^j \tag{6.8}$$

$$\sigma^c_{axial} = \sum_{j=0}^{3} d_j \cdot (\rho_{app})^j \tag{6.9}$$

$$\sigma^c_{trans} = \sum_{j=0}^{3} e_j \cdot (\rho_{app})^j \tag{6.10}$$

where the coefficients c_j, d_j and e_j are presented in Table 6.2. The elasticity modulus and the ultimate compression stress approximated by the curves presented in Eqs. (6.7), (6.8), (6.9) and (6.10) are expressed in MPa and the apparent density ρ_{app} in g/cm^3.

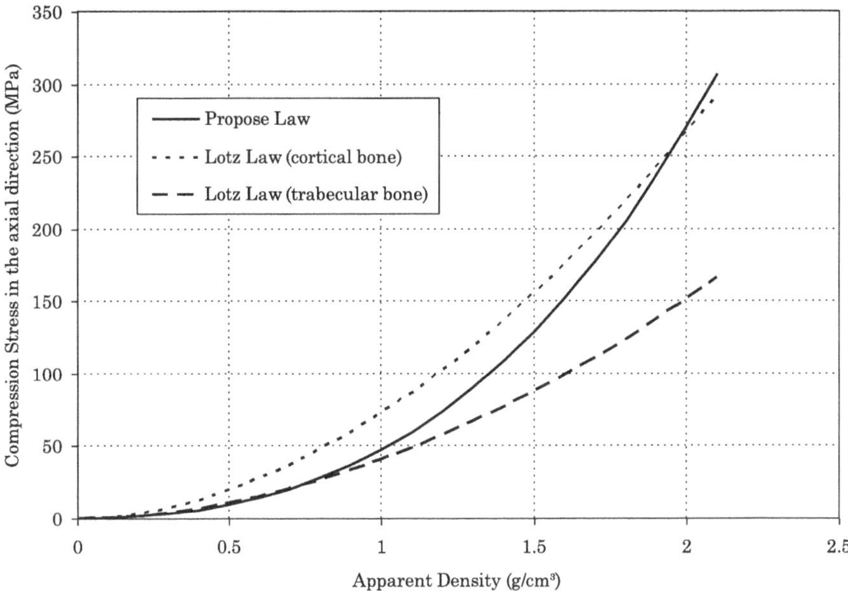

Fig. 6.7 Compression stress in the axial direction. Lotz law for cortical and trabecular bone compared with the mathematical model proposed in this book

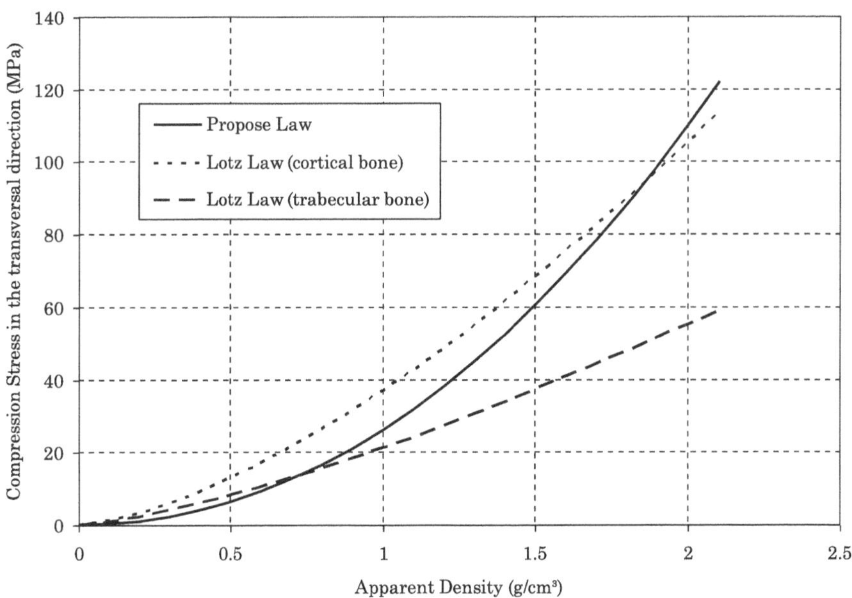

Fig. 6.8 Compression stress in the transversal direction. Lotz law for cortical and trabecular bone compared with the mathematical model proposed in this book

The plot of the approximation for the elasticity modulus in the transversal direction is presented in Fig. 6.6. For the compression stress in the axial direction and in the transversal direction the approximation curves are presented respectively in Figs. 6.7 and 6.8. The ultimate tension stress of the bone tissue is usually defined as a proportion of the ultimate compression stress, $\sigma_i^t = \alpha \cdot \sigma_i^c$. There are several studies on this subject, experimental [29–31] and numerical [32]. The value of α to be used is not consensual, in the various works available in the literature α varies between 0.33 and 1.00, as so in this work it will be considered $\alpha = 0.5$, once it is a conservative value. Although it will be not considered in the analyses on this book, the bone shear ultimate stress varies between 49 MPa and 69 MPa [31].

6.3 Bone Remodelling Algorithms

Recent experimental investigations shown the existence, in the bone remodelling process, of a strong correlation between the bone functional adaptations and the induced stress (or the strain). The strain distribution, the dynamic nature of the loads and the number of loading cycles seem to be the most significant external stimuli in the bone remodelling process. However, experimental research has shown that the bone remodelling and functional adaptation are quite complex and that both cannot be readily described in detail at the present moment.

In the pursue of explaining the nature of bone remodelling phenomena, numerous semi empirical laws, phenomenological based mathematical descriptions of the remodelling processes, were developed and proposed. With these semi empirical laws it is possible to predict the actual stress distribution and stress path and correlate them with the bone remodelling process. These mathematical formulations consider the bone tissue as a local adaptive material, directly dependent on the mechanical loading, mostly characterized by the strain or stress tensors, and aim to numerically predict the local remodelling reactions observed in experiments through appropriate bone growth laws. The basis of the computational tools simulating the natural bone adaptation, occurring under known stress states and changes in geometry and stiffness, are these semi empirical bone growth laws. Next, a brief overview over bone remodelling theories proposed by several authors is presented.

6.3.1 Pauwels's Model

Pauwels [33] was one of the firsts to suggest a mathematical formulation for the "Wolff's Law". In order to ensure a balanced state of bone reabsorption and deposition, it was presumed the existence of an optimal mechanical stimulus S_n.

Pauwels was only interested in the surface remodelling of long bones, in which the stress state can be approximately described as uniaxial, thus the remodelling mechanical stimulus S_n was assumed to be the axial stress. Therefore the optimal value for S_n corresponds to an optimal axial stress value, σ_n. Stress values exceeding σ_n lead to bone hypertrophy and values below σ_n indicate bone atrophy. This material law can be used iteratively, forcing the bone stress state to move in the direction of the optimal homeostatic value, σ_n. In Pauwels law one has to consider, as bone material properties, the maximum admissible stress, σ_{max}, and the minimum admissible stress, $\sigma_{\min}$, being $\sigma_n = (\sigma_{\max} - \sigma_{\min})/2$. The actual stress must be limited between the interval: $\sigma_{\min} \leq \sigma \leq \sigma_{\max}$. Kummer [34] summarized this model in a simple cubic expression,

$$\frac{\partial m}{\partial t} = \beta(\sigma - \sigma_{\min})(\sigma - \sigma_n)(\sigma - \sigma_{\max}) \tag{6.11}$$

In this expression $\partial m / \partial t$ is the variation of mass bone through time. The coefficient β is an empiric model parameter, which must be obtained experimentally [34].

6.3.2 Cowin's Model

The bone internal remodelling process (the growth and the reabsorption of bone material) was described as the result of chemical reactions between bone matrix and the extracellular fluids in a sophisticated continuum theory developed by Cowin [35], named adaptive elasticity model of Cowin. In order to account the reorientation and the changes in the anisotropic material behaviour of the trabecular design, Cowin et al. [36] used the fabric tensor to develop an anisotropic material model for trabecular bone. The fabric tensor is a symmetric second order tensor, which describes the stereological behaviour of the microstructural arrangement of trabeculae and pores [37]. The fabric tensor is directly related with the material elasticity tensor [38]. In this model it is assumed that the state of equilibrium remodelling is achieved when the stress (and strain) principal directions are coincident with the fabric tensor principal axes. This model presents a significant disadvantage, since it is necessary to define a high number of bone remodelling parameters in order to describe the remodelling behaviour.

In the pursue to overcome this parameter identification problem it was suggested [39] that, in the case of bone surface and core remodelling, the strain energy density (SED) could serve as a mechanical stimulus. In Huiskes work [39] it was proposed a SED 'lazy-zone' in which no bone remodelling take place. This modified version of the 'Adaptive Elasticity' approach was used to predict the proximal femur density distribution [39]. The obtained results converge to a solution very close with the actual density patterns observed in the real sample femur.

This model was improved by many other authors. The overstrain necrosis, with anisotropic material behaviour, was considered along with a SED approach by

Harrigan and Hamilton [40–42]. Based in the idea that the remodelling process is not just influenced by the momentary strain state, Bucháček [43] developed an extended version of the adaptive elasticity model of Cowin where bone density and material orientation are dependent on the whole bone strain history. It was assumed that the difference between the actual and optimal density and the variation between the actual and optimal material orientation, with both depending on the strain state, serve mutually as the adaptive stimuli and have a time-fading effect on the remodelling process. In this manner, at a given time t any past strain event will have some effect on the remodelling process, however because it is considered an exponential time-fading function the bone reactions are dominated mainly by the recent strain state.

6.3.3 Carter's Model

A mathematical formulation for the functional adaptation of trabecular bone based on a self-optimization concept was proposed by Carter and co-workers [44–46]. In this model, as in Pauwels's model, it was also assumed, within the bone tissue, that a mechanical stimulus S_n must be present in order to maintain a quasi-stationary state in the bone remodelling process. Carter suggested that this stimulus should be constant for the entire bone and proportional to the effective stress.

$$S_n \propto \sum_{i=1}^{l} m_i \, \bar{\sigma}_i^{\kappa} \tag{6.12}$$

In this formulation distinct load cases can be considered, $i = 1, \ldots, l$, and each one is weighted by the corresponding number of load cycles, m_i. The influence of the magnitude of the correspondent stress state is considered by the exponent κ. The effective stress $\bar{\sigma}_i^{\kappa}(\sigma_i, \rho_{app})$ depends on the local stress state σ_i of load case i and on the local apparent density ρ_{app}. Regardless the considered bone material law or the biological basis of bone remodelling algorithm, with this model it is assumed that the functional adaptation provides the bone with the ability to maximize its structural integrity with the least amount of bone mass. Which is the same as assuming that the induced stress from bone remodelling acts as an optimization tool, minimizing an objective function [47]. Instead of using the local effective stress, it is possible to use the strain energy density (SED) approach. The use of the strain energy is connected with the idea that the bone tissue is attempting to maximize its stiffness while the use of a failure stress criterion is related with the material strength optimization. The SED approaches lead to a correlation between the apparent density ρ_{app} and the local strain energy, U.

$$\rho_{app} \propto \left(\sum_{i=1}^{l} m_i \, U_i^{\kappa} \right)^{\frac{1}{\kappa}} \tag{6.13}$$

This expression permits to estimate the apparent density as a function of the strain energy in an equilibrium state of the bone remodelling process. It can also be used as an optimization criterion in an iterative optimization procedure. The finite element method (FEM) was used, combined with this formulation, to predict the apparent density distribution in the actual bone [44]. The proximal femur was the presented example, and the FEM was used to obtain the stress and strain distributions for each distinct typical loading cases. With the proposed algorithm, starting from a homogenous density distribution, it was possible to predict density distributions similar with those found in the real femur within only a few iterations. However, the convergence could not be obtained, and the iteration process led to non-physiologic states, such as a complete bone with zero density or a complete bone with cortical density.

Carter assumed that the trabeculae are oriented in the direction of the principal stresses, respecting the trajectorial hypothesis accepted by many authors. For a single load case it was shown that an alignment between the material principal directions and the stress principal directions results in an optimal configuration with respect to the local stiffness maximization [47]. If multiple load cases are considered, a weighted combination of the normal stress components, with respect to a normal-vector $\boldsymbol{n}$, was suggested to serve as a stimulus for trabecular growth in the corresponding direction. The effective global normal stress $\bar{\sigma}_n$ is calculated in analogy with Eq. (6.13).

$$\bar{\sigma}_n(\boldsymbol{n}) = \left(\sum_{i=1}^{l} \frac{m_i}{\sum_{j=1}^{l} m_j} \bar{\sigma}_{n_i}^{\kappa}(\boldsymbol{n}) \right)^{\frac{1}{\kappa}} \tag{6.14}$$

It was suggested [46] that the material stiffness in any direction $\boldsymbol{n}$ was directly dependent on the magnitude of the corresponding global normal stress $\bar{\sigma}_n$, however no practical implementation of this trabecular orientation approach was shown.

A modified version of Carter's algorithm was proposed in research works about the adaptive growth reactions of bone following total hip joint replacement [48, 49]. In order to reduce the number of independent material properties of the orthotropic case, the bone material was assumed to be transversely isotropic. This modified version respect the trajectorial hypotheses of Wolff's law, the directions of the material axes were aligned with the stress principal direction. The femur density distribution obtained numerically, in the case of the pre-surgery state, was very close with the actual density distribution.

Pettermann et al. [49] suggested an improved version of Carter's algorithm. It is based in the assumption that the adaptation of bone tissue can be described appropriately on the continuum level by using overall tissue material parameters and stress/strain measures. The specific mechanical stimuli act as driving forces in the remodelling processes. Any material parameter actually contributing to the local bone stiffness will be subjected to a specific remodelling process which tries to adapt the effective stiffness behaviour at the particular site under consideration according to the local stresses and strains. In this approach the essential material parameters governing the elastic material behaviour of bone tissue are given by the apparent density, ρ_{app}. The structural anisotropy is described by the orthotropic parameters and the orientation of the principal material axes, with respect to the global coordinate system, are described by a rotational vector. Negative values of the stimulus lead to bone reabsorption whereas positive ones give rise to local bone hypertrophy. Bone hypertrophy in the case of internal remodelling is interpreted as an increase in bone apparent density. Each remodelling stimulus has to exceed a specific threshold level to cause any actual adaptive changes at all, which means that bone material is assumed to show a 'lazy-zone' behaviour in the vicinity of its homeostatic state.

Within Carter's model, the effects of the nonlinearity in the equation of bone remodelling was also studied [50] and developed [51]. This nonlinear development of Carter's model begins with the work of Huiskes [39, 52], Weinans et al. [53] and Mullender et al. [54]. The Carter model was widely applied with the FEM [55], with diverse variations in order to obtain the most accurate density distribution [56, 57]. An interesting approach was proposed by Chen et al. [58], in which it is presented an alternative way to obtain the density, very useful for the case of meshless methods. The density is iteratively obtained in the nodes, and not in the integration points, reducing in this manner the computational time and also the accuracy of the process. More recently the Carter's model was extended to the micro-finite element method [59, 60], with stunning results. The bone adaptation and the inner remodelling processes were studied and the results show a very high correlation with the actual density of the human proximal femur. Another conclusion was that bone surface abnormalities contribute insignificantly to the bone global structural integrity and generally tend to disappear during the remodelling process.

6.3.4 Rodrigues' Model

Rodrigues and co-workers presented a model [61] considering a global–local hierarchical approach in which a global model of an entire bone supplies strain and density information to a series of local models characterizing the trabecular microstructure at each global model location. The process of bone adaptation is described for two levels of the bone structure: the macroscopic level, where the bone apparent density is determined; and a microscopic level, characterizing the trabecular structure. The law of bone remodelling is obtained assuming that bone adapts to functional demands in order to satisfy a multicriteria for structural stiffness and metabolic cost of bone formation [62]. More recently the model was successfully extended to the three-dimensional analysis [63], being the distribution of the bone apparent density, as well as the microstructural designs characterizing both anisotropy and bone surface area density, consistent with the real bone apparent density distribution.

6.3.5 Proposed Adaptation of Carter's Model

In the present book it is considered the remodelling algorithm presented by Belinha and co-workers [27, 28], which assumes that the mechanical stimulus, suitably described by stress and/or strain measures, acts as the principal driving force in the bone tissue remodelling process. The local density and material orientation is dependent on the stress/strain field caused by the mechanical load. The presented remodelling algorithm is an adaptation of Carter's model for meshless methods.

6.3.5.1 Model Description

In this subsection a bone remodelling nonlinear equation is introduced and combined with the meshless method procedure. The expression can be presented as a differential equation, in which a temporal-spatial based functional, $\rho_{app}(\boldsymbol{x}, t)$, is minimized with respect to time,

$$\frac{\partial \rho_{app}(\boldsymbol{x}, t)}{\partial t} \cong \frac{\Delta \rho_{app}(\boldsymbol{x}, t)}{\Delta t} = (\rho_{app}^{model})_{t_j} - (\rho_{app}^{model})_{t_{j+1}} = 0 \tag{6.15}$$

Being $\rho_{app}(\boldsymbol{x}, t) : \mathbb{R}^{d+1} \mapsto \mathbb{R}$ defined for the temporal one-dimension and the spatial d-dimensions. The analysed problem must be discretized in space and time. It is assumed that the d-dimensional spatial domain is discretized in N nodes: $\boldsymbol{X} = \{\boldsymbol{x}_1, \boldsymbol{x}_2, \ldots, \boldsymbol{x}_N\} \in \Omega$, leading to Q interest points: $\boldsymbol{Q} = \{\boldsymbol{x}_1, \boldsymbol{x}_2, \ldots, \boldsymbol{x}_Q\} \in \Omega$, being $\boldsymbol{x}_i \in \mathbb{R}^d$. The temporal domain is discretized in iterative fictitious time steps $t_j \in \mathbb{R}$, with $j \in \mathbb{N}$. The medium apparent density for the complete model domain is defined by $(\rho_{app}^{model})_{t_j}$ at a fictitious time t_j. Within the same iterative step, the medium apparent density of the model, ρ_{app}^{model}, can be determined with,

$$\rho_{app}^{model} = Q^{-1} \sum_{i=1}^{Q} (\rho_{app})_i \tag{6.16}$$

being and $(\rho_{app})_I$ the infinitesimal apparent density on interest point $\boldsymbol{x}_I$ defined by $\rho_I = g(\sigma_I)$. The functional $g(\sigma_I) : \mathbb{R}^3 \mapsto \mathbb{R}$ is defined by,

$$g(\sigma_I) = \max(\{\sigma_1^{-1}(\rho_I), \sigma_2^{-1}(\rho_I), \sigma_3^{-1}(\rho_I)\}) \tag{6.17}$$

Notice that σ_j are the three principal stresses obtained for the interest point $\boldsymbol{x}_I$ and are the inverse functions of $\sigma_j^{-1}(\rho_I)$ defined in Eqs. (6.9) and (6.10). Therefore it is possible to obtain the following expression for the principal stress σ_1,

$$\sigma_1^{-1}(\rho_I) = \sigma_{axial}^{-1}(\rho_I) = \rho_{axial} = 8.14 \times 10^{-4} f_1(\sigma_1) + \frac{235.3}{f_1(\sigma_1)} - 0.439 \quad (6.18)$$

and for the remaining principal stresses σ_j with $j = \{2, 3\}$,

$$\sigma_j^{-1}(\rho_I) = \sigma_{trans}^{-1}(\rho_I) = \rho_{trans} = 1.34 \times 10^{-3} f_2(\sigma_j) + \frac{3.34 \times 10^4}{f_2(\sigma_j)} - 0.669 \quad (6.19)$$

the functions $f_1(\sigma_1) : \mathbb{C} \mapsto \mathbb{R}$ and $f_2(\sigma_j) : \mathbb{C} \mapsto \mathbb{R}$ are defined as,

$$f_1(\sigma_1) = \mathrm{Re}\left(-1.54 \times 10^8 + 4.47 \times 10^7 \sigma_1 + 2.44 \times 10^3 \sqrt{-2.31 \times 10^9 \sigma_1 + 3.35 \times 10^9 \sigma_1^2}\right) \quad (6.20)$$

$$f_2(\sigma_j) = \mathrm{Re}\left(-1.25 \times 10^{11} + 1.68 \times 10^8 \sigma_j + 1.49 \times 10^4 \sqrt{-1.88 \times 10^{11} \sigma_j + 1.26 \times 10^8 \sigma_j^2}\right) \quad (6.21)$$

The expressions presented in Eqs. (6.18) and (6.19) are applied to the interest points with SED values belonging to the following interval,

$$U(\boldsymbol{x}_I) \in [U_m, U_m + \alpha \cdot \Delta U[\cup]U_M - \beta \cdot \Delta U, U_M], \quad \forall U(\boldsymbol{x}_I) \in \mathbb{R} \quad (6.22)$$

being $U_m = \min(\boldsymbol{U})$, $U_M = \max(\boldsymbol{U})$ and $\Delta U = U_M - U_m$. It is possible to define the SED field of the problem domain by: $\boldsymbol{U} = \{U(\boldsymbol{x}_1) \quad U(\boldsymbol{x}_2) \quad \cdots \quad U(\boldsymbol{x}_Q)\}$. The parameters α and β define the growth rate and the decay rate of the apparent density. The remodelling equilibrium is achieved when,

$$\frac{\Delta\rho}{\Delta t} = 0 \quad \vee \quad (\rho_{app}^{model})_{t_j} = \rho_{app}^{control} \quad (6.23)$$

The values of parameters α and β and the value of the control apparent density $\rho_{app}^{control}$ vary with the analysed problem.

6.3.5.2 Remodelling Algorithm

In this subsection the implemented iterative remodelling process, a forward Euler scheme, is described with detail. The presented algorithm is shown in Fig. 6.9. First the NNRPIM pre-processing phase is initiated, in which the problem domain is discretized in a nodal mesh and the respective Voronoï Diagram is constructed.

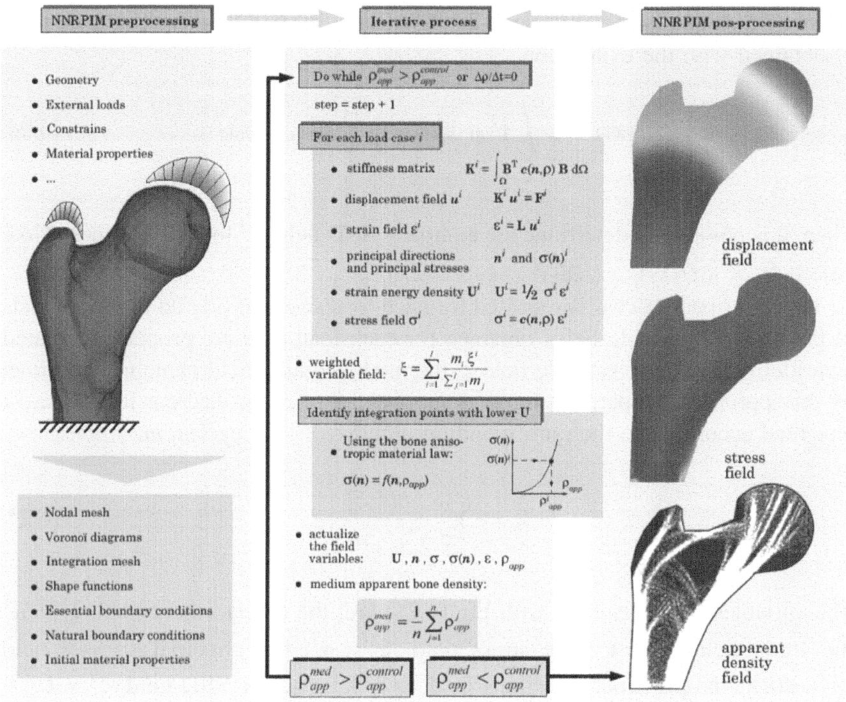

Fig. 6.9 Proposed bone remodelling algorithm

Then, it is possible to establish the nodal connectivity (the influence-cells), construct the integration mesh and determine the interpolation functions. In this early phase the essential and natural boundary conditions are enforced and the material properties are allocated to the respective domain areas. Afterwards, the iterative remodelling algorithm is initiated.

In the first step of the iterative loop a trial linear analysis of the problem is performed in order to obtain the principal directions of the stress field. In this first step no remodelling takes place, the only purpose of this first step is to align the material constitutive matrix of each interest point with the principal direction of the respective maximum principal stress obtained in each interest point. Therefore in the first step of the iterative loop, $i = 0$, it is considered an initial isotropic constitutive elastic matrix, obtained from the compliance matrix presented in Sect. 2.1.2, the local stiffness matrix for each integration point is constructed and afterwards assembled into a global stiffness matrix $\boldsymbol{K}_i$. Since the presented remodelling algorithm permits to consider simultaneously several load cases, for each load case j the correspondent displacement field is obtained with $\boldsymbol{u}_i^j = \boldsymbol{K}_i^{-1} \boldsymbol{f}^j$ and subsequently the respective strain field $\boldsymbol{\varepsilon}_i^j$ and stress field $\boldsymbol{\sigma}_i^j$ can be determined. The principal stresses, $\boldsymbol{\sigma}(\boldsymbol{n})_i^j$, and directions, $\boldsymbol{n}_i^j$, are obtained using the stress field $\boldsymbol{\sigma}_i^j$ as described in Sect. 2.1.1.5. Being the SED for each interest point $\boldsymbol{x}_I$ obtained with the expression,

$$U(\boldsymbol{x}_I) = \frac{1}{2} \int_{\Omega_I} \boldsymbol{\varepsilon}^T \boldsymbol{\sigma} \, d\Omega_I = \widehat{w}_I \left[\frac{1}{2} \boldsymbol{\sigma}(\boldsymbol{x}_I)^T \boldsymbol{\varepsilon}(\boldsymbol{x}_I) \right] \tag{6.24}$$

Then, it is possible to determine the complete SED field $\boldsymbol{U}_i^j$ for the considered load case j: $\boldsymbol{U}_i^j = \{ U(\boldsymbol{x}_1)_i^j \quad U(\boldsymbol{x}_2)_i^j \quad \cdots \quad U(\boldsymbol{x}_Q)_i^j \}$.

The described process is repeated for each load case considered in the analysis. In the end, the variable fields obtained for each load case are properly weighted. Considering a generic variable field ξ, the final weighted field value is determined by an appropriate superposition of a number of relevant discrete load cases, l, weighted according to the corresponding number of load cycles, m.

$$\xi_i = \sum_{j=1}^{l} \frac{m^{(j)} \, \xi_i^j}{\sum_{k=1}^{l} m^{(k)}} \tag{6.25}$$

The variables field weighted with Eq. (6.25) are: the displacement field, $\xi_i^j = \boldsymbol{u}_i^j$; the strain field, $\xi_i^j = \boldsymbol{\varepsilon}_i^j$; the stress field, $\xi_i^j = \boldsymbol{\sigma}_i^j$; the principal stresses field, $\xi_i^j = \boldsymbol{\sigma}(\boldsymbol{n})_i^j$; the principal directions field, $\xi_i^j = \boldsymbol{n}_i^j$; and the SED field, $\xi_i^j = \boldsymbol{U}_i^j$.

Afterwards, the interest points presenting the SED values indicated in the Eq. (6.22) are identified and subjected to a density remodelling process, all the other interest points maintain the previous density. The weighted principal stress field of the interest points presenting lower SED values is determined using Eq. (6.25), then the individual new apparent density of these interest points is obtained with Eq. (6.17). With the new apparent density defined, the process moves forward the next iteration step.

Using the new apparent density field, in each interest point the material properties are actualized with Eqs. (6.7) and (6.8). Then, the constitutive elasticity matrix, defined for each interest point, is oriented using the principal directions obtained in the previous iteration step. Thus, the constitutive matrix is obtained with (2.45). This procedure permits to align iteratively the material properties with the actualized load path. The process stops when the medium apparent density of the model, ρ_{app}^{model}, reaches a control value, $\rho_{app}^{control}$, or if two consecutive iteration steps present the same medium apparent density, $\Delta\rho/\Delta t = 0$. The control value can be determined by the user, based on clinical observations.

Notice that the presented remodelling algorithm is in fact a topology optimization based model for bone adaptation. In each step not all the interest points optimize the density in relation to the mechanical stimulus. Although for every interest point the material is oriented with the principal directions obtained in each iteration step, the density on each interest point is not. Only the interest points with lower SED or higher SED, Eq. (6.22), are subject to the density remodelling

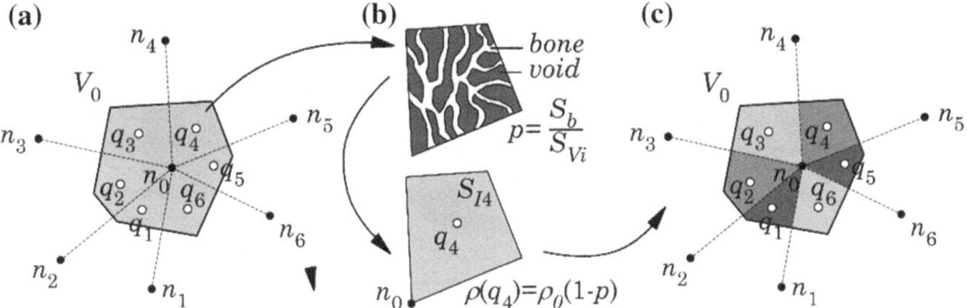

Fig. 6.10 **a** Voronoï cell with the quadrature points. **b** Theoretical trabecular architecture of the sub-cell and homogenized apparent density. **c** Voronoï cell with the integration points homogenized apparent densities

process, all the others maintain the previous density. With this approach the material properties orientation is continuously optimized and only a small fraction of bone material have its density actualized each time.

The presented iterative process follows the forward Euler scheme with some particular adaptations to suit the bone internal remodelling analysis. The inclusion of the NNRPIM meshless method in the remodelling analysis is an asset and not just another way to obtain the stress and the strain field, since the accuracy of the remodelling algorithm depends on the accuracy of the used numerical method.

6.3.5.3 Interpretation of the NNRPIM Results

Since all the bone tissue analyses presented in this book are performed using only the NNRPIM, first the analysed problem domain, $\Omega \subset \mathbb{R}^d$, is discretized by a nodal distribution, $\boldsymbol{X} = \{\boldsymbol{x}_1, \boldsymbol{x}_2, \ldots, \boldsymbol{x}_N\} \in \Omega$ with $\boldsymbol{x}_i \in \mathbb{R}^d$ and then the Voronoï diagram, $\boldsymbol{V} = \{V_1, V_2, \ldots, V_N\}$, is obtained, being $\Omega = \bigcup_{i=1}^{N} V_i$. Using the Voronoï diagram the integration points are determined, $\boldsymbol{Q} = \{\boldsymbol{q}_1, \boldsymbol{q}_2, \ldots, \boldsymbol{q}_Q\} \in \Omega$ with $\boldsymbol{q}_j \in \mathbb{R}^d$. Recall that the integrations points are sequentially obtained for each Voronoï cell V_i, therefore each Voronoï cell V_i produces k integration points, being: $\{\boldsymbol{q}_1, \boldsymbol{q}_2, \ldots, \boldsymbol{q}_k\} \in \boldsymbol{Q}$ and $\{\boldsymbol{q}_1, \boldsymbol{q}_2, \ldots, \boldsymbol{q}_k\} \subset V_i$, Fig. 6.10a.

In the end of each iteration step the local apparent density of each integration point $\boldsymbol{q}_j$ is obtained, $\rho(\boldsymbol{q}_j)_{app}$. The infinitesimal subdivisions of each Voronoï cell (the sub-cells) are the smallest dimensional partition of the domain. Each Voronoï cell infinitesimal subdivision, S_{V_i}, is numerically represented by the respective integration point. Therefore, it is not possible to obtain the detailed microscale trabecular arrangement represented in Fig. 6.10b, in which the bone volume, S_b, and the void volume, S_V, are clearly defined. For the infinitesimal subdivision area represented by the integration point it is only possible to obtain the volume porosity, $p(\boldsymbol{q}_j)$, and then the local apparent density $\rho(\boldsymbol{q}_j)_{app} = \rho_0(1 - p(\boldsymbol{q}_j))$, being $\rho_0 = 2.1\ \text{g/cm}^3$ the compact bone density, Fig. 6.10b.

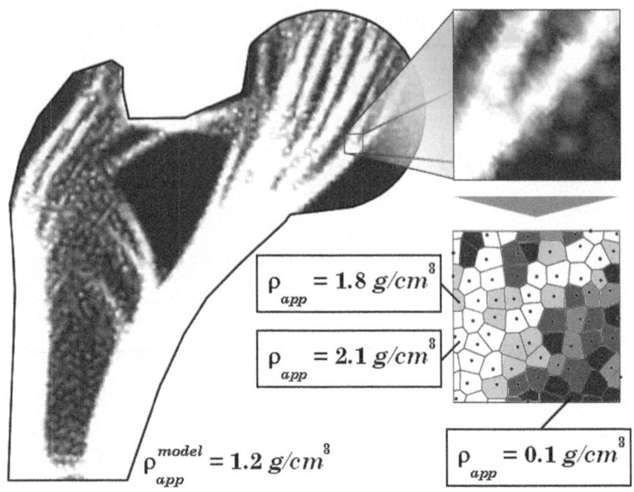

Fig. 6.11 Isomap representing the trabecular architecture of the femoral bone

As it is understandable, decreasing the size of the infinitesimal subdivisions permits to increase the detail of the analysis, however it will increase also the computational cost of the analysis. As represented in Fig. 6.10c each integration point will probably present a distinct local apparent densitiy.

Similarly it is possible to obtain the local apparent density of each field node $\boldsymbol{x}_i$, which can be determined using the following expression,

$$\rho(\boldsymbol{x}_i)_{app} = \frac{\sum_{j=1}^{k} \widehat{w}_j \cdot \rho(\boldsymbol{q}_j)_{app}}{\sum_{j=1}^{k} \widehat{w}_j} \tag{6.26}$$

where $\widehat{w}_j$ is the integration weight of an integration point $\boldsymbol{q}_j$ belonging to the Voronoï cell V_i of the field node $\boldsymbol{x}_i$. With Eq. (6.26) the local apparent density field can be defined, which can be represented in isomaps. Therefore, the results regarding the evolution of the trabecular architecture, in all numerical examples studied in this book, are presented as grey tone isomaps. In those grey tone isomaps the white colour represents the considered maximum apparent density $\rho_0 = 2.1\,\text{g/cm}^3$ and the dark-grey colour represents the minimum apparent density $\rho_0 = 0.1\,\text{g/cm}^3$ admitted in the analysis. All the other gray tones in the middle represent transitional apparent densities. In each isomap presented in this book it is also indicated the domain medium apparent density, which is obtained applying Eq. (6.16).

In Fig. 6.11 it presented an isomap example. This isomap was obtained with the proposed remodelling algorithm combined with the NNRPIM [27, 28]. The analysed femur domain was discretized in 5,991 nodes and the three load cases suggested in the literature were considered [64, 65]. Apparently the result obtained for a medium apparent density $\rho_{app}^{model} = 1.2\,\mathrm{g/cm^3}$ indicates a well-defined trabecular arrangement. However as it is perceptible the presented trabecular arrangement is dependent on the domain discretization. If more nodes were used in the analysis it will be possible to obtain a more accurate trabecular architecture.

References

1. Netter FH (1994) Musculoskeletal system—Netter collection of medical illustrations, 1st edn, vol 8. Saunders, USA
2. Wolff J (1986) The law of bone remodeling (Das Gesetzder Transformationder Knochen, Hirschwald, 1892). Springer, Berlin, New York
3. Aubin JE, Heersche JNM (2000) Osteoprogenitor cell differentiation to mature bone-forming osteoblasts. Drug Dev Res 49(3):206–215
4. D'Ippolito G, Schiller PC, Ricordi C, Roos BA, Howard GA (1999) Age-related osteogenic potential of mesenchymal stromal stem cells from human vertebral bone marrow. J Bone Miner Res 14(7):1115–1122
5. Miller SC and Jee WSS (1992) Bone lining cells. In Bone. CRC Press, Boca Raton, pp 1–19
6. Lanyon LE (1993) Osteocytes, strain detection, bone modelling and remodelling. Calcif Tissue Int 53:102–107
7. Cowin SC, Moss-Salentijin L, Moss ML (1991) Candidates for the mechanosensory system in bone. J Biomech Eng 113(2):191–197
8. Lee TC, Staines A, Taylor D (2002) Bone adaptation to load: microdamage as a stimulus for bone remodelling. J Anat 201(6):437–446
9. Taylor D, Kuiper JH (2001) The prediction of stress fractures using a 'stressed volume' concept. J Orthop Res 19(5):919–926
10. Prendergast PJ, Taylor D (1994) Prediction of bone adaptation using damage accumulation. J Biomech 27(8):1067–1076
11. Carter DR, Hayes WC (1976) Bone compressive strength: the influence of density and strain rate. Science 194(4270):1174–1176
12. Carter DR, Hayes WC (1977) The compressive behaviour of bone as a two-phase porous structure. J Bone Joint Surg 59(A):954–962
13. Carter DR, Spengler DM (1978) Mechanical properties and composition of cortical bone. Clin Orthop Relat Res 135:192–217
14. Gibson LJ (1985) The mechanical behaviour of cancellous bone. J Biomech 18:317–328
15. Goldstein SA (1987) The mechanical properties of trabecular bone: dependence on anatomic location and function. J Biomech 20:1055–1061
16. Rice JC, Cowin SC, Bowman JA (1988) On the dependence of the elasticity and strength of cancellous bone on apparent density. J Biomech 21(2):155–168
17. Martin RB (1991) Determinants of the mechanical properties of bones. J Biomech 24(1):79–88
18. Pietruszczak S, Jiang J, Mirza FA (1988) An elastoplastic constitutive model for concrete. Int J Solids Struct 24(7):705–722
19. Currey JD (1969) The mechanical consequences of variation in the mineral content of bone. J Biomech 2(1):1–11
20. Currey JD (1988) The effect of porosity and mineral content on the young s modulus of elasticity of compact bone. J Biomech 21:131–139
21. Schaffler MB, Burr DB (1988) Stiffness of compact bone: effects of porosity and density. J Biomech 21:13–16
22. Keller TS (1994) Predicting the compressive mechanical behaviour of bone. J Biomech 27:1159–1168

23. Keyak JH, Lee I, Skinner HB (1994) Correlations between orthogonal mechanical properties and density of trabecular bone: use of different densitometric measures. J Biomed Mater Res 28(11):1329–1336
24. Hernández CJ, Beaupré GS, Keller TS, Carter DR (2001) The influence of bone volume fraction and ash fraction on bone strength and modulus. Bone 29(1):74–78
25. Lotz JC, Gerhart TN, Hayes WC (1991) Mechanical properties of metaphyseal bone in the proximal femur. J Biomech 24(5):317–329
26. Zioupos P, Cook RB, Hutchinsonc JR (2008) Some basic relationships between density values in cancellous and cortical bone. J Biomech 41:1961–1968
27. Belinha J, Jorge RMN, Dinis LMJS (2013) A meshless microscale bone tissue trabecular remodelling analysis considering a new anisotropic bone tissue material law. Comput Methods Biomech Biomed Eng 16(11):1170–1184
28. Belinha J, Jorge RMN, Dinis LMJS (2012) Bone tissue remodelling analysis considering a radial point interpolator meshless method. Eng Anal Boundary Elem 36(11):1660–1670
29. Keaveny TM, Wachtel EF, Ford CM, Hayes WC (1994) Differences between the tensile and compressive strengths of bovine tibial trabecular bone depend on modulus. J Biomech 27(9):1137–1146
30. Stone JL, Beaupre GS, Hayes WC (1983) Multiaxial strength characteristics of trabecular bone. J Biomech 16(9):743–752
31. Reilly DT, Burstein AH (1975) The elastic and ultimate properties of compact bone tissue. J Biomech 8(6):393–405
32. Keyak JH, Rossi SA (2000) Prediction of femoral fracture load using finite element models: an examination of stress- and strain-based failure theories. J Biomech 33(2):209–214
33. Pauwels F (1956) Gesammelte Abhandlungen zur Funktionellen Anatomie des Bewegungsapparates. Springer, Berlin
34. Kummer BKF (1972) Biomechanics of bone: mechanical properties, functional structure, functional adaptation. In: Fung YC, Perrone N, Anliker M (eds) Biomechanics its foundation and objectives. Englewood Cliffs, Prentice-Hall, pp 237–271
35. Cowin SC, Hegedus DH (1976) Bone remodeling I: a theory of adaptive elasticity. J Elast 6:313–326
36. Cowin SC, Sadegh AM, Luo GM (1992) An evolutionary wolff's law for trabecular architecture. J Biomech Eng 114(1):129–136
37. Turner CH, Cowin SC, Rho JY, Ashman RB, Rice JC (1990) The fabric dependence of the orthotropic elastic constants of cancellous bone. J Biomech 23(6):549–561
38. Cowin SC (1985) The relationship between the elasticity tensor and the fabric tensor. Mech Mater 4(2):137–147
39. Huiskes R, Weinans H, Grootenboer HJ, Dalstra M, Fudala B, Slooff TJ (1987) Adaptive bone-remodelling theory applied to prosthetic-design analysis. J Biomech 20(11–12):1135–1150
40. Harrigan TP, Hamilton JJ (1992) An analytical and numerical study of the stability of bone remodelling theories: dependence on microstructural stimulus. J Biomech 25(5):477–488
41. Harrigan TP, Hamilton JJ (1992) Optimality conditions for finite element simulation of adaptive bone remodelling. Int J Solids Struct 29(3):2897–2906
42. Harrigan TP, Hamilton JJ (1993) Finite element simulation of adaptive bone remodelling: a stability criterion and a time stepping method. Int J Numer Meth Eng 36(5):837–854
43. Buchácek K (1990) Non equilibrium bone remodelling: changes of mass density and of the axes of anisotropy. Int J Eng Sci 28(10):1039–1044
44. Carter DR, Fyhrie DP, Whalen RT (1987) Trabecular bone density and loading history: regulation of connective tissue biology by mechanical energy. J Biomech 20(8):785–794
45. Whalen RT, Carter DR, Steele CR (1988) Influence of physical activity on the regulation of bone density. J Biomech 21(10):825–837
46. Carter DR, Orr TE, Fyhrie DP (1989) Relationship between loading history and femoral cancellous bone architecture. J Biomech 22(3):231–244

47. Fyhrie DP, Carter D (1986) A unifying principle relating stress to trabecular bone morphology. J Orthop Res 4:304–317
48. Starke GR, Mercer CD, Spirakis A, Martin JB, Learmonth ID (1992) Some aspects of adaptive bone growth as a result of total hip joint replacement. In ABACUS user's conference, Providence
49. Pettermann H, Reiter T, Rammerstorfer FG (1997) Computational simulation of internal bone remodeling. Arch Comput Methods Eng 4(4):295–323
50. Xinghua Z, He G, Dong Z, Bingzhao G (2002) A study of the effect of non-linearities in the equation of bone remodeling. J Biomech 35:951–960
51. He G, Xinghua Z (2006) The numerical simulation of osteophyte formation on the edge of the vertebral body using quantitative bone remodelling theory. Jt Bone Spine 73:95–101
52. Huiskes R, Weinans H, Dalstra M (1989) Adaptive bone remodeling and biomechanical design considerations for noncemented total hip arthroplasty. Orthopedics 12:1255–1267
53. Weinans H, Huiskes R, Grootenboer HJ (1992) The behaviour of adaptive bone-remodelling simulation models. J Biomech 25(12):1425–1441
54. Mullender MG, Huiskes R, Weinans H (1994) A physiological approach to the simulation of bone remodeling as a self organizational control process. J Biomech 27(11):1389–1394
55. Prendergast PJ (1997) Finite element models in tissue mechanics and orthopaedic implant design. Clin Biomech 12(6):343–366
56. Jacobs CR, Levenston ME, Beaupre GS, Simo JC, Carter DR (1995) Numerical instabilities in bone remodelling simulations: the advantages of a node-based finite element approach. J Biomech 28(4):449–459
57. Jacobs CR, Simo JC, Beaupré GS, Carter DR (1997) Adaptive bone remodeling incorporating simultaneous density and anisotropy considerations. J Biomech 30(6):603–613
58. Chen G, Pettet G, Pearcy M, McElwain DLS (2007) Comparison of two numerical approaches for bone remodelling. Med Eng Phys 29:134–139
59. Jang IG, Kim IY (2010) Computational simulation of simultaneous cortical and trabecular bone change in human proximal femur during bone remodeling. J Biomech 43:294–301
60. Jang IG, Kim IY (2010) Application of design space optimization to bone remodeling simulation of trabecular architecture in human proximal femur for higher computational efficiency. Finite Elem Anal Des 46(4):311–319
61. Rodrigues H, Jacobs C, Guedes J, Bendsøe M (1999) Global and local material optimization applied to anisotropic bone adaptation. In: Perdersen P, Bendsøe M (eds) Synthesis in bio solid mechanics. Kluwer Academic Publishers, Berlin, pp 209–220
62. Coelho PG, Fernandes P, Guedes J, Rodrigues H (2008) A hierarchical model for concurrent material and topology optimization of three-dimensional structures. Struct Multi Optim 35:107–115
63. Coelho PG, Fernandes PR, Rodrigues HC, Cardoso JB, Guedes JM (2009) Numerical modeling of bone tissue adaptation—a hierarchical approach for bone apparent density and trabecular structure. J Biomech 42(7):830–837
64. Beaupré GS, Orr TE, Carter DR (1990) An approach for time dependent bone modelling and remodelling theoretical development. J Orthop Res 8(5):651–661
65. Beaupré GS, Orr TE, Carter DR (1990) An approach for time dependent bone modelling and remodelling. A preliminary remodelling simulation. J Orthop Res 8(5):662–670

Chapter 7
Bone Tissue Remodelling Analysis

Abstract In this chapter the proposed bone tissue remodelling algorithm using the NNRPIM is applied to several problems. First a two-dimensional benchmark example is used to validate the bone trabecular remodelling. In this example distinct material laws are studied as well as the influence of the model nodal discretization and the anisotropy of the biomaterial. Next, the study is extended to the three-dimensional analysis, where a test problem based in another benchmark example is presented. Afterwards, it is numerically simulated the bone tissue remodelling occurring in natural bones. Thus, the calcaneus bone is simulated using a two-dimensional approach, for this example the obtained trabecular bone architecture is in very good agreement with the one that can be found in calcaneus bone X-ray images. The same quality results were found in the two-dimensional approach of the femur example. Additionally, a three-dimensional analysis of the femur is presented. Ending this section, it is studied a two-dimensional model of the maxillary central incisor constructed using the available data in clinical literature. The complete elasto-static analysis of the incisor/maxillary structure, using the NNRPIM, is evaluated and then the nonlinear iterative local NNRPIM analysis of the maxillary bone tissue remodelling is performed. The last section of the present chapter shows the bone tissue remodelling due to the insertion of implants. First it is studied the bone tissue remodelling process of the premolar region of the mandible due to the inclusion of an implant system. Then, the bone tissue remodelling response to the insertion of a femoral implant, after an idealized subcapital or transcervical neck fracture, is studied.

7.1 Bone Patch Analysis

In this section the proposed bone remodelling algorithm, using the NNRPIM, is applied to two-dimensional and three-dimensional bone micro-patches. The purpose is to validate the biomechanical numerical model, comparing the obtain

J. Belinha, *Meshless Methods in Biomechanics*, Lecture Notes in Computational Vision and Biomechanics 16, DOI: 10.1007/978-3-319-06400-0_7,

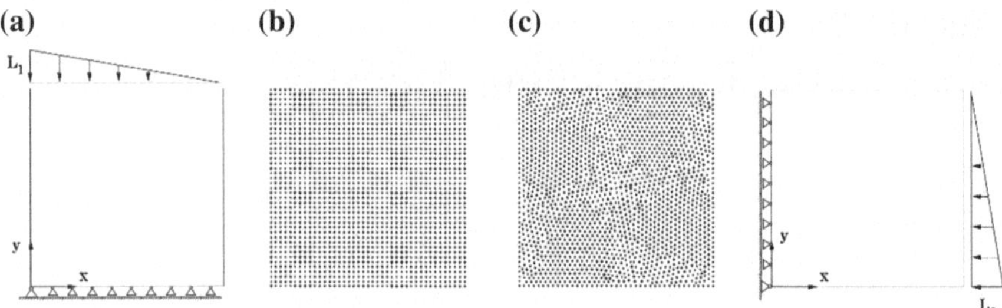

Fig. 7.1 **a** Plate model geometry and essential and natural boundary conditions. **b** Regular nodal distribution (1,681 nodes). **c** Irregular nodal distribution (1,952 nodes). **d** Second load case considered in the analysis [1]

solutions with available solutions in literature. Both the Lotz material law and the material law proposed in this book [1, 2] are studied and compared.

7.1.1 2D Bone Patch

In this example the unit square two-dimensional patch indicated in Fig. 7.1a is studied. This benchmark example [3–5] is used to validate the bone trabecular remodelling algorithms. Two types of nodal discretizations were used, a regular mesh (RM) and an irregular mesh (IM), both indicated in Fig. 7.1b, c. The two-dimensional patch is subjected to a compressive stress distribution, decreasing linearly over the top edge. The node displacement is constrained in the y direction along the line $y = 0$ and on the axis' origin is constrained in both x and y directions. The natural and essential boundary conditions are presented in Fig. 7.1a. For all studied patch examples, an uniform initial density distribution $\rho_{app}^{max} = 2.1\ \text{g/cm}^2$ is assumed, with a Poisson ratio $\upsilon = 0.3$, regardless the material direction, and a control medium apparent density $\rho_{app}^{control} = 0.4\ \text{g/cm}^2$.

Firstly it was studied the influence of the parameters α and β Eq. (6.22). These parameters define respectively the growth rate and the decay rate of the bone tissue apparent density. Thus, using the regular discretization indicated in Fig. 7.1b, and considering the bone tissue as an isotropic material, the bone tissue square patch was analysed. The material law used to obtain the material properties was the one proposed in Sect. 6.2.2. However, in order to respect the isotropic material assumption, the material properties in the transverse direction assume the correspondent axial values.

The square patch was analysed considering four distinct α parameters: $\alpha = \{0.1, 0.05, 0.02, 0.01\}$. The β parameter was assumed as $\beta = \alpha$. In Fig. 7.2 it is possible to observe the evolution of bone tissue remodelling process for each one of the considered α and β parameters values.

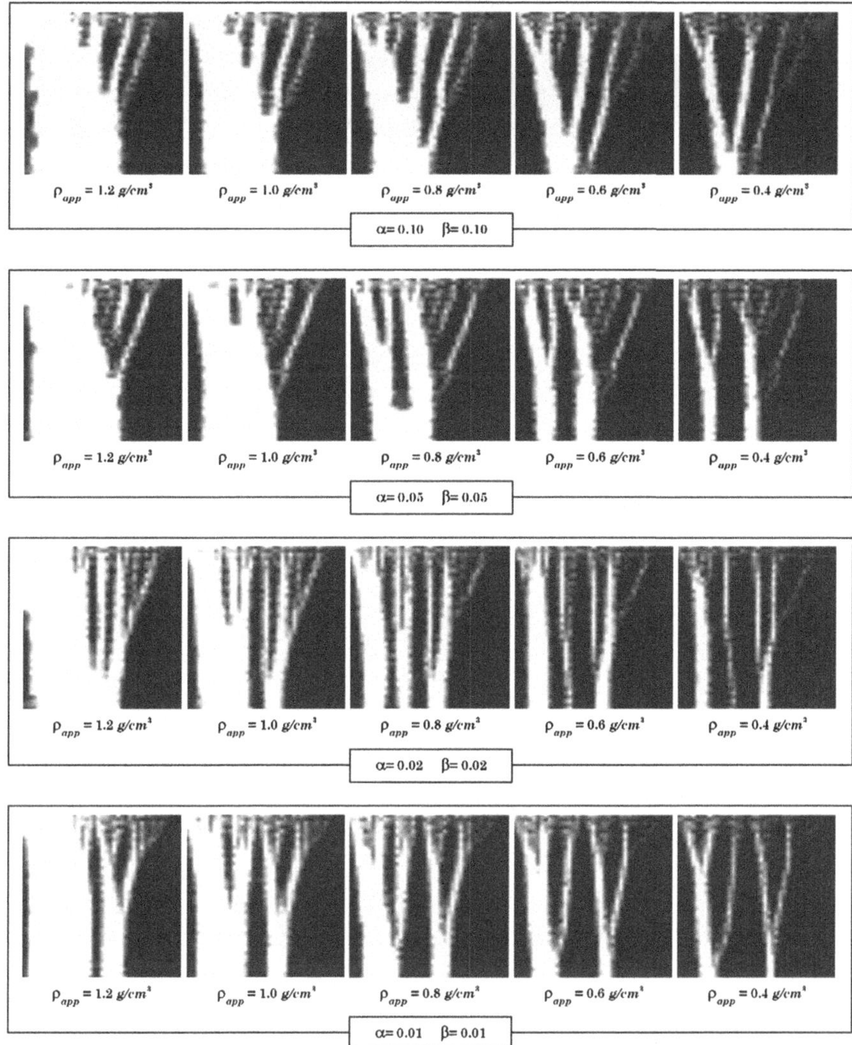

Fig. 7.2 Evolution of the medium apparent density in the square bone patch (proposed material law, regular nodal distribution and isotropic material properties) [1]

The results obtained with $\alpha = \beta = 0.01$, Fig. 7.2, are very similar with other numerical solutions available in the literature, also obtained using isotropic materials [3–5]. Nevertheless, the results obtained assuming $\alpha = \beta = 0.10$ are not very different from the expected trabecular architecture for the considered load case. Therefore, in further studies the same four distinct α parameters will be considered until a consistent conclusion can be obtained.

Next, the influence of the material anisotropy is analysed. Thus, in this study the bone tissue is considered as an anisotropic material. The material law proposed in

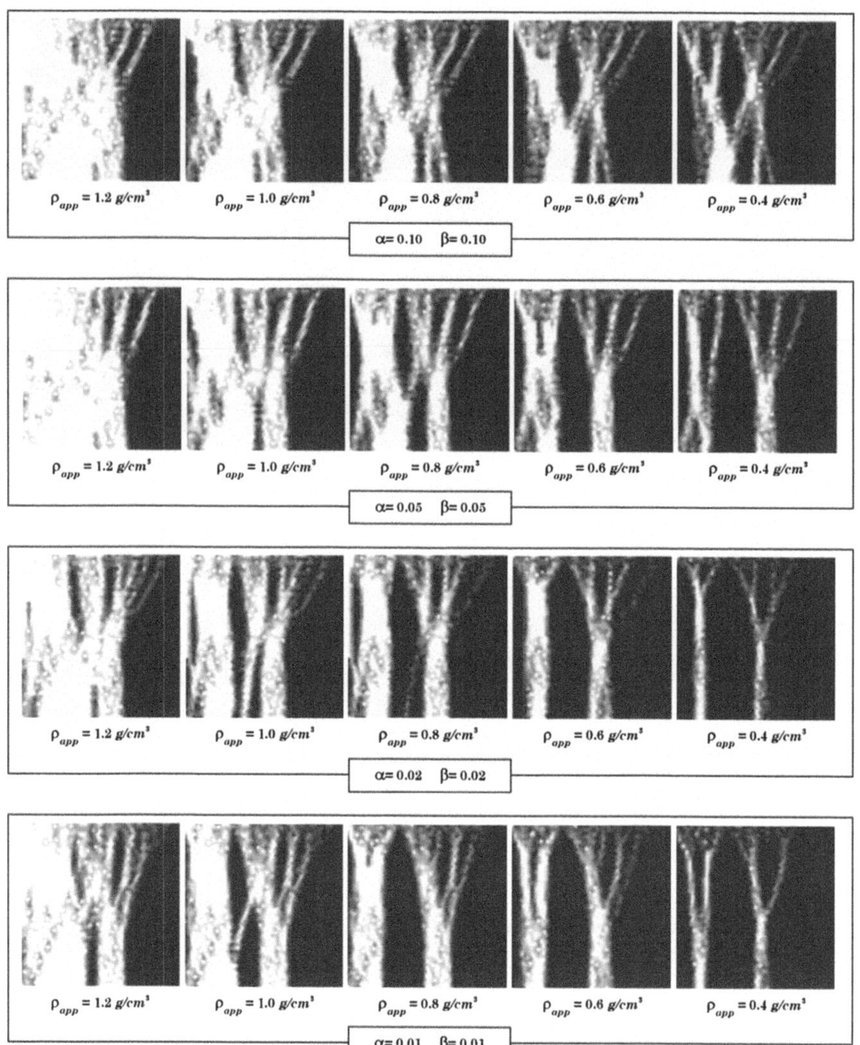

Fig. 7.3 Evolution of the medium apparent density in the square bone patch (proposed material law, regular nodal distribution and anisotropic material properties) [1]

Sect. 6.2.2 is fully tested, being the axial and the transverse directions distinctly considered. The same four distinct α parameters: $\alpha = \{0.1, 0.05, 0.02, 0.01\}$ are assumed and $\beta = \alpha$. In this study two distinct nodal distributions are considered in the analysis, a uniform nodal distribution, Fig. 7.1b, and an irregular nodal distribution Fig. 7.1c. The evolution of bone tissue trabecular architecture for each set of parameters is presented in Fig. 7.3 for the regular nodal distribution and for the irregular nodal distribution the evolution of trabecular architecture is presented in Fig. 7.4.

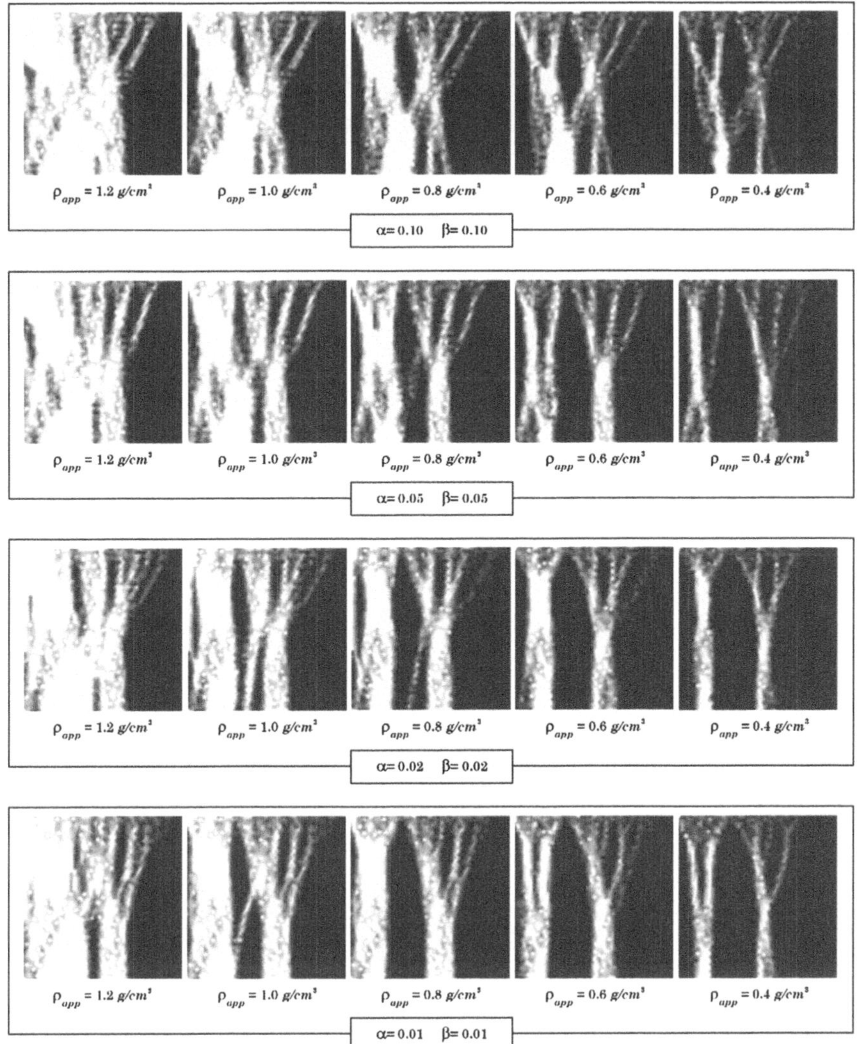

Fig. 7.4 Evolution of the medium apparent density in the square bone patch (proposed material law, irregular nodal distribution and anisotropic material properties) [1]

The results show that the obtained trabecular distribution is different from the achieved trabecular architecture when an isotropic material is considered. It is possible to observe that the obtained solution does not depend strongly on the used nodal distribution, since the results presented in Figs. 7.3, 7.4 are very similar. This observation shows that the nodal distribution discretizing the problem domain does not influence significantly the bone remodelling process.

In order to validate the material law proposed in this book, the same problem was analysed using the Lotz material law considering the bone as an anisotropic

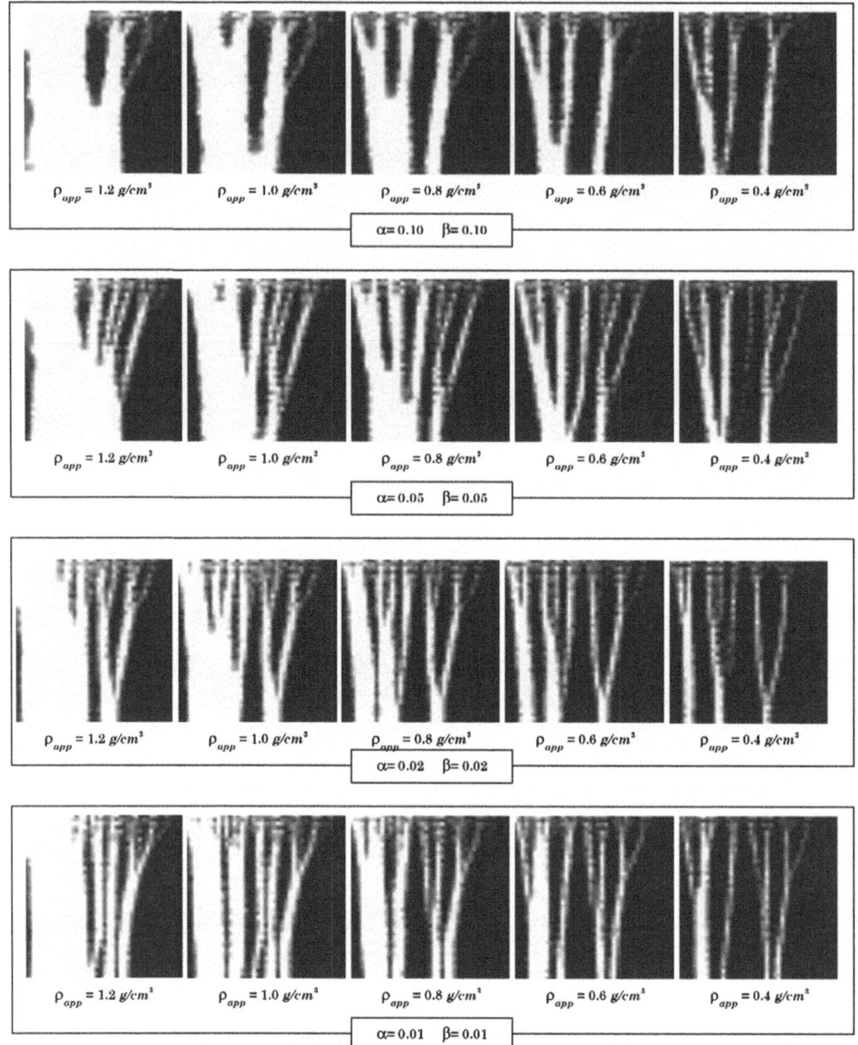

Fig. 7.5 Evolution of the medium apparent density in the square bone patch (Lotz material law, regular nodal distribution and anisotropic material properties) [1]

material, Sect. 6.2.1. The two distinct nodal distributions presented in Fig. 7.1b, c are considered and the same remodelling criterion expressed by Eq. 6.22 is assumed. Therefore, the same four distinct α parameters: $\alpha = \{0.1, 0.05, 0.02, 0.01\}$ are considered, being $\beta = \alpha$. The results obtained with the uniform nodal distribution are presented in Fig. 7.5. In Fig. 7.6 the results regarding the irregular nodal distribution are presented.

Using the material law of Lotz it is visible a slight difference between the results obtained considering the regular nodal distribution and the solution

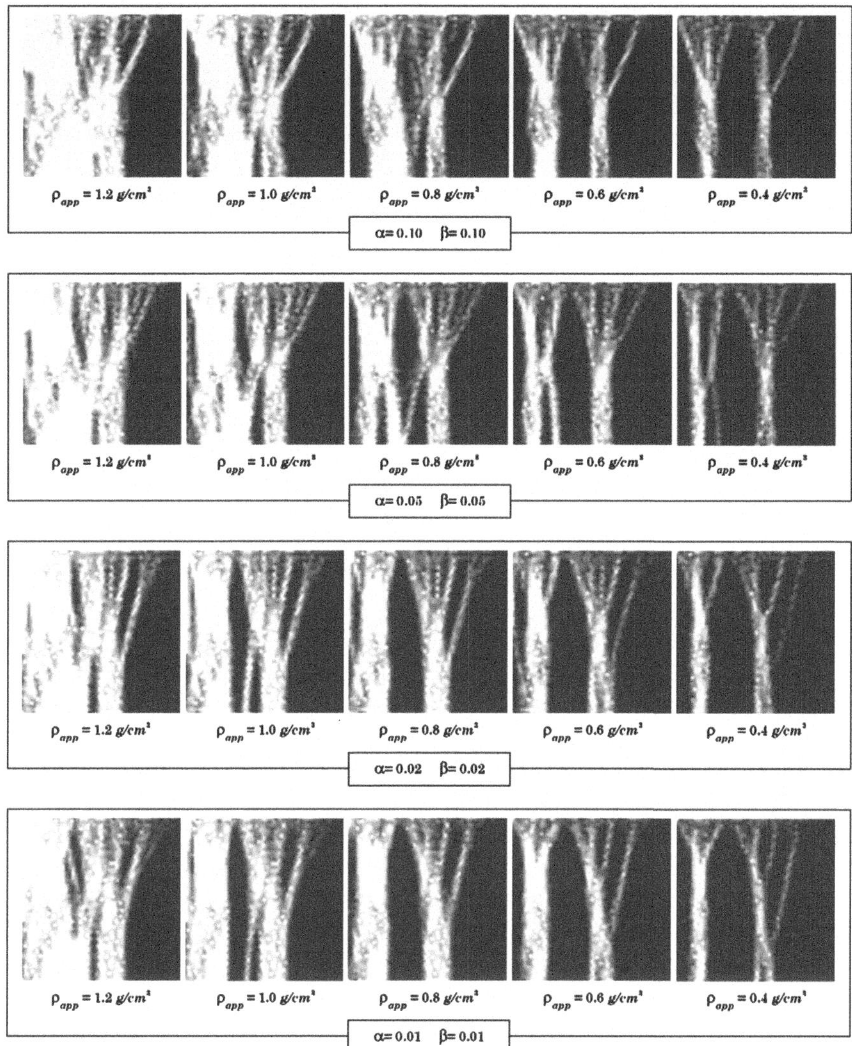

Fig. 7.6 Evolution of the medium apparent density in the square bone patch (Lotz material law, irregular nodal distribution and anisotropic material properties) [1]

obtained with the irregular nodal distribution. Nevertheless, the trabecular architecture obtained with the irregular nodal distribution considering the Lotz material law is very similar with the trabecular arrangement obtained with the proposed material law. The phenomenological curves proposed by Lotz describe separately the material properties of the cortical bone tissue and the trabecular bone tissue, Sect. 6.2.1. Whenever an interest point shows an apparent density $\rho_{app} > 1.3\,\mathrm{g/cm^3}$, in the analysis it is considered the cortical material curve,

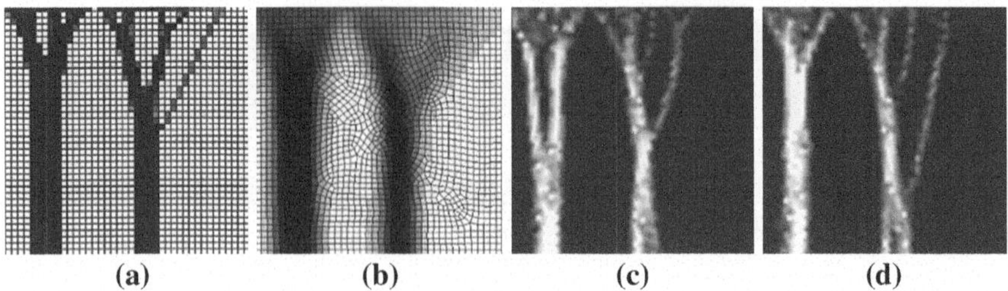

Fig. 7.7 Evolution of the trabecular architecture in the square bone patch. **a** FEM analysis [4]. **b** FEM analysis [5]. **c** Meshless analysis using the proposed material law. **d** Meshless analysis using Lotz material law [1]

however if the same interest point shows $\rho_{app} > 1.3\,\mathrm{g/cm^3}$ the trabecular material curve is assumed. It is not a continuous process, as it is in the proposed material law. In Lotz material law when an interest point leaves the cortical curves and passes to the trabecular curve a hug leap in the material laws occurs (notice, for the Lotz material law curves shown in Figs. 6.5–6.8, the difference between the material properties of the cortical bone and the trabecular bone). This maybe in the origin of the difference between the results obtained with the regular and the irregular nodal distributions.

In Fig. 7.7 the results obtained with the NNRPIM are compared with the results obtained with the FEM [4, 5]. Being this example a benchmark problem, the nodal distribution used in both FEM studies respects the same nodal density used in the meshless analysis. It is visible that the FEM results available in the literature [4, 5], Fig. 7.7a, b, resemble the results of a simple topological structural problem, in opposition the results obtained with the meshless method, Fig. 7.7c, d, look like the real trabecular architecture which can be seen in X-ray plates. The results of the meshless method are obtained considering $\alpha = \beta = 0.01$. As it is visible in Figs. 7.4, 7.6, both $\alpha = \beta = 0.01$ and $\alpha = \beta = 0.02$ lead to good results when compared with the FEM solution, Fig. 7.7a, b. Therefore in the following studies presented in this book the parameters α and β are assumed as: $\alpha = \beta = 0.01$.

In order to validate the remodelling algorithm when distinct mechanical load cases are consider, the bone patch was subjected to two individual loads; a load case $L1$, Fig. 7.1a, and a load case $L2$, Fig. 7.1d. In both cases the applied load has the same magnitude. In this study only the proposed anisotropic material law was considered and the problem was analysed with the same nodal distributions presented in Fig. 7.1b, c.

In a first approach, both loads are applied with the same number of cycles. Therefore, load case $L1$ was applied with 1,000 cycles and load case $L2$ was also applied with 1,000 cycles. The results are shown in Fig. 7.8. As it was expected the trabecular remodelling resembles in both directions (x and y).

A final test regarding the square bone patch example was conducted. The two load cases already referred were now applied with the following condition: load case $L1$ was applied with 1,000 cycles and load case $L2$ with 5,000 cycles. All the

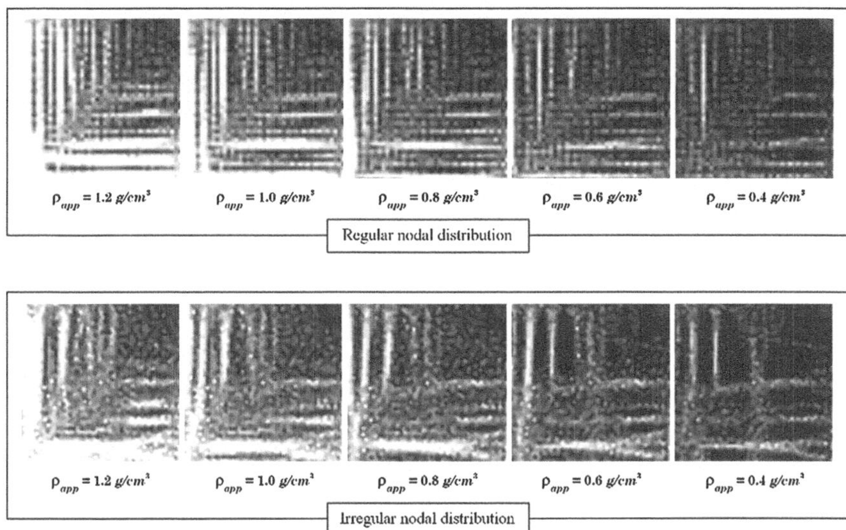

Fig. 7.8 Evolution of the apparent density in the square bone patch considering the two load cases: L1 with 1,000 cycles and L2 with 1,000 cycles [1]

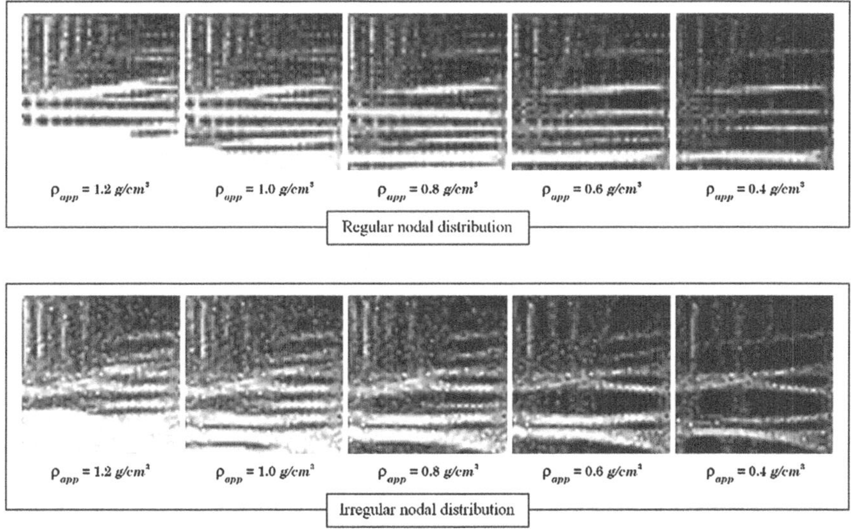

Fig. 7.9 Evolution of the apparent density in the square bone patch considering the two load cases: L1 with 1,000 cycles and L2 with 5,000 cycles [1]

other variables remain the same as in previous study. The results are shown in Fig. 7.9. It is clear the trabeculae structural design preference. The trabeculae developed towards the higher load, in the x direction.

The results of this benchmark example have shown the ability of the proposed remodelling algorithm, combined with the proposed new material law and the NNRPIM accuracy, to predict the principal and secondary trabecular structures for the two-dimensional analysis.

7.1.2 3D Bone Patch

It is possible to find in the literature three-dimensional benchmark examples developed to validate bone tissue remodelling algorithms [6]. Generally, those benchmark examples consist on cubic bone patches submitted to localized loads, which originate the formation of well-known trabeculae structures. In this book similar hexahedron bone patches are studied. Firstly consider the 3D patch presented in Fig. 7.10a, with a volume $2 \times 1 \times 2\,\text{mm}^3$. A surface load $F = 1.0\,\text{N/mm}^2$, with the direction indicated in Fig. 7.10a, is applied in two square areas on the top of the hexahedron patch. On the patch bottom another two square areas locally constrain the patch movement in all directions. The problem is analysed considering the regular mesh, with 2,681 nodes, presented in Fig. 7.10c. In order to present the results the cubic patch is sectioned by the section presented in Fig. 7.10d.

In this example only the proposed material law is considered in the bone remodelling algorithm. For all studied examples, an uniformly initial density distribution $\rho_{app}^{max} = 2.1\,\text{g/cm}^3$ is assumed for the 3D patch, with a Poisson ratio $\upsilon = 0.3$, regardless the material direction. In the remodelling algorithm it is assumed $\alpha = \beta = 0.01$ and a control medium apparent density $\rho_{app}^{control} = 0.4\,\text{g/cm}^3$.

In Fig. 7.11 is presented the evolution of the trabecular bone remodelling process until the control apparent density $\rho_{app}^{control} = 0.4\,\text{g/cm}^3$ is reached. As expected, the applied loads lead the bone to build vertical trabeculae.

In order to verify the influence of the nodal discretization, a single diagonal load is considered, as Fig. 7.10b illustrates. This example is capable to analyse the influence of the nodal discretization because now, in opposition with the previous example, the load path is unable to travel from the application point to the constrain point following a trivial linear string of field nodes.

The same essential boundary conditions and material properties are assumed. Again, for the remodelling algorithm, it is assumed $\alpha = \beta = 0.01$ and a control medium apparent density $\rho_{app}^{control} = 0.4\,\text{g/cm}^3$. The obtained results are presented in Fig. 7.12. It is visible that evolution of the trabecular bone remodelling process leads to a single diagonal trabecula.

In Fig. 7.13 are presented the three-dimensional section views for both analyses when the apparent medium density $\rho_{app} = 0.4\,\text{g/cm}^3$ is achieved. Notice that, as

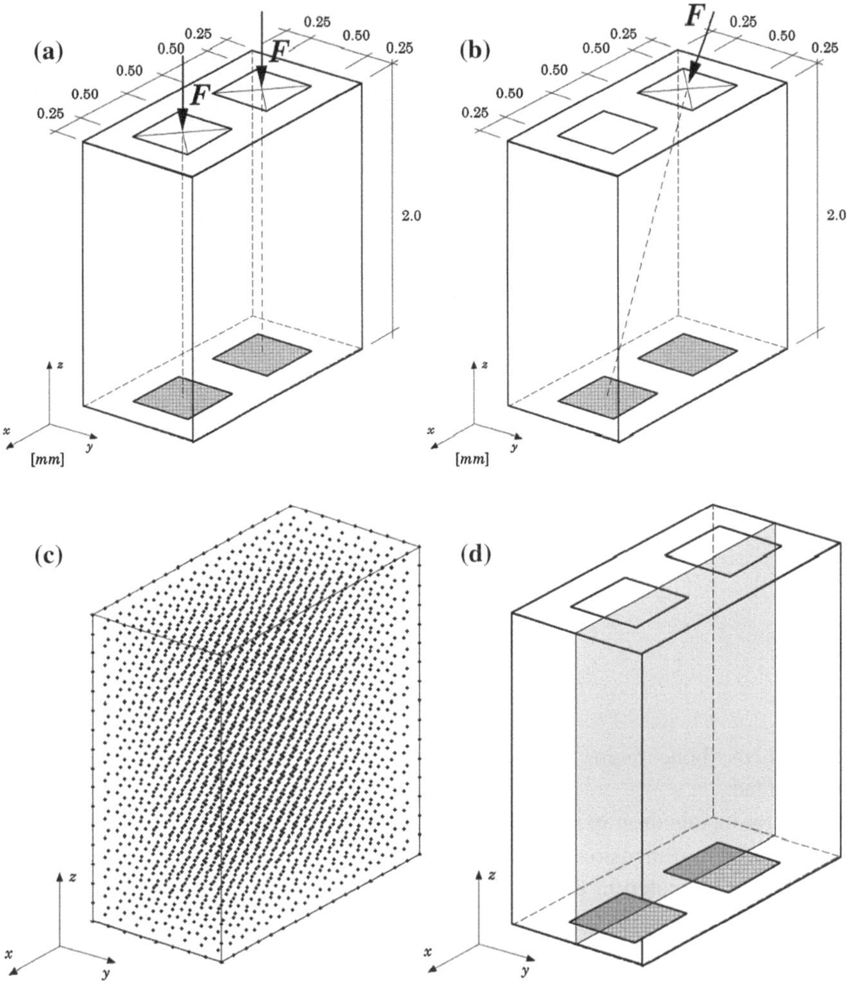

Fig. 7.10 **a** Hexahedron bone patch model submitted to vertical loads. **b** Hexahedron bone patch model submitted to a diagonal load. **c** Regular nodal discretization (2,681 nodes). **d** Patch section cut [1]

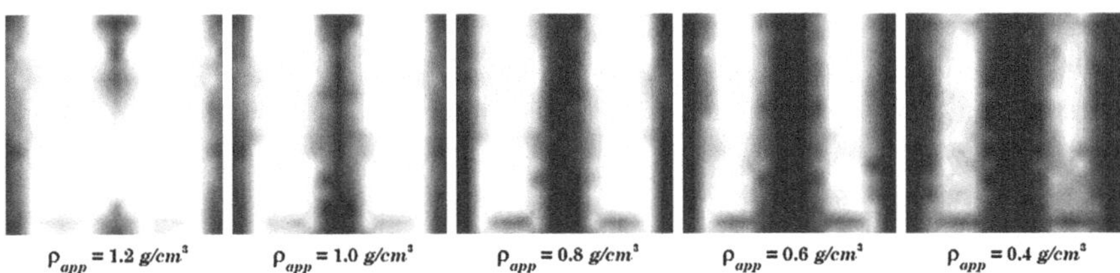

Fig. 7.11 Evolution of the trabecular architecture in the bone hexahedron patch for vertical loads [1]

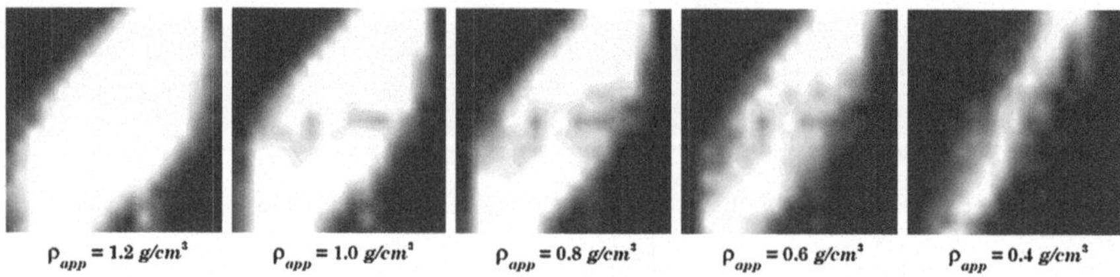

Fig. 7.12 Evolution of the trabecular architecture in the bone hexahedron patch for the diagonal load [1]

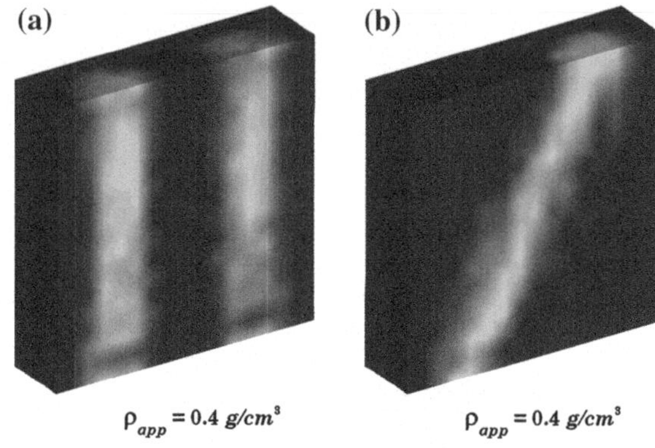

Fig. 7.13 Three-dimensional view of the section cut. **a** Vertical loads. **b** Diagonal load [1]

expected, the bone seems to form cylindrical trabeculae in the direction of the applied load.

Another hexahedron example is studied. Consider the cubic patch presented in Fig. 7.14a, with dimensions $2 \times 2 \times 2\,\text{mm}^3$. A surface load $F = 1.0\,\text{N/mm}^2$, with the direction indicated in the figure, is diagonally applied in a square area on the top of the cubic patch. On the patch bottom four square areas locally constrain the patch movement in all directions. The problem is analysed considering a regular nodal distribution with 2,744 nodes, Fig. 7.14b. In the analysis are assumed the same material properties as in previous example. For the remodelling algorithm, it is considered $\alpha = \beta = 0.01$ and a control medium apparent density $\rho_{app}^{control} = 0.4\ \text{g/cm}^3$. In order to present the results the cubic patch is sectioned by the division presented in Fig. 7.14c. The results regarding the evolution of the trabecular bone remodelling process are presented in Fig. 7.15.

Once again, despite the non-collinearity between the applied load and the nodal distribution, the formed trabecula is perfectly oriented in the load direction. Notice that for apparent densities $\rho_{app} > 0.6\,\text{g/cm}^3$ a secondary trabecula remains in the bone patch. The structural function of this secondary trabecula is to stabilize the principal diagonal trabecula, stopping a possible buckling phenomenon. The manifestation of the secondary trabecula proves that the remodelling algorithm and the proposed material law can predict additional secondary trabecular structures in the bone tissue potential three-dimensional domain.

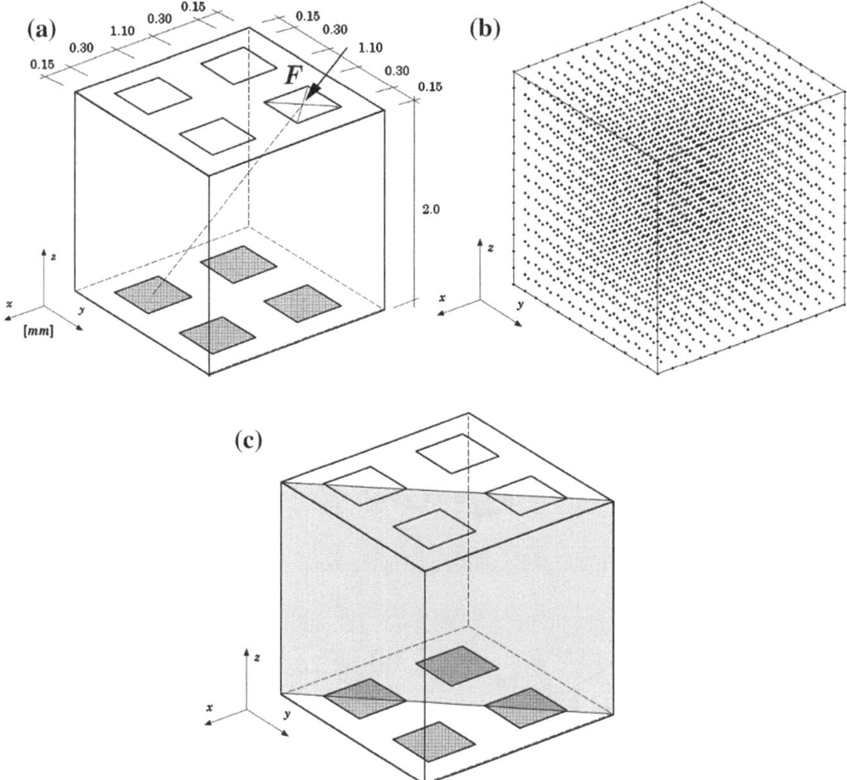

Fig. 7.14 **a** Cubic patch model submitted to a diagonal load. **b** Regular nodal discretization (2,744 nodes). **c** Cubic patch section cut [1]

To end this section dedicated to the three-dimensional bone patch analysis, the same cubic bone patch with dimensions $2 \times 2 \times 2\,\text{mm}^3$ is analysed, however in this case surface loads $F = 1.0\,\text{N/mm}^2$ are cross diagonally applied in square areas on the top of the cubic patch, Fig. 7.16a. The essential boundary conditions are the same as in previous analysis, the bottom four square areas locally constrain the cubic patch movement in all directions. The problem is analysed considering the same regular nodal distribution presented in Fig. 7.14b. With this example it is expected to stimulate torsion effects in the bone patch, and in response the bone should resist remodelling into a suitable trabecular structure. In order to observe the internal bone reorganization, four sections are made in the cubic patch, Fig. 7.16b.

The present analysis was performed using three distint bone tissue material laws. For the three performed analyses, the remodelling algorithm assumed $\alpha = \beta = 0.01$ and a control medium apparent density $\rho_{app}^{control} = 0.4\,\text{g/cm}^3$.

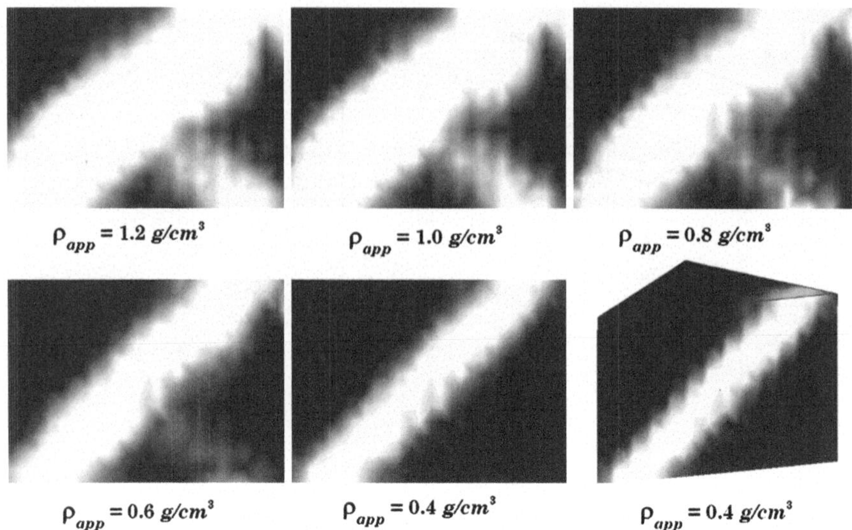

Fig. 7.15 Evolution of the trabecular architecture in the bone cubic patch for diagonal load [1]

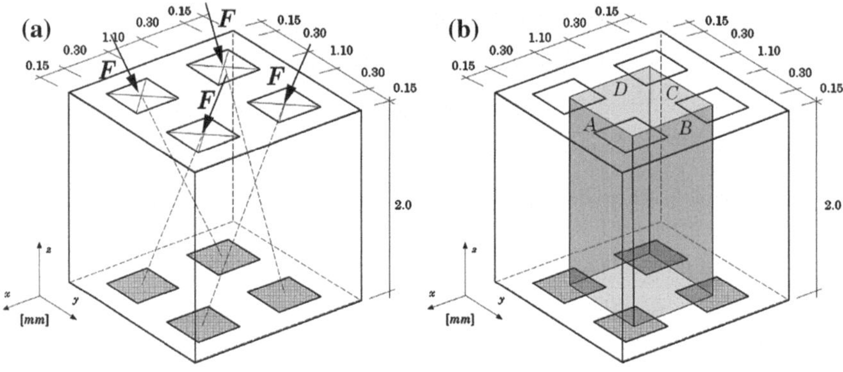

Fig. 7.16 **a** Cubic patch model submitted to cross diagonal loads. **b** Patch sections [1]

First the cubic bone patch was analysed considering the bone as an isotropic material. The isotropic phenomenological mathematical law was obtained from the proposed anisotropic material law, Sect. 6.2.2, considering for the transverse direction the correspondent axial values in order to respect the isotropic material assumption. The results on the evolution of the medium apparent density in the bone cubic patch for the cross load are presented in Fig. 7.17.

The results of Fig. 7.17 show that the bone forms standardized diagonal trabeculae to resist the torsion effect produced by the applied load. Secondary trabeculae do not appear as evident structures. Therefore the study continues and the cubic patch was analysed considering the bone as an anisotropic material, using

Fig. 7.17 Evolution of the trabecular architecture in the bone cubic patch for the cross diagonal loads (isotropic material considering the proposed material law) [1]

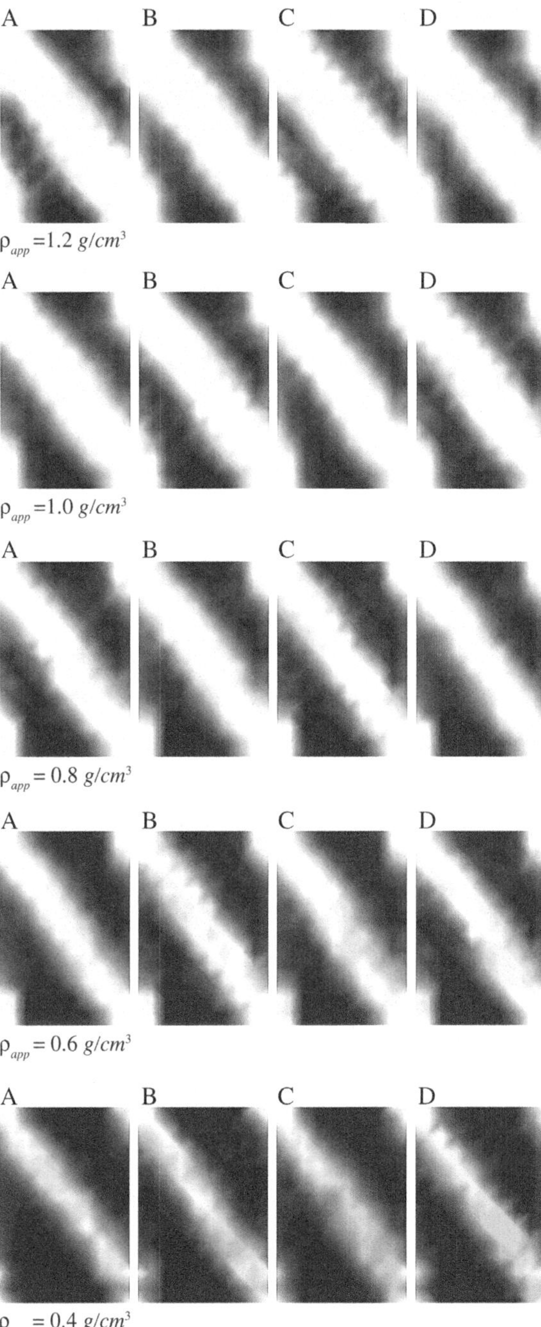

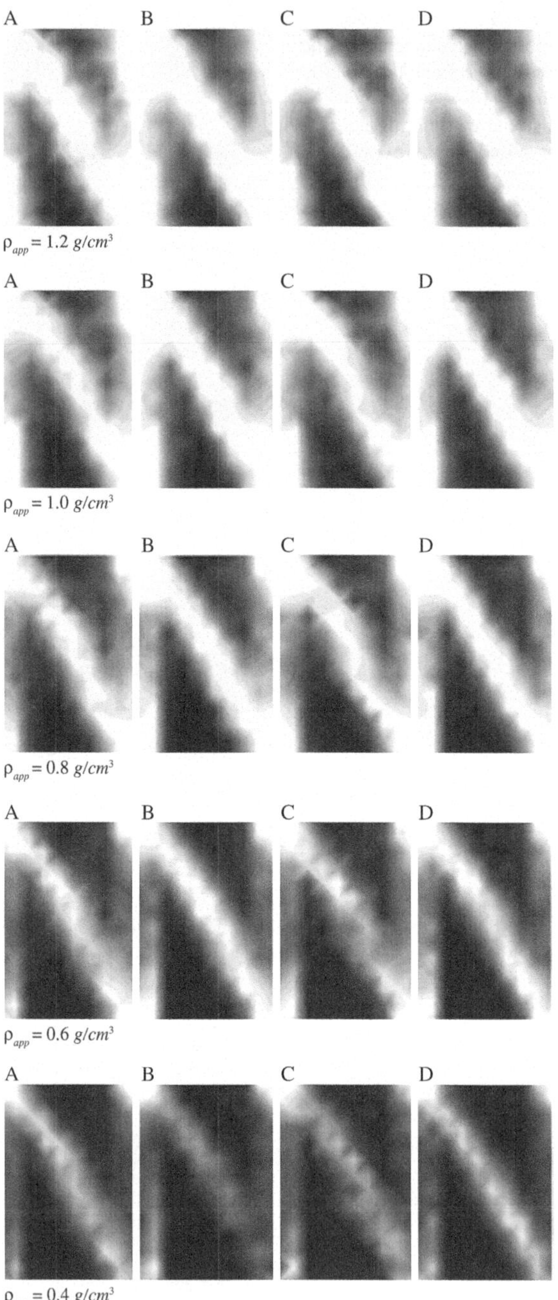

Fig. 7.18 Evolution of the trabecular architecture in the bone cubic patch for the cross diagonal loads (anisotropic material considering the proposed material law) [1]

the anisotropic material law proposed by Belinha et al. [1, 2], Sect. 6.2.2. The results are shown in Fig. 7.18 and it is possible to observe the growth of secondary trabecular structures along the vertical direction. These results are corroborated

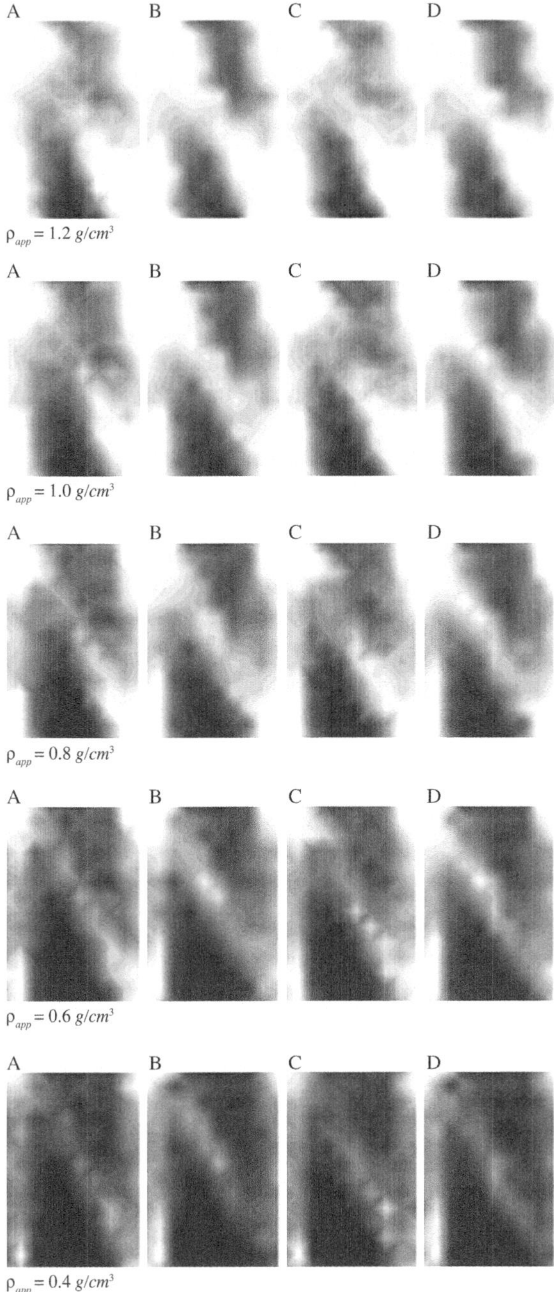

Fig. 7.19 Evolution of the trabecular architecture in the bone cubic patch for the cross diagonal loads (anisotropic material considering Lotz material law) [1]

with the solution obtained when the Lotz anisotropic material law is considered, Fig. 7.19. The same vertical secondary structures appear when the Lotz anisotropic material law is considered. This example shows the importance of considering the bone as an anisotropic material. The secondary trabecular structures confer to the trabecular bone mesh a higher stability and resistance. The manifestation of such structures indicates that the present approach on bone remodelling is moving forward the right path.

7.2 Bone Structures

The objective of the present section is to obtain numerically trabecular distributions resembling the real trabecular architecture that can be found in natural bones. Therefore, in this sections three distinct bone structures are studied: the maxillary bone; the calcaneus bone; and the femoral bone.

To numerically obtain the accurate trabecular distributions three components are required: a suitable bone tissue phenomenological law; an efficient iterative remodelling algorithm; and an accurate numerical method. However, these three components are not enough, the use of the correct numerical model is very important. The geometry of the model must be precise and the essential and natural boundary conditions must be truthfully determined. Since the remodelling model proposed in this book is driven by the mechanical stimulus, if the essential and natural boundary conditions are incorrect it will be impossible to achieve the correct trabecular architecture. Therefore, a preliminary detailed research should be performed to determine the correct essential and natural boundary conditions for each bone model.

7.2.1 Incisor

In this example the used computational model of the maxillary central incisor was based in the computerized axial tomography (CAT) scan performed by Poiate et al. [7]. The problem domain and the main biologic structure, shown in Fig. 7.20a, were discretized in an irregular nodal discretization with 4,245 nodes, represented in Fig. 7.20b, respecting the domain differentiation between the biologic structures indicated in Fig. 7.20a.

As suggested in the literature [7], in the model upper domain boundary the nodal displacements are constrained in both directions, Fig. 7.20b. Regarding the natural boundary conditions other research works [7, 8] suggest the five loads F_i presented in Fig. 7.20b, oriented with an angle α_i in relation to the incisor longitudinal middle axis.

In order to compare the obtained NNRPIM results for the distinct studies, eight interest regions of the model were considered, Fig. 7.20b. In the indicated regions are expected stress concentrations caused by the applied load and the model essential boundary conditions. Notice that in regions 1, 4, 6 and 8 are expected compressive stresses and tensile stresses in regions 2, 3, 5 and 7.

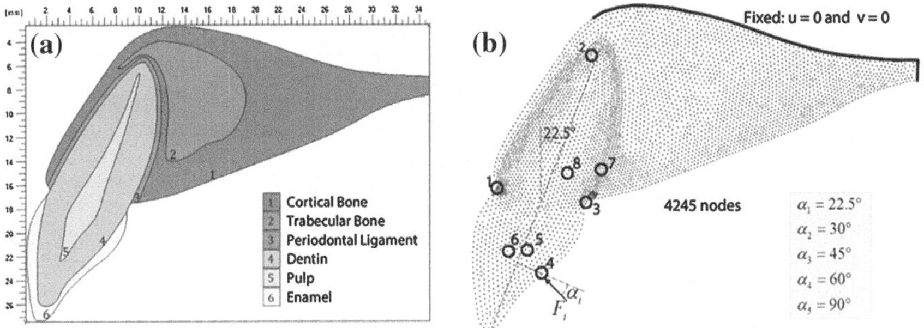

Fig. 7.20 **a** Maxillary central incisive for the two dimensional analysis. **b** Nodal discretization and considered essential and natural boundary conditions

Table 7.1 Mechanical properties of the anatomical structures

Anatomical structure	Young modulus (GPa)	Poisson's ratio
Pulp	0.02	0.45
Dentin	18.60	0.31
Enamel	41.00	0.30
Trabecular bone	1.37	0.30
Cortical bone	13.70	0.30
Periodontal ligament	0.0689	0.45

7.2.1.1 NNRPIM/FEM Comparison

This first presented analysis regards a comparison study between the NNRPIM and the FEM. The previously described 2D model is analysed considering the material properties suggested in the literature [7], which are presented in Table 7.1 for the biologic structures indicated in Fig. 7.20a.

In this comparison study the same load suggested in the FEM study [7] is applied: a localized load $\boldsymbol{F}_1 = F_0 \cdot \{-\cos\theta, \sin\theta\}$, being $F_0 = 100\,\mathrm{N}$ and the angle $\theta = 22.5 + \alpha_1$, as indicated in Fig. 7.20b. In this example, in order to eliminate local stress concentrations, the localized load $\boldsymbol{F}_1$ was distributed along 5 boundary nodes.

In the end of the elastostatic analysis the meshless results were compared with the FEM results [7]. The von Mises effective stress distribution map obtained with the NNRPIM is presented in Fig. 7.21. Comparing the obtained NNRPIM stress distribution, Fig. 7.21, with the FEM results available in the literature [7] it is visible that the NNRPIM results are considerably smoother.

In Table 7.2 are presented the maximum principal stresses obtained in regions 1 to 8 with the NNRPIM. The meshless results are compared with the results obtained with a 2D linear triangular finite element (CTRIA3) and a quadratic quadrilateral finite element (CQUAD8), [7].

The results presented in Table 7.2 show that the NNRPIM solution is very close to both FEM solutions. Additionally it is important to refer that the NNRPIM solution was obtained using a discretization with 4,245 nodes and the CTRIA3 FEM and

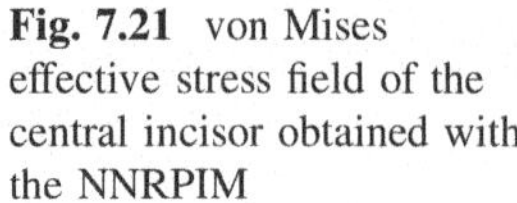
Fig. 7.21 von Mises effective stress field of the central incisor obtained with the NNRPIM

Table 7.2 Maximum principal stress results obtained for the analysis of the central incisor

σ_1 (MPa)	Region1	Region2	Region3	Region4	Region5	Region6	Region7	Region8
CTRIA3	−9.80	17.00	57.10	−103.00	97.10	–	–	–
CQUAD8	−18.40	33.80	68.50	−123.00	138.00	–	–	–
NNRPIM	−12.84	27.67	65.74	−106.69	140.23	−86.75	160.37	−112.63

CQUAD8 FEM solutions were obtained with computational meshes with 9,259 nodes and 24,868 nodes respectively, showing that the NNRPIM is capable to obtain good results with a lower discretization level.

7.2.1.2 Maxillary Bone Tissue Remodelling

The model presented in Fig. 7.20 is used to study the bone tissue remodelling process of the maxillary bone supporting the central incisive. The same computational mesh with 4,245 nodes, Fig. 7.20b, is used to discretize the problem domain and the considered mechanical properties for the pulp, dentin, enamel and periodontal ligament are indicated in Table 7.1. As in previous examples, in the model upper domain boundary the nodal displacements are constrained in both directions, Fig. 7.20b. The four load cases suggested in the literature [8], corresponding to the normal solicitation of the incisor due to the daily mastication activity, are considered in the present analysis.

The four load cases consist in localized loads $\boldsymbol{F}_i = F_0 \cdot \{-\cos\theta_i, \sin\theta_i\}$, being the global force $F_0 = 100\,\text{N}$ and the total angle $\theta_i = 22.5 + \alpha_i$, Fig. 7.20b. Load case 1 is obtained considering $i = 5$, load case 2 considers $i = 4$ and load cases 3 and 4 are obtained considering $i = 3$ and $i = 2$ respectively. In this work, to eliminate local stress concentration, all the localized loads $\boldsymbol{F}_i$ were distributed along 5 boundary nodes.

For all studied examples, as required by the proposed remodelling algorithm, it is considered an initial uniform density distribution $\rho_{app}^{max} = 2.1\,\text{g/cm}^3$ and a

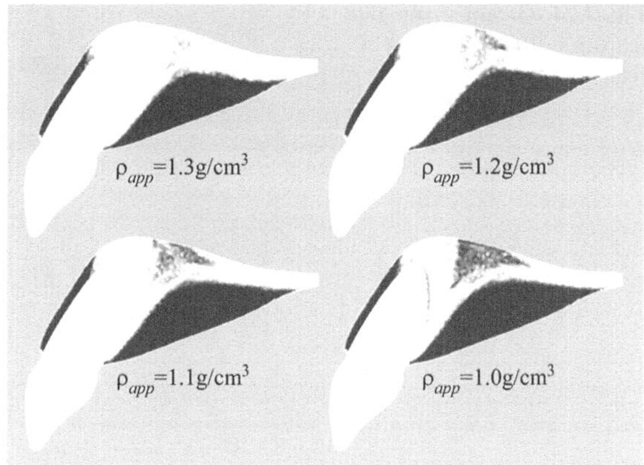

Fig. 7.22 Obtained apparent densities distributions for load case 1 (i = 5)

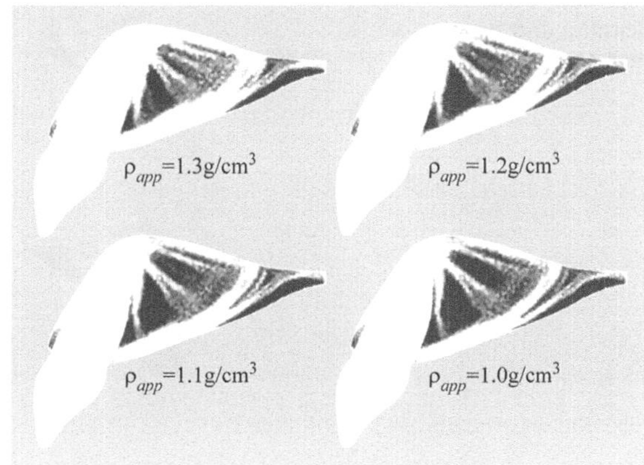

Fig. 7.23 Obtained apparent densities distributions for load case 2 (i = 4)

Poisson ratio $\upsilon = 0.3$. It is assumed $\rho_{app}^{control} = 1.0\,\text{g/cm}^3$ as the remodelling algorithm medium bone density control value. For the α and β parameters ruling the growth and the decay of the bone tissue it is considered: $\alpha = \beta = 0.01$.

The bone tissue remodelling results are presented with a grey scale as usual and in each presented figure it is indicated the bone model medium apparent density.

Firstly each one of the load cases are independently analysed. The results obtained for load case 1 are presented in Fig. 7.22. It is possible to observe the achieved apparent density distribution for four distinct medium bone densities.

The results regarding load cases 2, 3 and 4 are respectively presented in Figs. 7.23, 7.24 and 7.25. In each figure are shown four distinct medium bone densities obtained with the respective load case.

In order to obtain a trabecular architecture similar to the real trabecular distribution of the maxillary bone on the surroundings of the central incisor, it is necessary to consider simultaneously the four load cases. For each load case are considered 2,500 cycles per day, totalising 10,000 masticating movements per day. Recall that the considered remodelling algorithm weights each load case in the

Fig. 7.24 Obtained apparent densities distributions for load case 3 (i = 3)

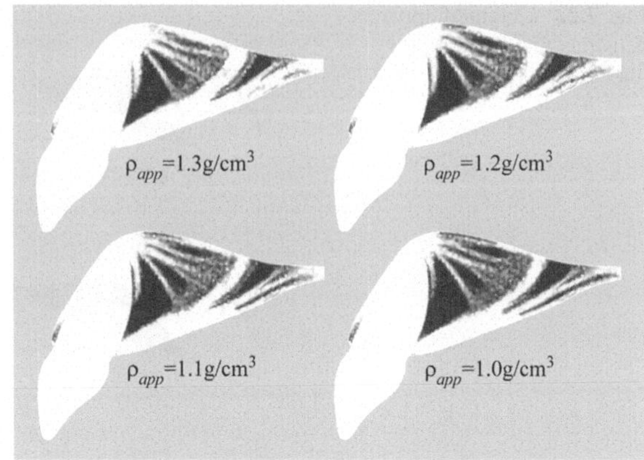

Fig. 7.25 Obtained apparent densities distributions for load case 4 (i = 2)

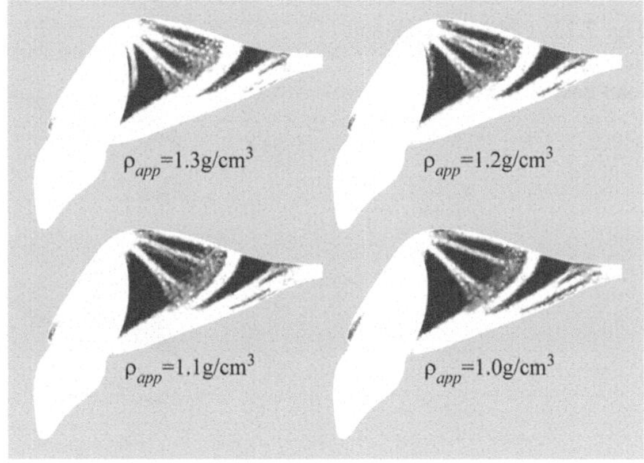

same proportion with the number of associated cycles. The results for the four load cases simultaneously applied are presented in Fig. 7.26.

The results in Fig. 7.26 show an apparent similarity with the central incisor sagittal plane X-ray plates, which can be found in the literature [9]. With the proposed numerical approach it is possible to predict the same trabecular triangular area in the incisor posterior zone and the superior and inferior maxillary cortical layer. These results indicate that the combination of the NNRPIM with the remodelling algorithm permits to achieve the internal trabecular bone structure if the correct mechanical cases are known.

7.2.2 Calcaneus Bone

In the human species the foot is the first mechanical contact with the ground. The calcaneus is one of the many bones composing the foot structure. Its function is to distribute the load from above. The body weight is transmitted by the talus bone,

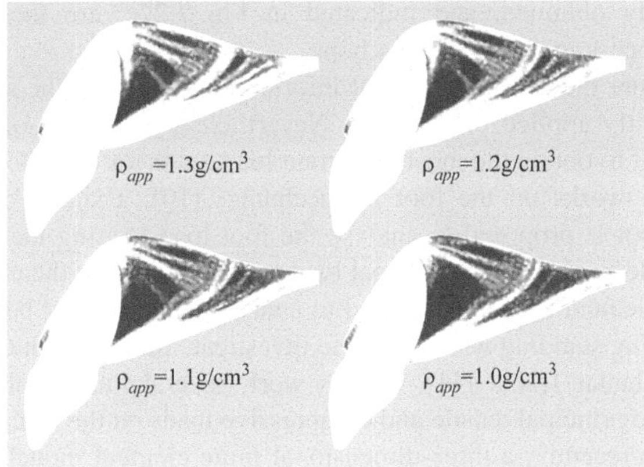

Fig. 7.26 Obtained apparent densities distributions for combination of all load cases

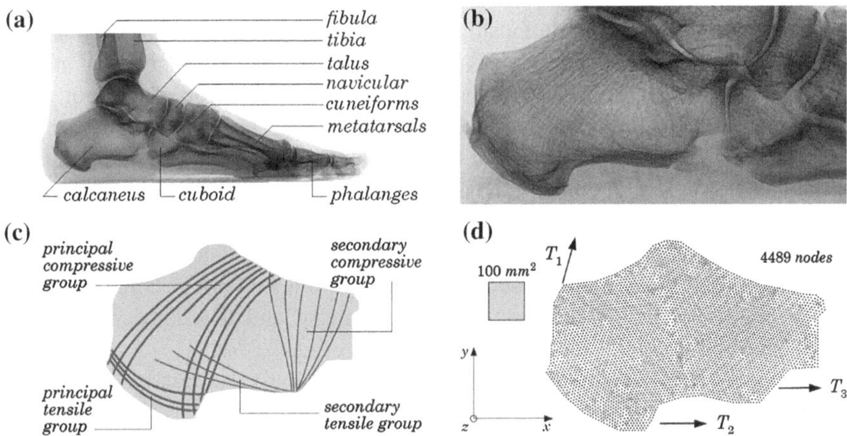

Fig. 7.27 **a** Foot X-ray plate. **b** Calcaneus X-ray detail. **c** Calcaneus internal principal trabecular structures. **d** Geometry and nodal discretization used in the analysis of the calcaneus bone. Tendonal forces: Achilles tendon force (*T1*), plantar fascia fan-tendon force (*T2*) and calcaneus-metatarsal tendon force (*T3*) [2]

which is directly loaded by the tibia bone. The cuboid bone articulates with the calcaneus bone anterior side, the Achilles tendon is inserted into a roughened area on the superior side and the plantar fascia fan-tendon is inserted on the bottom of the calcaneus bone.

In Fig. 7.27a it is possible to visualize a foot X-ray plate in which foot bones are indicated. A closer look on the calcaneus bone, Fig. 7.27b, permit to empirically determine the compressive and tensile lines indicated in Fig. 7.27c. The foot is a dynamic structure, the several foot bones suffer a constant shift of load cases. Therefore the foot bones are forced to achieve an internal trabecular architecture which permits to resist the several applied load cases. The compressive and tensile lines empirically obtained, and indicated in Fig. 7.27c, are the result of the remodelling process in the calcaneus bone.

In the literature there is no sufficient information regarding the several loading conditions directly applied on the foot. Nevertheless there are some works that indirectly permit to obtain the most important load cases for the calcaneus bone. In one of the first works on the foot biomechanics [10], a simple biomechanical model of the foot is proposed to analyse the foot load bearing mechanism, from which is possible to obtain the principal load contact points on the calcaneus bone. Later, a finite element model developed to analyse the structural behaviour of the human foot during standing was utilized to investigate the biomechanical effects of releasing the plantar fascia [11]. In this work several numerical models were analysed and the principal tensile and compressive loads on the calcaneus bone are proposed. More recently, a three-dimensional finite element model of the human foot and ankle, incorporating geometrical and material nonlinearity, was employed to investigate the loading response of the plantar fascia in the standing foot with different magnitudes of Achilles tendon loading [12]. This work supplies important information regarding the relationship between the Achilles tendon tensile force and the plantar fascia tensile force. The same relationship was obtained in another recent research work [13], in which it was recreated the position of the foot when stretch is introduced on the plantar fascia (foot movement). All these authors agree that the principal tendons inserted in the calcaneus bone are the Achilles tendon, the plantar fascia fan-tendon and the calcaneus-metatarsal tendon. In Fig. 7.27d it is possible to schematically visualize the respective tendon tensile forces in a two-dimensional representation.

Based on the X-ray plate presented in Fig. 7.27b a two-dimensional model of the calcaneus bone was constructed. The geometry and the nodal distribution discretizing the problem domain are presented in Fig. 7.27d.

The human foot is a structure which possesses a robust and relative wide angle articulation, permitting this way diverse load cases. However, two structural mechanical conditions are recurrent: standing and walking (gait). In this book three mechanical cases, based in the previous referred articles [10–13], are proposed. Although the human species possesses two foots it is common to stand only in one of them, and during normal walking merely one foot is standing in the ground. Therefore, in this example it is considered a human with a body weight of $P = 100\,kg \cong 1{,}000\,\mathrm{N}$ standing always in one foot. In Fig. 7.28 it is presented the first mechanical case. This case is due to the arrival of the foot to the ground during the walking movement (gait). When the foot arrives to the floor the first contact is with the calcaneus bone. In this mechanical case the compressive forces are the most relevant. The pressures applied in the distinct natural boundaries have the magnitude, distribution and orientation indicated. The essential boundary condition indicated as ebc_1 constrains the movement on the direction normal to the surface and the ebc_2 essential boundary condition constrains the movement on the x direction.

The second mechanical case is presented in Fig. 7.29. This case corresponds to the foot lifting movement. This movement is used to jump or simply to walk. In this example the magnitudes of the forces involved correspond to the walking

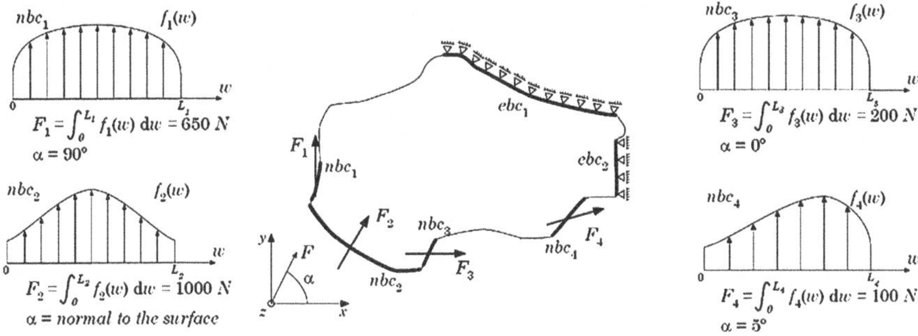

Fig. 7.28 Loads and constrains of the first mechanical case of the calcaneus bone [2]

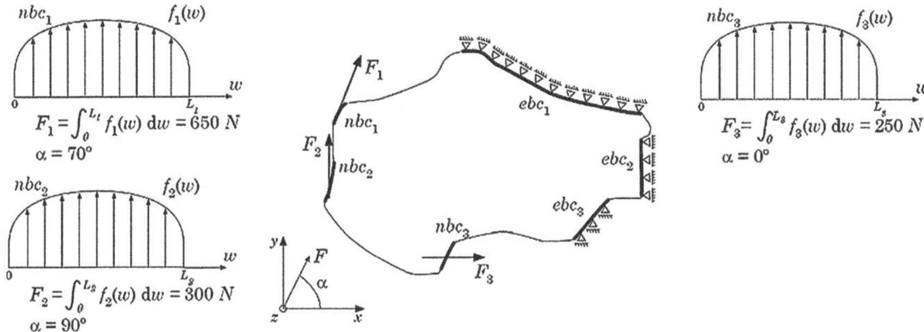

Fig. 7.29 Loads and constrains of the second mechanical case of the calcaneus bone [2]

movement (gait). The calcaneus bone is submitted to high tensile forces from the Achilles tendon and the plantar fascia tendon. The talus bone distributes most of the compressive forces to the navicular bone and the cuboide bone, constraining the calcaneus bone movements. The pressures applied in the distinct natural boundaries have the magnitude, distribution and orientation indicated in Fig. 7.29. The essential boundary condition ebc_1 constrains the movement on the direction normal to the surface, the ebc_2 essential boundary condition constrains the movement on the x direction and ebc_3 on the y direction.

In Fig. 7.30 the third mechanical case is presented. This case corresponds to the standing position. During the standing position the Achilles tendon is always in tension, providing a dynamic equilibrium for the body. In this mechanical case there is a balance between the compressive forces and the tensile forces applied in the calcaneus. In Fig. 7.30 are presented the pressures applied in the distinct natural boundaries and the respective magnitudes, distributions and orientations. The essential boundary conditions ebc_1 and ebc_3 constrain the movement on the direction normal to the surface and the ebc_2 essential boundary condition constrains the movement on the y direction.

Firstly each one of the load cases is independently analysed. In all studied examples it is imposed an initial uniform density distribution $\rho_{app}^{max} = 2.1$ g/cm^3,

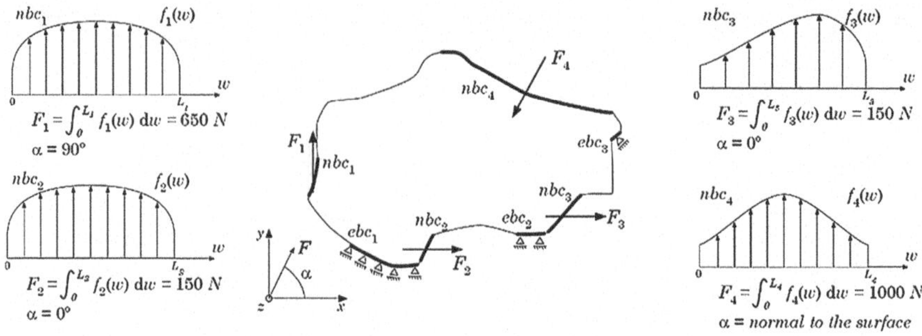

Fig. 7.30 Loads and constrains of the third mechanical case of the calcaneus bone [2]

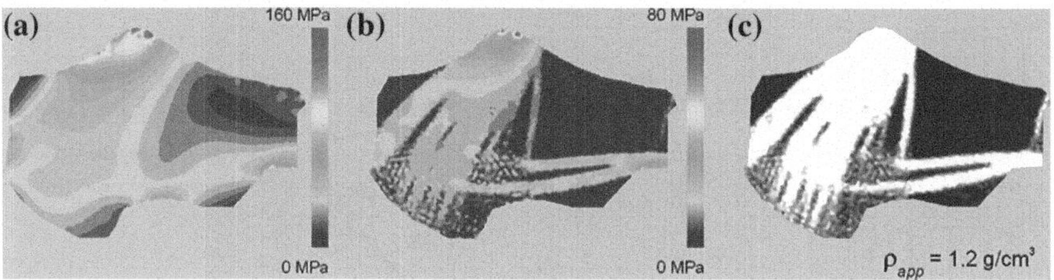

Fig. 7.31 First mechanical case. **a** Initial von Mises effective stress isomap. **b** Final von Mises effective stress isomap. **c** Final obtained trabecular architecture

from which it is possible to determine the initial bone tissue material properties using the proposed phenomenological material law. It is considered a Poisson ratio $\upsilon = 0.3$ and $\rho_{app}^{control} = 1.2\,\text{g/cm}^3$ as the remodelling algorithm medium bone density control value. The α and β parameters, governing the growth and the decay of the bone tissue, are assumed as: $\alpha = \beta = 0.01$.

The results obtained regarding the proposed first mechanical case are presented in Fig. 7.31. The von Mises effective stress distribution obtained for the first step and the final step of the iterative remodelling analysis are presented respectively in Fig. 7.31a, b. The final trabecular architecture obtained in the analysis is shown in the apparent density isomap presented in Fig. 7.31c. The results regarding the second and third calcaneus mechanical cases are presented in Figs. 7.32 and 7.33. The results are in accordance with the internal trabecular structures indicated in Fig. 7.27c, demonstrating that it is possible to individually achieve the internal trabecular calcaneus bone structure.

In Fig. 7.34 it is presented evolution of bone tissue trabecular architecture for each one of the studied mechanical cases considering $\rho_{app}^{control} = 0.4\,\text{g/cm}^3$ as the remodelling algorithm medium bone density control value.

The next phase is to mix the three mechanical cases in order to achieve a closer solution to the X-ray plate initially presented. Therefore it was considered the following combination: The three mechanical cases are simultaneously applied with 6,000 cycles per day for the first mechanical case, 6,000 cycles per day for the

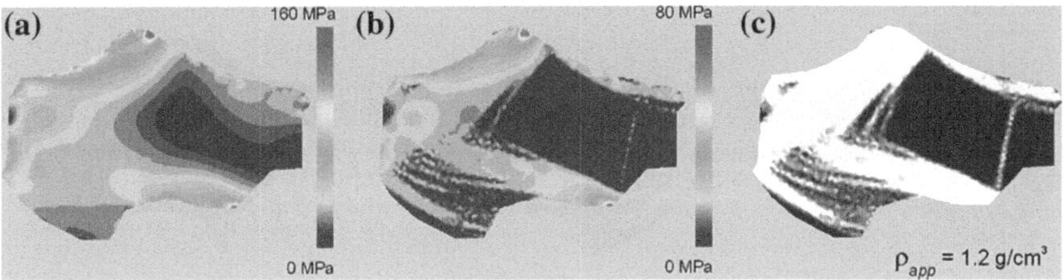

Fig. 7.32 Second mechanical case. **a** Initial von Mises effective stress isomap. **b** Final von Mises effective stress isomap. **c** Final obtained trabecular architecture

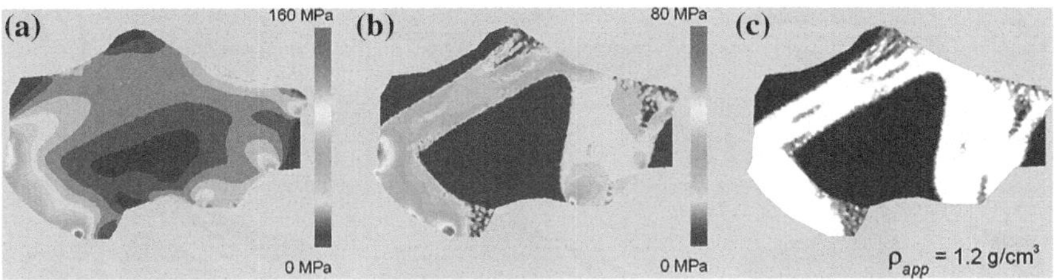

Fig. 7.33 Third mechanical case. **a** Initial von Mises effective stress isomap. **b** Final von Mises effective stress isomap. **c** Final obtained trabecular architecture

second mechanical case and 4,200 cycles per day for the third mechanical case. The results are presented in Fig. 7.35.

Despite the good results obtained with the previously described load combination, another load combination was analysed, corresponding to the simultaneous application of the three mechanical cases with 6,000 cycles per day for all the three mechanical cases. In Fig. 7.36 it is possible to compare the evolution of bone tissue trabecular architecture obtained for each one of the load combinations for $\rho_{app}^{control} = 0.4\,\text{g/cm}^3$.

The results in Fig. 7.36 show a clear similitude with the X-ray plate presented in Fig. 7.27b and corroborate the empirical trabecular architecture suggested in Fig. 7.27c. These results indicate that it is possible with the proposed bone tissue remodelling algorithm to achieve the internal trabecular bone structure if the correct mechanical cases are known.

7.2.3 Femur

The femur bone is probably the most studied bone example available in the literature. The remodelling process of this long bone was analysed by several authors using the two-dimensional analysis [14–22] and the three-dimensional analysis [23–25]). In this section the bone tissue material law proposed in Sect. 6.2.2 and the bone tissue remodelling algorithm presented in Sect. 6.3.5 are combined with

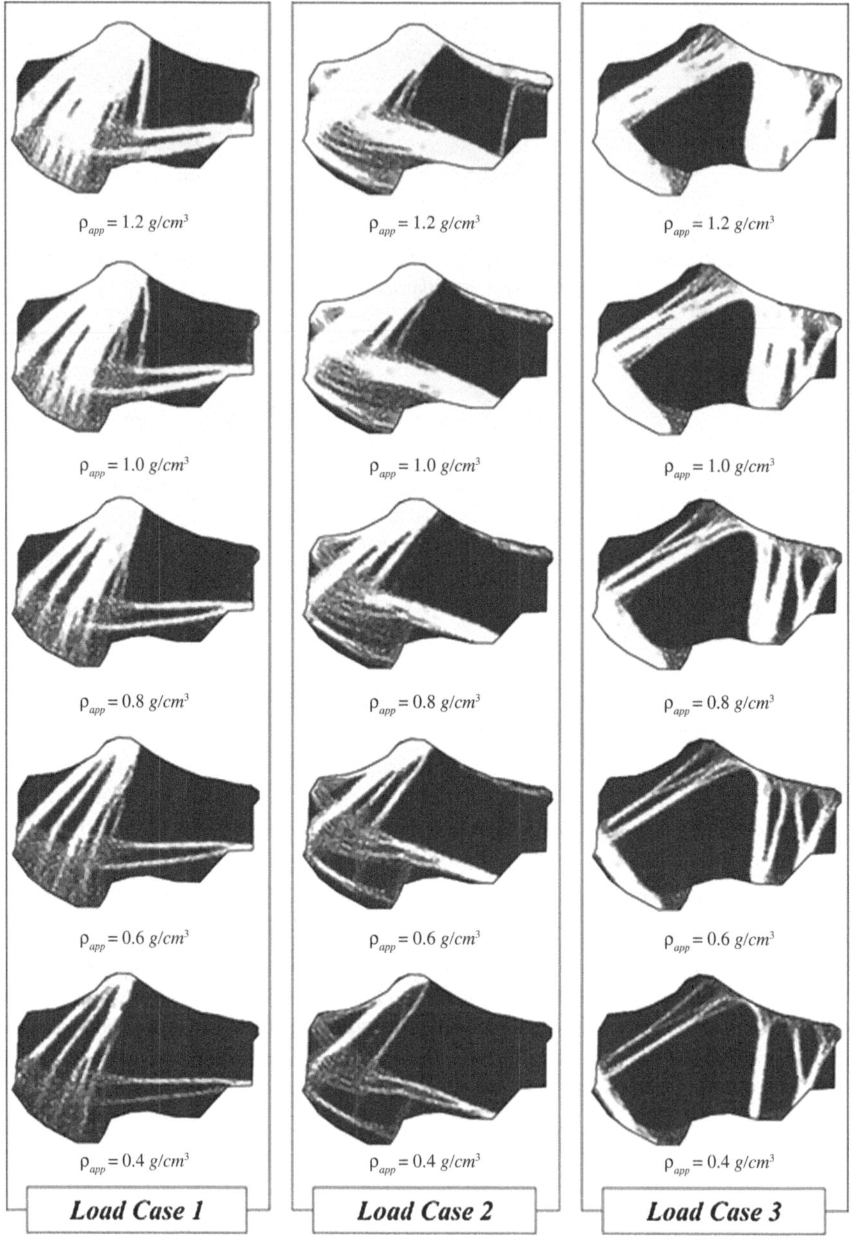

Fig. 7.34 Evolution of bone tissue trabecular architecture for each one of the studied mechanical cases

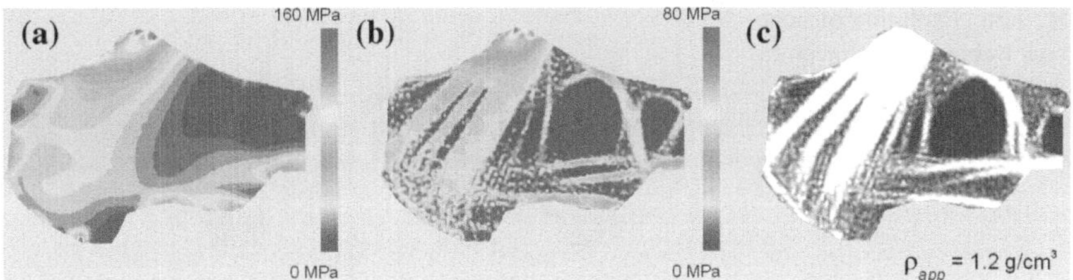

Fig. 7.35 Combination of all mechanical cases. **a** Initial von Mises effective stress isomap. **b** Final von Mises effective stress isomap. **c** Final obtained trabecular architecture

the NNRPIM to achieve a femoral internal trabecular bone architecture similar with the real density distribution found in the femur bone.

7.2.3.1 Femoral Two Dimensional Analysis

The femur is a long bone articulated in the hip-bone. The body weight is directly applied in the femur head. This long bone is a natural choice to validate an anisotropic remodelling algorithm, since it is a well-studied bone in biomechanics and the trabecular structure in the proximal femur is relatively well oriented. An example of a X-ray plate of the proximal femur is presented in Fig. 7.37a. From Fig. 7.37a it is possible to empirically obtain the compressive and tensile lines indicated in Fig. 7.37b. It was used the geometry of a two-dimensional proximal femur model proposed in the literature [26]. The domain was discretized with the nodal distribution presented in Fig. 7.38.

The femur loading history was approximated by the three-load cases used by Beaupré et al. [14, 15], each consisting of one parabolic distributed load over the joint surface, nbc_1, and another parabolic distributed load on the trochanter, nbc_2, representing the abductor muscle attachment. In Fig. 7.39 it is possible to observe the resultant of each applied parabolic distributed load and the correspondent direction. For the three considered mechanical cases, all degrees of freedom are constrained in the basis ebc_1.

As in previous example, in a first step, each one of the load cases presented in Fig. 7.39 are separately analysed. In order to determine the initial bone tissue material properties using the proposed phenomenological material law, in each analysis it is imposed an initial uniform density distribution $\rho_{app}^{max} = 2.1\ \text{g/cm}^3$. The Poisson ratio is assumed as $\upsilon = 0.3$ and the apparent density control value is considered as $\rho_{app}^{control} = 1.2\ \text{g/cm}^3$. The α and β parameters, governing the growth and the decay of the bone tissue, are assumed as: $\alpha = \beta = 0.01$.

The results obtained regarding the first mechanical case proposed by Beaupré et al. [14, 15] are presented in Fig. 7.40. In Fig. 7.40a it is presented the von Mises effective stress isomap obtained for the first step of the iterative remodelling

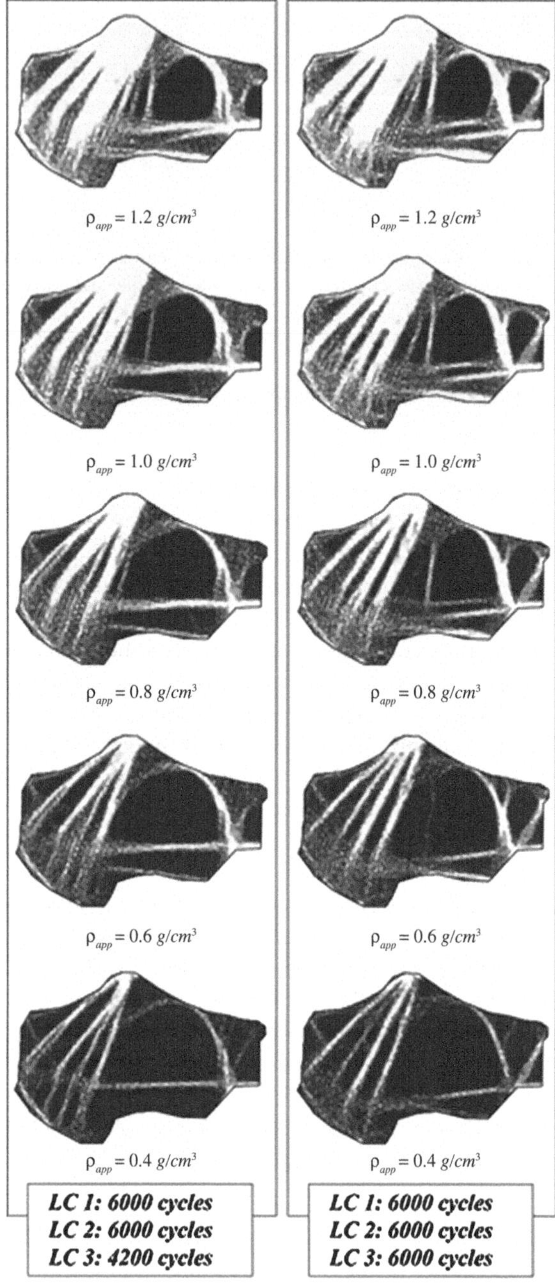

Fig. 7.36 Evolution of bone tissue trabecular architecture for the studied load cases combinations

analysis and in Fig. 7.40b it is shown the final von Mises effective stress distribution obtained in the analysis. The final trabecular architecture obtained in the analysis is presented in Fig. 7.40c. In Figs. 7.41 and 7.42 are respectively

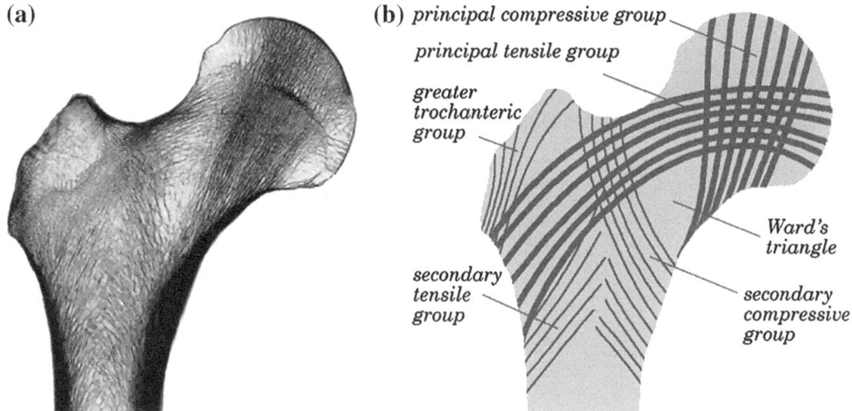

Fig. 7.37 **a** Femoral X-ray plate. **b** Internal principal trabecular structures found in the femur bone

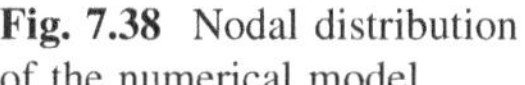
Fig. 7.38 Nodal distribution of the numerical model

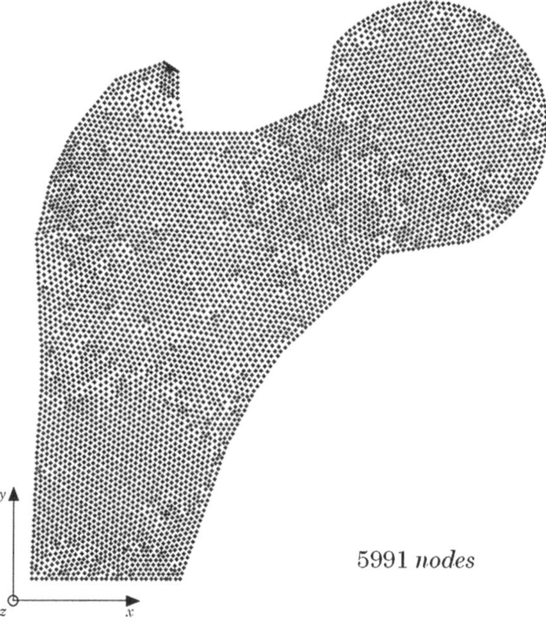

presented the results regarding the second and third mechanical cases. Notice, in all femoral mechanical cases examples, that the formation of principal and secondary trabecular structures are coherent with the ones presented in Fig. 7.37a, b.

In order to obtain a closer result to the real proximal femur X-ray plate it is necessary to simultaneously apply all mechanical cases. In the work of Beaupré et al. [14, 15] it was originally suggested that the first mechanical case should be applied assuming 6,000 cycles per day and 2,000 cycles per day for the second and

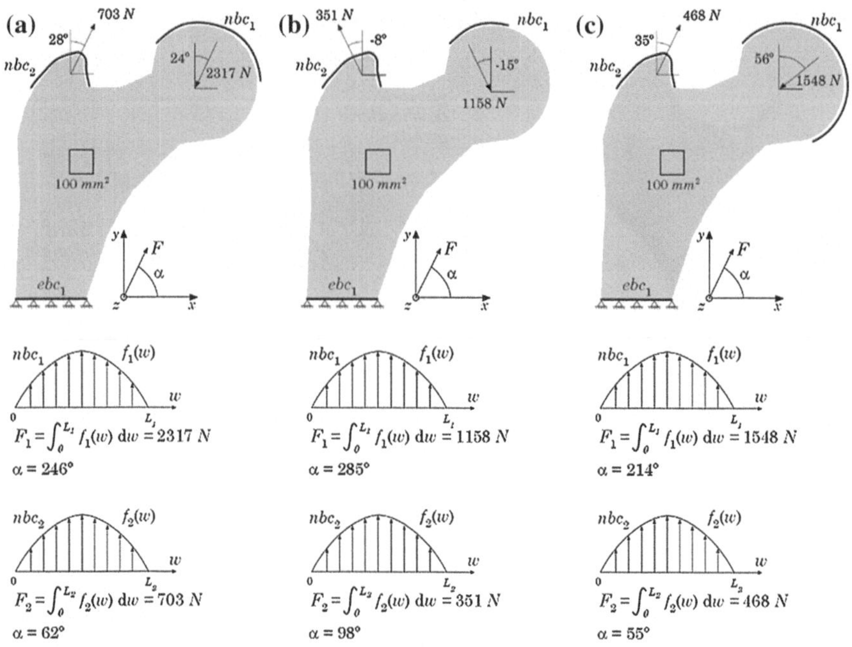

Fig. 7.39 Loads and constrains of the various mechanical case of the femur bone. **a** First mechanical case, **b** second mechanical case and **c** third mechanical case

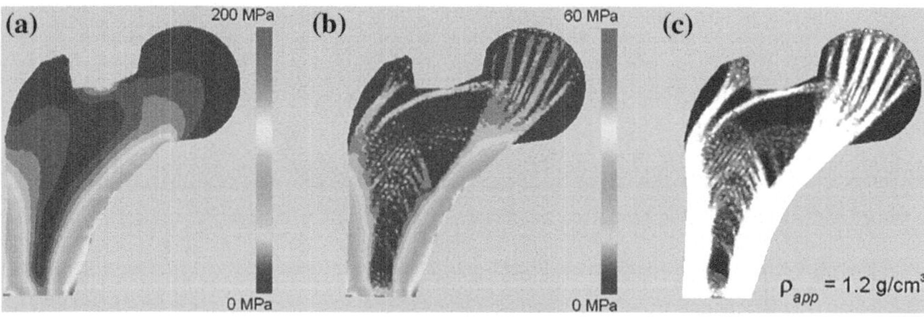

Fig. 7.40 First mechanical case. **a** Initial von Mises effective stress isomap. **b** Final von Mises effective stress isomap. **c** Final obtained trabecular architecture

the third mechanical cases. The results of the three mechanical cases simultaneously applied are presented in Fig. 7.43. It is possible to observe in Fig. 7.43 all the internal trabecular structures indicated in Fig. 7.37b and the remaining internal formations, such the Ward's triangle and the greater trochanteric group. Notice also that the secondary structures can be accurately predicted with the proposed remodelling algorithm and the bone anisotropic material law.

In order to broadly understand the remodelling process followed by the proposed algorithm, it is presented in Fig. 7.44 the evolution of bone tissue trabecular

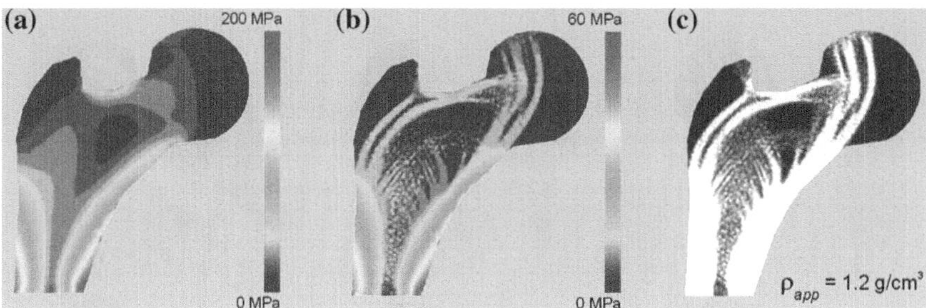

Fig. 7.41 Second mechanical case. **a** Initial von Mises effective stress isomap. **b** Final von Mises effective stress isomap. **c** Final obtained trabecular architecture

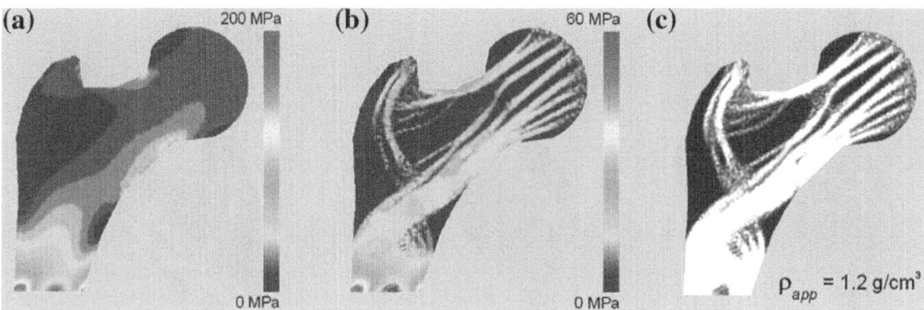

Fig. 7.42 Third mechanical case. **a** Initial von Mises effective stress isomap. **b** Final von Mises effective stress isomap. **c** Final obtained trabecular architecture

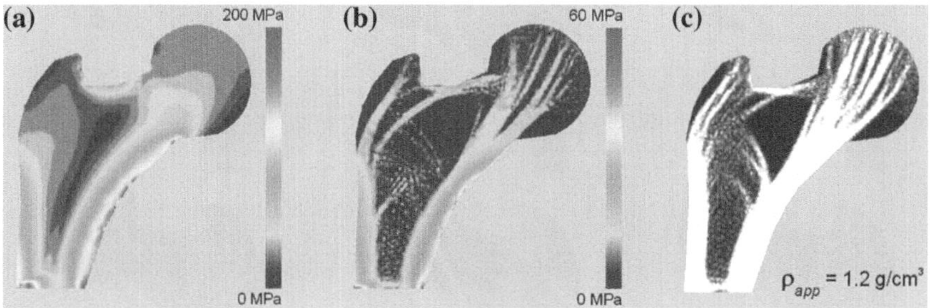

Fig. 7.43 Combination of all mechanical cases. **a** Initial von Mises effective stress isomap. **b** Final von Mises effective stress isomap. **c** Final obtained trabecular architecture

architecture for each one of the studied mechanical cases considering $\rho_{app}^{control} = 0.4\,\text{g/cm}^3$ as the remodelling algorithm medium bone density control value. In Fig. 7.45 it is shown the evolution of bone tissue trabecular architecture for the combination of the three mechanical cases assuming $\rho_{app}^{control} = 1.0\,\text{g/cm}^3$.

There are only a few works in the literature exploring the potential of meshless methods in the bone tissue remodelling analysis. Besides the research work

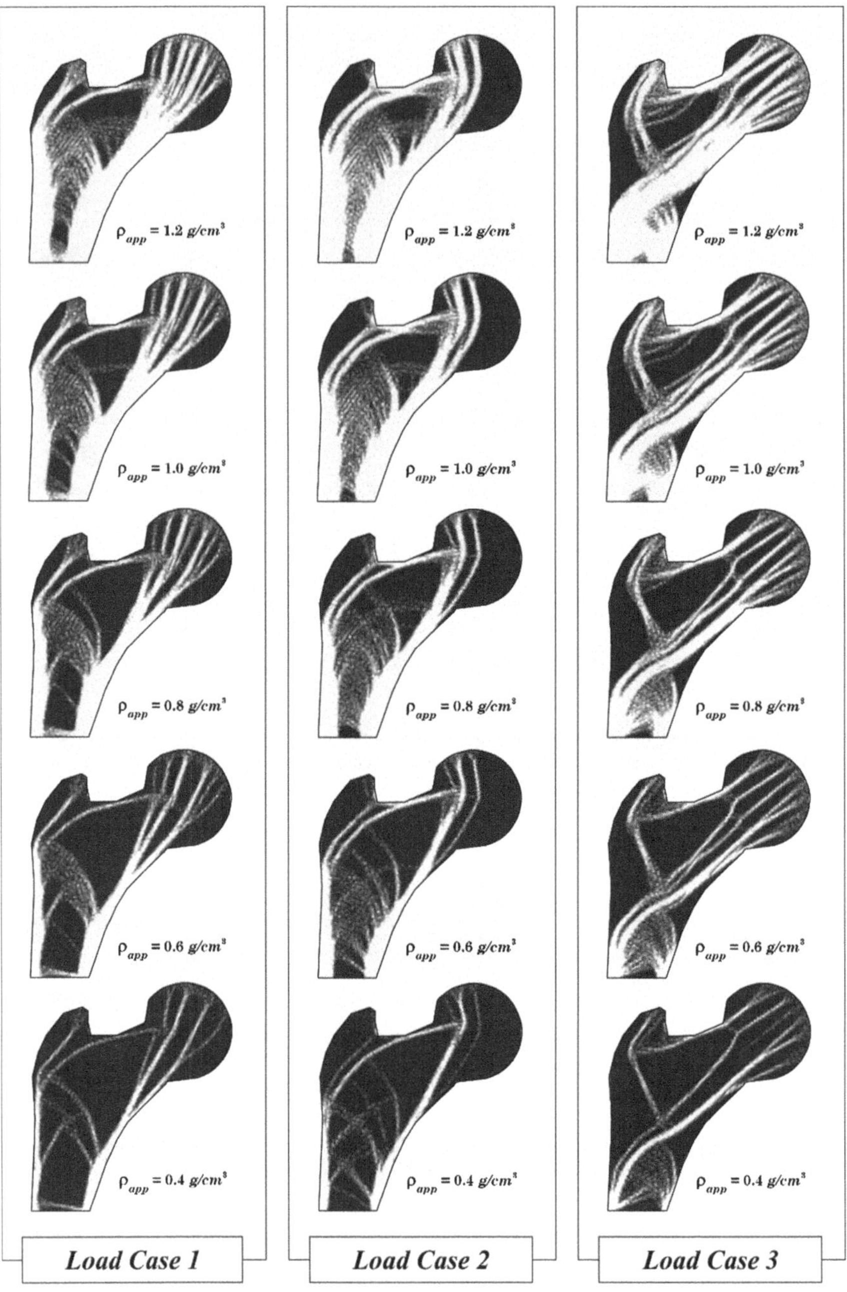

Fig. 7.44 Evolution of bone tissue trabecular architecture for each one of the studied mechanical cases

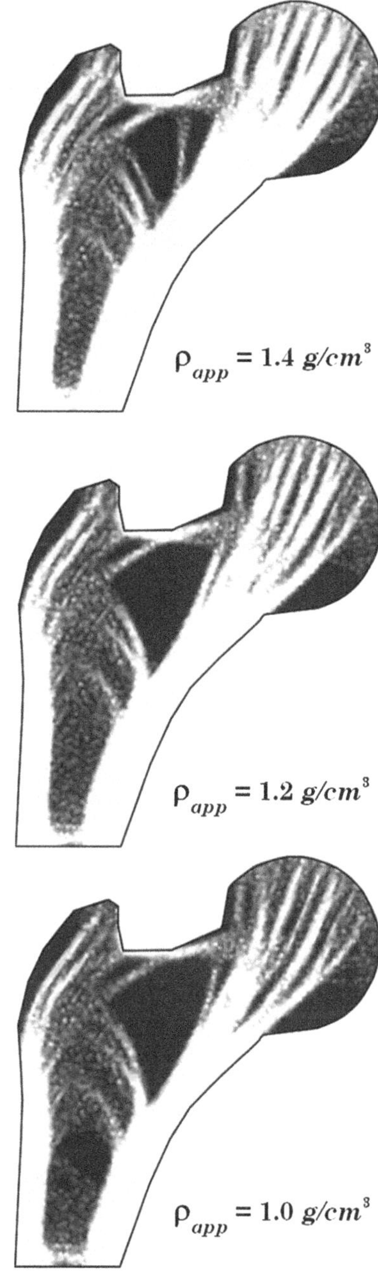

Fig. 7.45 Evolution of bone tissue trabecular architecture for the studied load cases combinations

published by Belinha et al. [1, 2], the other most relevant research work on the subject [24] does not show a trabecular arrangement as accurate as the one presented in Fig. 7.45.

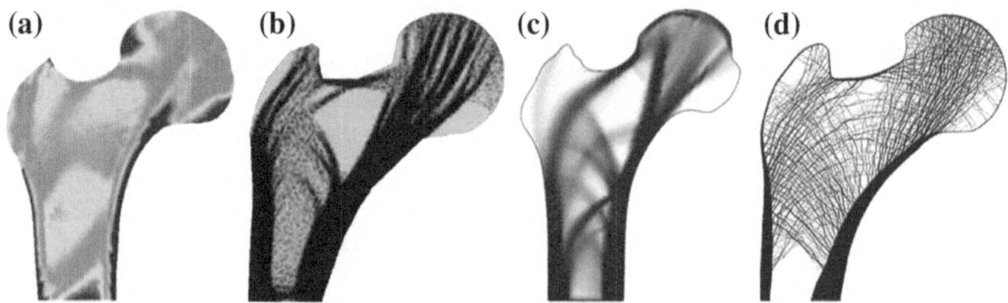

Fig. 7.46 Bone tissue density distribution. **a** Meshless analysis using a 1,500 nodal distribution [24]. **b** Results obtained with the NNRPIM using 6,000 nodes. **c** FEM analysis using 6,334 nodes (2033 8-node elements) [20]. **d** FEM analysis using the micro-finite element (with 0.76 million quadratic elements) [22]

Figure 7.46 shows the difference between the meshless results presented by other authors [24] and the results obtained with NNRPIM (previously shown in Fig. 7.45). The comparison between the results is clear: The present analysis is much more similar with the femur X-ray plate, presented in Fig. 7.37a. Notice that in Fig. 7.46 the colour-scale used in the isomap was inverted, the objective is to present isomaps with the same colour-scale of the X-ray plate presented in Fig. 7.37a.

In order to prove the accuracy and the efficiency of the proposed analysis, in Fig. 7.46c, d are presented some results from other works using the Finite Element Method [20, 22]. In the FEM analysis presented in [20], it was used an element mesh with 6,334 nodes, which is similar with the nodal distribution used in the present meshless analysis. It is clear the accordance between Fig. 7.46b, c. In the author opinion Fig. 7.46b resembles much more the real trabecular architecture, which can be seen in X-ray plates such as Fig. 7.37a, than the results on Fig. 7.46c. At the moment it is not possible to achieve the results obtained in Fig. 7.46d, because the author do not possess the computational power to proceed with the analysis of a nodal distribution with approximately 1.0 million nodes.

7.2.3.2 Femoral Three Dimensional Analysis

The three-dimensional analysis of the femur bone is similar with the two-dimensional analysis. For this analysis a three-dimensional model of the femur was constructed, Fig. 7.47. The model was discretized in 3,318 nodes, which is a very coarse three-dimensional nodal mesh. However, for the three-dimensional analysis computational limitations impose a limit around the 3,500 nodes. In order to have a three-dimensional nodal discretization with the same density as the one used in the two-dimensional analysis, the problem domain should be discretized with nearly 300,000 nodes.

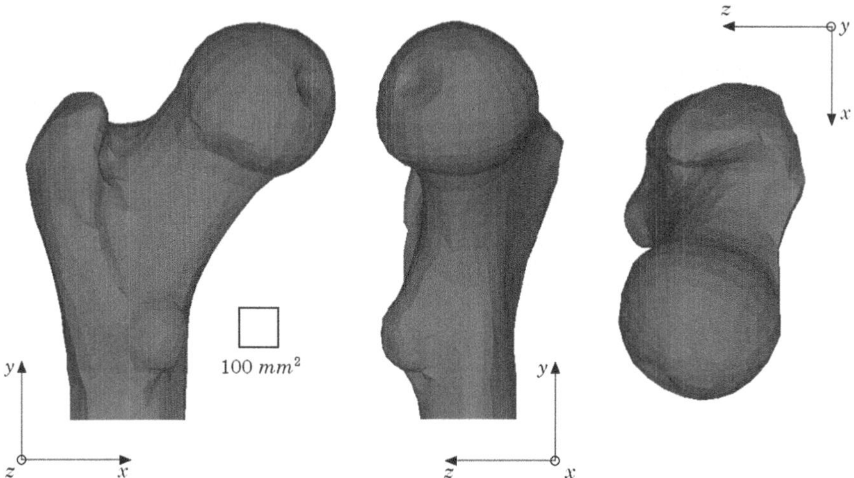

Fig. 7.47 Three-dimensional model of the femur bone

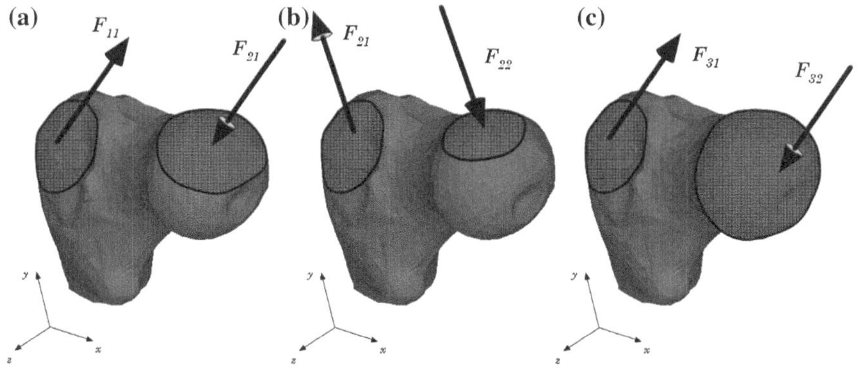

Fig. 7.48 **a** First mechanical case. **b** Second mechanical case. **c** Third mechanical case

For the three-dimensional analysis three-load cases were considered, adapting the load cases suggested by Beaupré et al. [14, 15], Fig. 7.48. The loads respect the direction and magnitude of the ones presented in Fig. 7.39.

By choise, the load vectors possess a null z component, $F_{ij} = \{f_x \quad f_y \quad 0\}$. The loads are, as in the two-dimensional analysis, parabolic distributed along the surfaces indicated in Fig. 7.48. In all three load cases the proximal femur model is clamped in the basis. In order to visualize the evolution of the internal apparent density in the femur bone, a sectional cut is made in the model, Fig. 7.49.

Following the same procedure of the two-dimensional study, each one of the load cases presented in Fig. 7.48 are independently analysed. Again, an initial

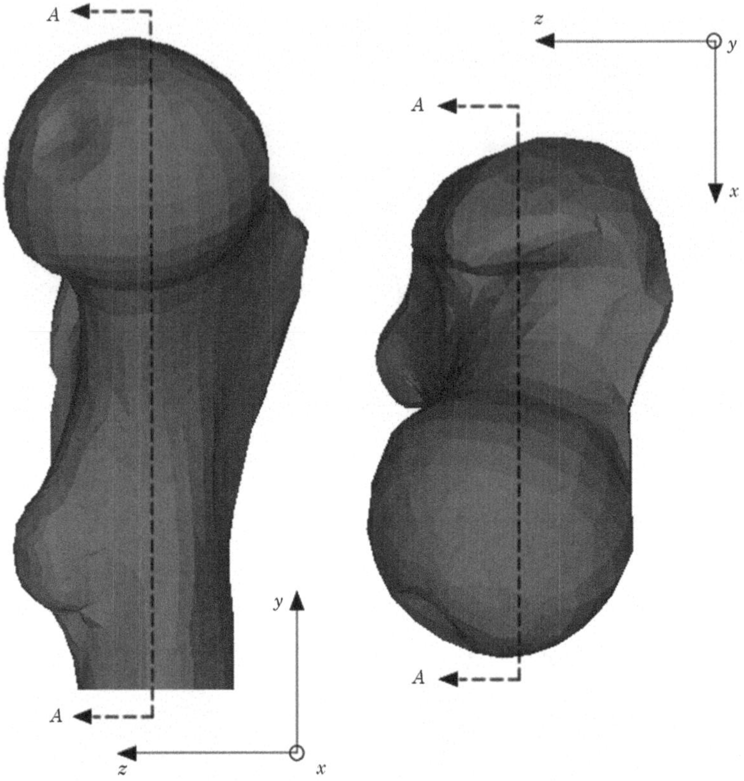

Fig. 7.49 Three-Dimensional model sectional cut AA

uniform density distribution $\rho_{app}^{max} = 2.1\,\text{g/cm}^3$ is imposed to define the initial bone tissue material properties using the proposed material law. The Poisson ratio is assumed as $\upsilon = 0.3$ and the apparent density control value is considered as $\rho_{app}^{control} = 0.4\,\text{g/cm}^3$. The α and β parameters are defined as $\alpha = \beta = 0.01$.

In Fig. 7.50 it is presented the evolution of the trabecular bone architecture in the femur for the distinct mechanical cases.

Notice that it is not possible to visualize clearly the trabecular structure of the inner bone. The used three-dimensional model was created with a very coarse nodal mesh, which does not permit the formation of trabecular structures as in the two-dimensional example. However, it is possible to observe clear density areas, which are in agreement with the two-dimensional results.

In the following analysis the three-load cases were also applied simultaneously. As in the two-dimensional example, the first mechanical case was applied with 6,000 cycles per day and the remaining mechanical cases with 2,000 cycles per day. The results are presented in Fig. 7.50d.

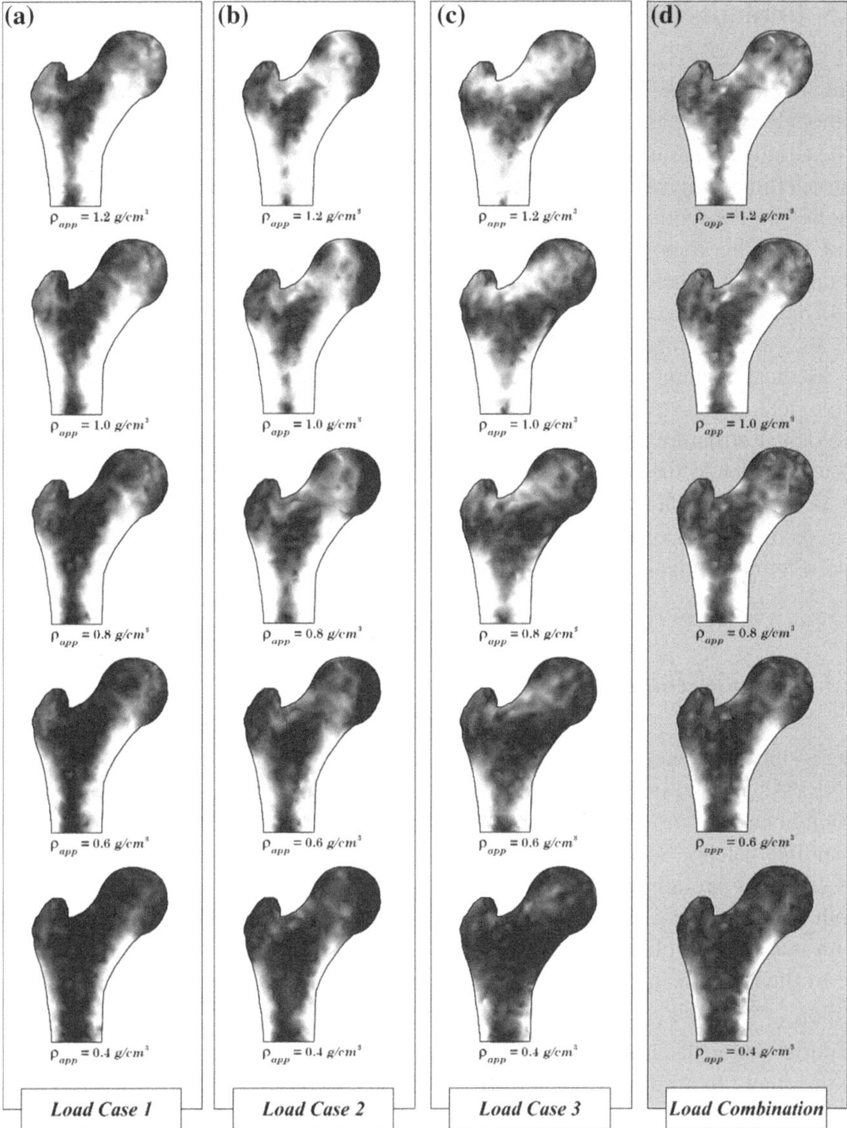

Fig. 7.50 Evolution of the trabecular architecture in the femoral bone for the **a** first mechanical case, **b** second mechanical case and **c** third mechanical case. **d** Three mechanical cases applied simultaneously

In Fig. 7.50d it is possible to observe the principal internal structures empirically suggested in Fig. 7.37b. For medium apparent densities $\rho_{app}^{control} = 0.8\,\text{g/cm}^3$ the principal compressive and tensile group are detectable and also the Ward's triangle and the greater trochanteric group.

7.3 Implants

The main purpose of this section is to verify the efficiency of the proposed remodelling approach in the prediction of the correct final trabecular architecture when rigid implants are inserted in the bone structure. As in the previous bone remodelling analyses, the initial bone apparent density is not relevant. All the studies in this section will start considering an initial maximum apparent density, and then the remodelling process will continue until a final configuration is obtained. Nevertheless, the proposed remodelling algorithm is capable of starting with any initial apparent density. The final result, regardless the initial values, must be always the same.

Medical implants are devices manufactured to: functionally improve existing biological structures; to replace lost biological structures; or to reinforce damaged biological structures. Generally, biocompatible materials cover the implant surface, in order to increase the success rate of the implant osseointegration.

In this section it is studied the bone tissue remodelling due to the insertion of two distinct types of implants: a dental implant, inserted in the mandibular bone; and a femoral prosthesis.

7.3.1 Dental Implants

Here, the presented bone tissue remodelling algorithm, combined with the NNRPIM, is used in the analysis of a single dental implant inserted in a mandible patch bone, Fig. 7.51, which corresponds to the position of the first premolar. The mandible patch is sectioned in two distinct analysis planes, Fig. 7.52. Each one of these planes are analysed separately considering a two-dimensional approach. In both analyses, the obtained trabecular bone architecture shows a good agreement with mandible/implant X-ray images.

In this section, for all the studied examples, an initial uniform density distribution $\rho_{app}^{max} = 2.1\,\mathrm{g/cm^3}$ is imposed in order to define the initial bone tissue material properties using the phenomenological bone tissue material law proposed in this book. In all the presented studies it is considered that the Poisson ratio does not depend on the apparent density, being a constant value: $\upsilon = 0.3$. The α and β parameters, required by the remodelling algorithm to govern the growth and the decay of the bone tissue, are considered: $\alpha = \beta = 0.01$.

7.3.1.1 NNRPIM/FEM Comparison

Considering the section plane indicated in Fig. 7.52b, a two-dimensional model of the mandible bone patch was developed, Fig. 7.53a. The proposed model is based on a computational model presented in the literature [27], which was obtained

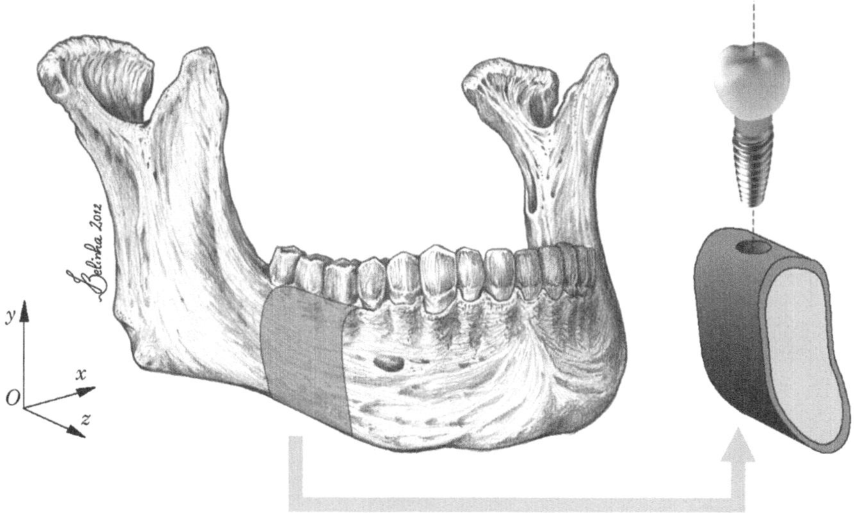

Fig. 7.51 Complete mandible bone and mandible bone patch

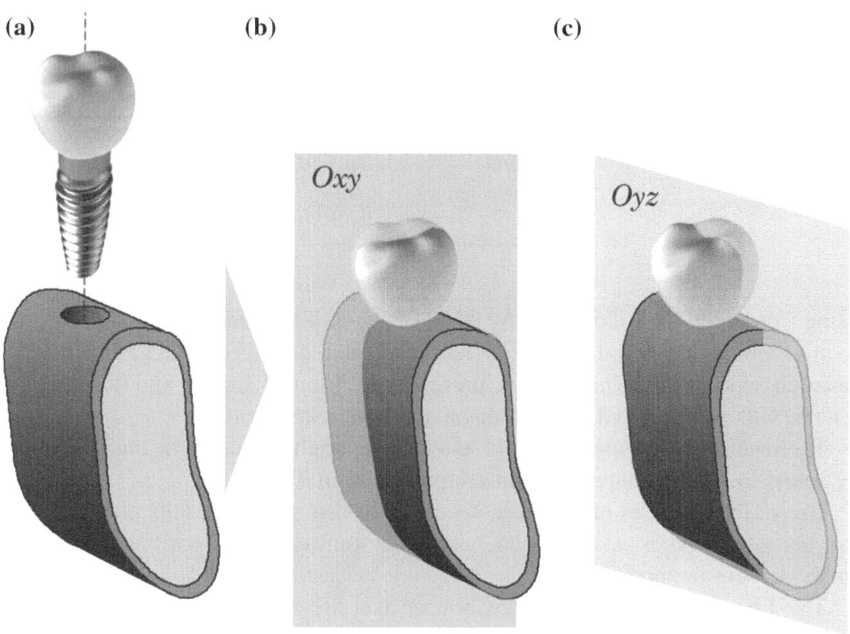

Fig. 7.52 **a** Mandible patch model with a dental implant. **b** Model *Oxy* section. **c** Model *Oyz* section

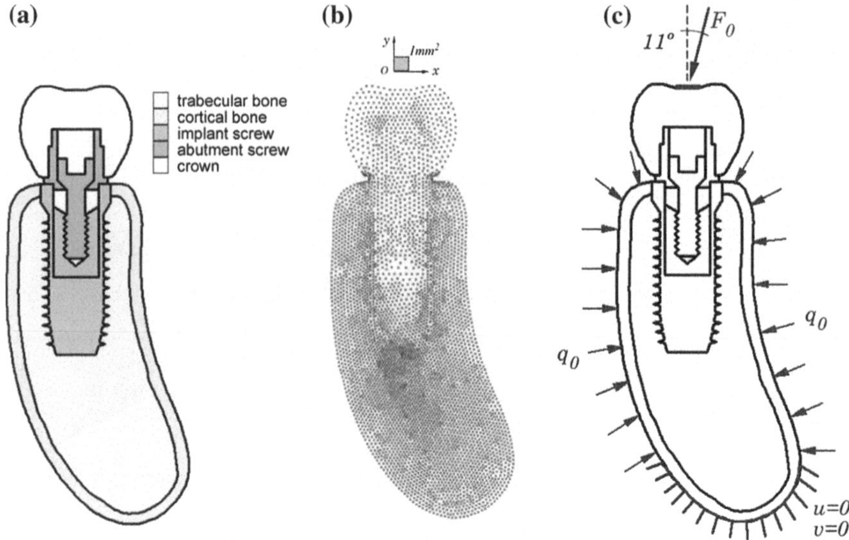

Fig. 7.53 **a** Distinct materials considered in the numerical model. **b** Geometry and nodal distribution used in the analysis (4,674 nodes). **c** Considered essential and natural boundary conditions

Table 7.3 Mechanical properties of the implant structures

Anatomical structure	Young modulus (GPa)	Poisson's ratio
Crown	172	0.30
Abutment	105	0.30
Abutment screw	92	0.30
Implant	105	0.30

using Computer Axial Tomography (CAT) images. The numerical model analysed in this work uses a dental implant system respecting dimensions suggested in other research works [27]. The implant presents a 4.5 mm diameter and 11.0 mm in length, with a 2° tapered conical shape and a helical thread.

The mechanical properties of the distinct materials indicated in Fig. 7.53a can be found in the literature [27] and are presented in Table 7.3.

One of the purposes of the present study is to compare the results obtained with the proposed numerical approach with other remodelling algorithms using the finite element method [27, 28].

The FEM model proposed by Chou et al. [28] considers a two-dimensional four-node finite element and discretize the problem domain with 12,800 elements, which corresponds approximately to 6,600 nodes. More recently Lian et al. [27] proposed another FEM model considering a two-dimensional three-node finite element with 7,132 elements, corresponding approximately to 3,800 nodes. For the present meshless analysis an irregular nodal distribution with 4,674 nodes is used to discretize the problem domain, Fig. 7.53b.

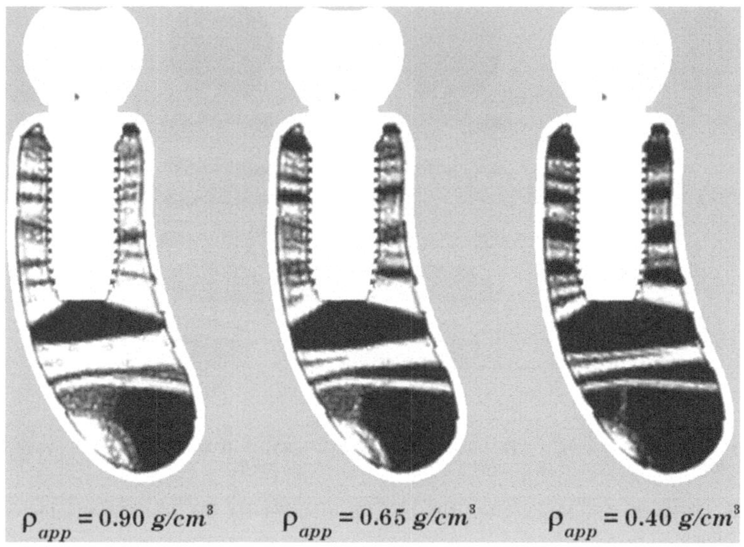

Fig. 7.54 Obtained trabecular architecture for the selected apparent density control values

The essential and natural boundary conditions are suggested in the work of Chou et al. [28]. The implant system is submitted to an occlusal load $F_0 = 100$ N applied directly in the crow, inclined 11° in relation to the implant longitudinal axis. A second load condition is suggested: an uniform distributed pressure along the outer surface of the cortical bone, $q_0 = 500$ kPa, intended to simulate the effect of the mandibular flexure [28] and a more realistic overall boundary condition [27]. Regarding the essential boundary conditions, the model is constrained in the basis, along x and y directions. In Fig. 7.53c it is possible to observe the referred boundary conditions.

Three distinct medium bone density control values were assumed: $\rho_{app}^{control} = 0.90\,\text{g/cm}^3$, $\rho_{app}^{control} = 0.65\,\text{g/cm}^3$ and $\rho_{app}^{control} = 0.40\,\text{g/cm}^3$. Notice that the process stops when the medium bone density reaches the control value.

The obtained trabecular distributions for each one of the control values are presented in Fig. 7.54 and the respective von Mises effective stress distributions is presented in Fig. 7.55. A permanent cortical bone perimeter with a 0.5 mm–0.8 mm thickness was considered as suggested in the literature [27, 28].

The results show that the proposed remodelling algorithm, combined with the NNRPIM, is capable of reproducing trabecular distributions very close to the FEM results, Fig. 7.56. The remodelling algorithm was able to predict the same higher apparent density regions on the mandible patch. Under the implant it is possible to observe high-density horizontally oriented regions connecting the cortical layers on the bone periphery. Also the bone resorption, immediately below the implant, and the diagonal trabeculae connecting the implant structure to the peripheric cortical layers are predicted by the bone tissue remodelling algorithm. A closer

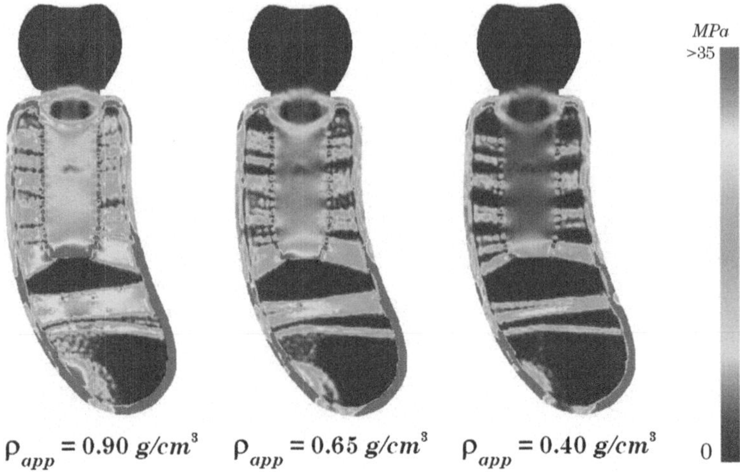

Fig. 7.55 Obtained von Mises effective stress distribution for the selected apparent density control values

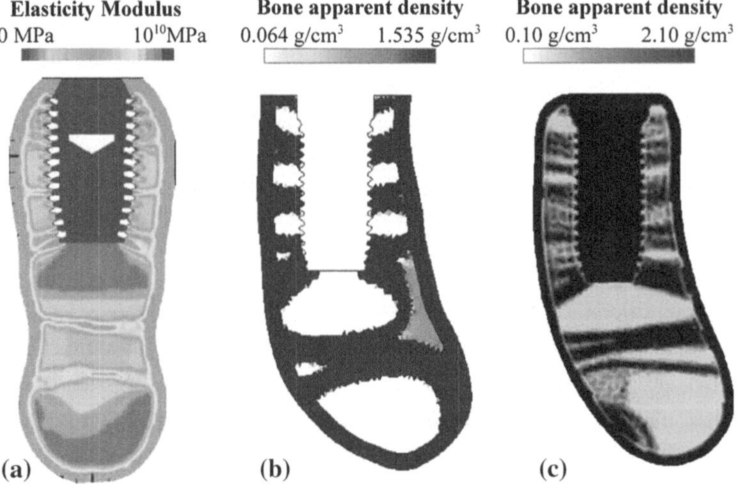

Fig. 7.56 Bone tissue apparent density distribution. **a** FEM solution [28]. **b** FEM solution [27]. **c** NNRPIM solution

look permits to verify that the meshless approach permits to achieve results that resembles much more real trabecular structures when compared with the FEM results [27, 28], Fig. 7.56.

Table 7.4 Considered load cases

	Load cycles	
Load case	F_0 (100 N)	q_0 (500 kPa)
LC1	100	900
LC2	250	750
LC3	500	500
LC4	750	250
LC5	900	100
LC6	999	1

7.3.1.2 Load System Analysis

In the pursuit of a suitable numerical load system resembling the real biological load system, in this example are proposed six load cases, combining the two loads previously presented, Fig. 7.53c. Each load, $F_0 = 100$ N and $q_0 = 500$ kPa, is weighted according with the number or cycles presented in Table 7.4 and combined using Eq. (6.25).

Assuming the same numerical model, Fig. 7.53, and the same material properties, Table 7.3, for the implant system presented in previous subsection, the problem was analysed considering each one of the six load cases presented in Table 7.4. Again an initial uniform apparent density distribution $\rho_{app}^{max} = 2.1$ g/cm^3 was considered, as well as the three distinct medium bone density control values assumed previously: $\rho_{app}^{control} = 0.90\,\text{g/cm}^3$, $\rho_{app}^{control} = 0.65\,\text{g/cm}^3$ and $\rho_{app}^{control} = 0.40\,\text{g/cm}^3$.

The results obtained with the suggested bone tissue remodelling algorithm, for each one of the 18 numerical analyses, are presented in Fig. 7.57. The von Mises effective stress distributions obtained are presented in Fig. 7.58. In the first row of Figs. 7.57 and 7.58 are presented the results regarding the analysis considering $\rho_{app}^{control} = 0.90\,\text{g/cm}^3$ as the medium bone density control value, the second and third row show the results for the medium bone density control values $\rho_{app}^{control} = 0.65\,\text{g/cm}^3$ and $\rho_{app}^{control} = 0.40\,\text{g/cm}^3$ respectively. The load cases are indicated with the acronym *LC#*.

From Figs. 7.57 and 7.58 it is possible to visualize that the uniform distributed pressure along the outer surface of the cortical bone, with an initial magnitude of $q_0 = 500\,\text{kPa}$, is extremely severe. The first five load cases reduce gradually the importance of q_0 in relation to the occlusal load $F_0 = 100\,\text{N}$, to a minimum of 100 cycles, which correspond roughly to 10 % of the initial value (load case LC5). Notice that despite the assumed reduction, the obtained results for the first five load cases are very close to each other, which indicates that the occlusal load does not contributes significantly to the mandible bone tissue remodelling.

Load case 6 (LC6) was introduced in the analysis to understand how the mandible remodelling evolves with the drastic decrease of the uniform distributed pressure weight, Table 7.4. This reduction corresponds approximately to consider a simple load set: $F_0 = 100\,\text{N}$ and $q_0 = 0.5\,\text{kPa}$.

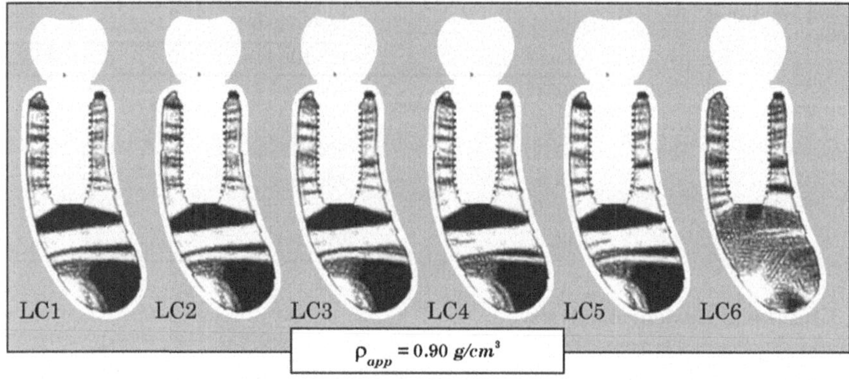

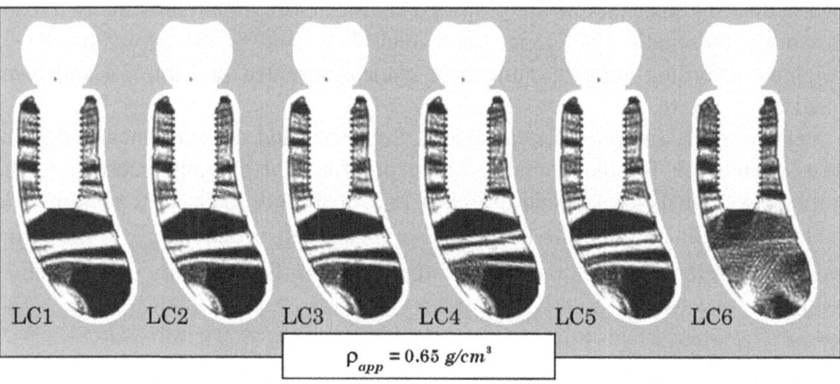

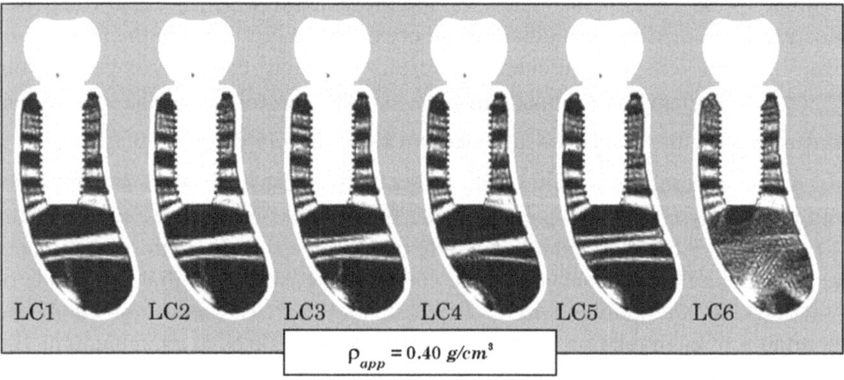

Fig. 7.57 Obtained trabecular architecture for the selected control values and for the distinct load combinations

The results presented in Figs. 7.57 and 7.58 for the load case LC6 resemble much more a mandible bone X-ray plate. Under the implant the high-density horizontally oriented trabecular bone structures, which appear on the other load cases connecting

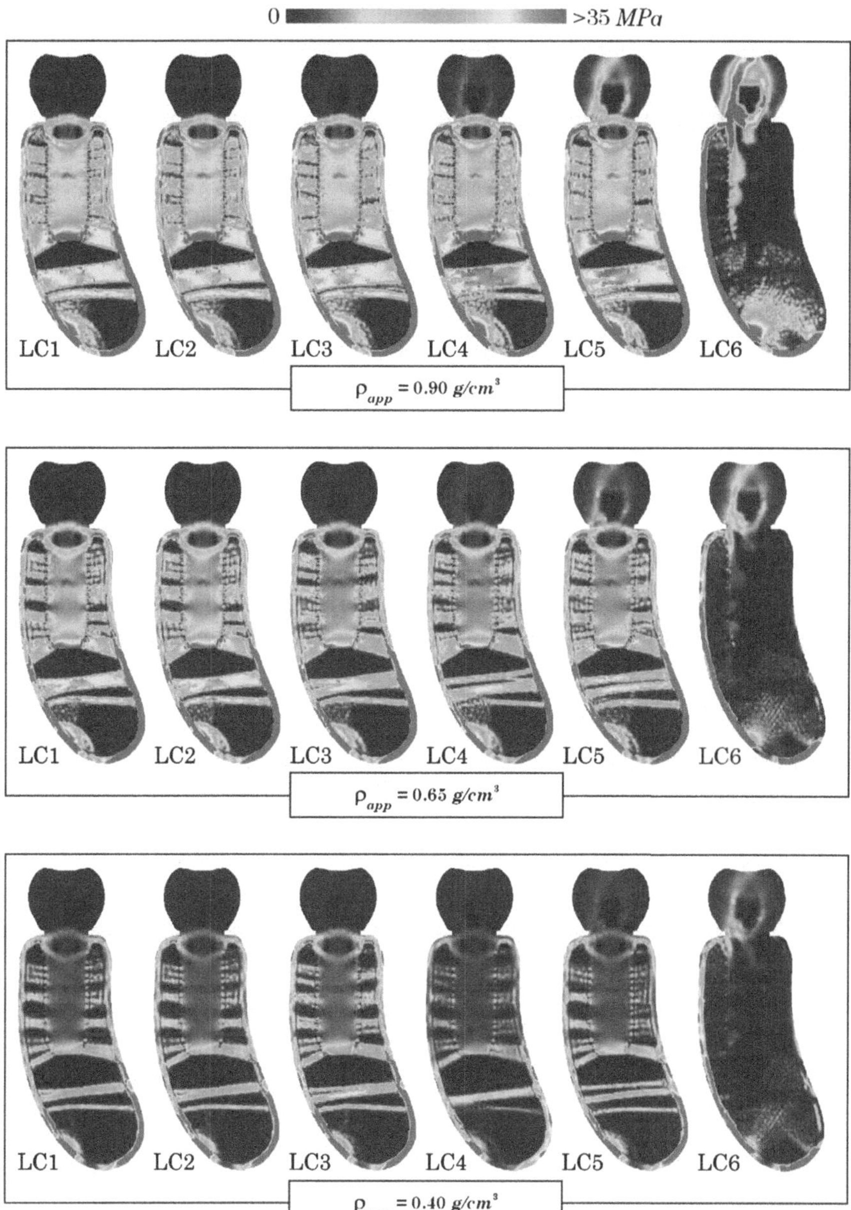

Fig. 7.58 Obtained von Mises effective stress distribution for the selected control values and for the distinct load combinations

the cortical layers on the bone periphery, have disappear and were substituted by a trabecular net much more realistic. The bone resorption, immediately below the implant, continues to be predicted by the algorithm, showing with LC6 a much more

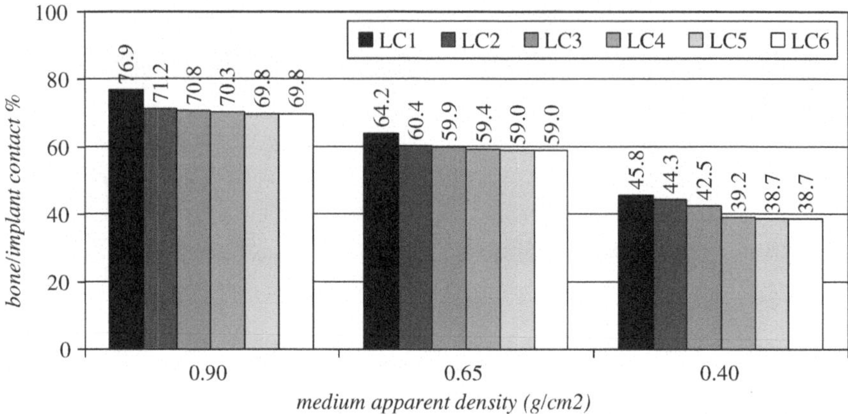

Fig. 7.59 Bone-implant contact percentages for each analysis

organic form. The diagonal trabeculae, connecting the implant structure to the peripheric cortical layers, present slight differences when compared with the other studied load cases. It is important to refer that the modification of the uniform distributed pressure magnitude does not introduces significant changes in the bone tissue remodelling in the implant vicinity.

Clinical experience considers that an implant is successfully inserted when in the end of the bone tissue remodelling phase 50–80 % bone-implant contact is verified [27]. In the work developed by Lian et al. [27] it is studied the bone-implant contact percentage due to the bone remodelling surrounding a dental implant. It was found that regardless the considered initial percentage of bone-implant contact, the numerical final result obtained is approximately a 60 % of contact.

Therefore, for each one of the 18 remodelling analyses previously described, it were obtained the percentage of bone-implant contact in the end of each analysis. The results are presented in Fig. 7.59. It is possible to observe that the obtained results with the NNRPIM corroborate the results obtained with the FEM [27]. For load case 3 (LC3), considering a medium bone density control value $\rho_{app}^{control} = 0.65\ \text{g/cm}^3$, it is obtained a bone-implant contact of 59.90 %. For the same analysis in [27] it was obtained a 59.58 % contact. It is also possible to observe that the uniform distributed pressure q_0 magnitude does not influence significantly the results, indicating that the bone-implant contact depends mainly on the occlusal load magnitude.

7.3.1.3 Mandible Plane Section Oyz Analysis

This last mandible bone example considers the section plane indicated in Fig. 7.52c. The developed two-dimensional model of the mandible bone section is presented in Fig. 7.60a.

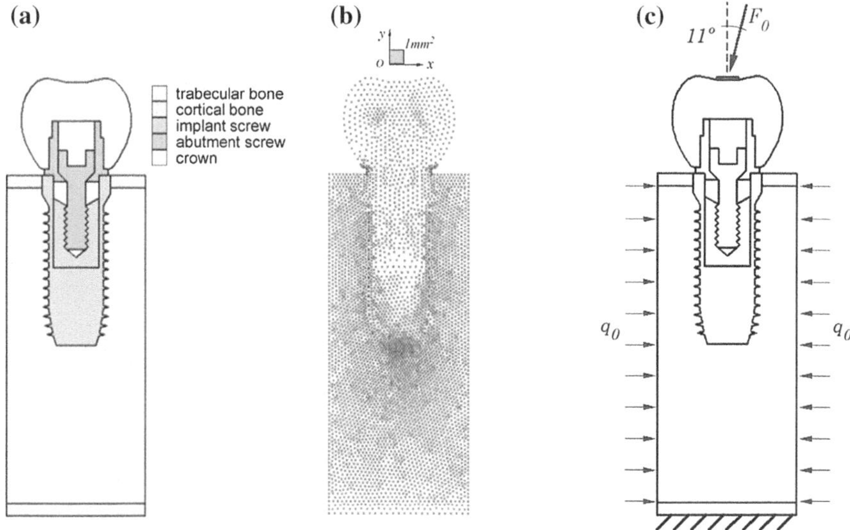

Fig. 7.60 **a** Distinct materials considered in the numerical model. **b** Geometry and nodal distribution used in the analysis (4,723 nodes). **c** Considered essential and natural boundary conditions

The implant system studied in this example is the same as in the previous mandible examples. The mechanical properties of the individual structures indicated in Fig. 7.60a are presented in Table 7.3. The solid domain was discretized in an irregular nodal distribution with 4,723 nodes, Fig. 7.60b. As in the previous examples, two distinct loads acting simultaneously are considered: an occlusal load $F_0 = 100$ N oriented 11° in relation to the implant longitudinal axis, once more applied directly in the crow; and a uniform distributed pressure q_0 acting in the vertical boundaries of the model. The uniform distributed pressure aims to simulate the stress induced by the mandibular flexure [28] and the internal fluid pressure.

Considering the results and the conclusion of the previous example, it is considered a much lower magnitude for the uniform distributed pressure: $q_0 = 0.5$ kPa. The schematic representation of the applied force system is presented in Fig. 7.60c. It is also possible to visualize in Fig. 7.60c that the model is constrained in the basis along x and y directions.

It was considered an initial uniform apparent density distribution $\rho_{app}^{max} = 2.1\,\mathrm{g/cm^3}$ and, as in previous examples, the same three distinct medium bone density control values were assumed: $\rho_{app}^{control} = 0.90\,\mathrm{g/cm^3}$, $\rho_{app}^{control} = 0.65\,\mathrm{g/cm^3}$ and $\rho_{app}^{control} = 0.40\,\mathrm{g/cm^3}$. A permanent cortical bone perimeter with a 0.5 mm thickness was considered in the model top and bottom bone surface.

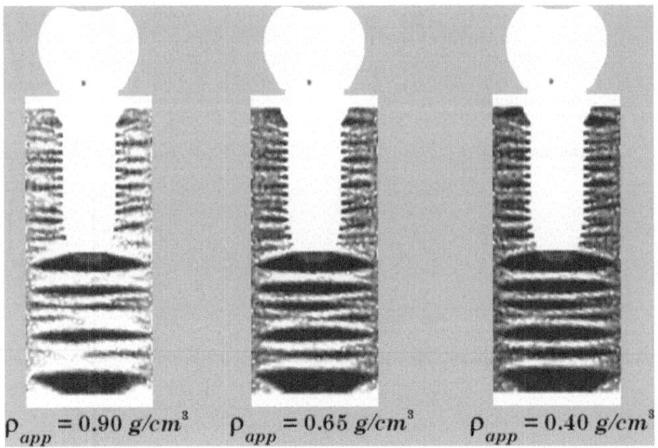

Fig. 7.61 Obtained von Mises effective stress distribution for the selected apparent density control values

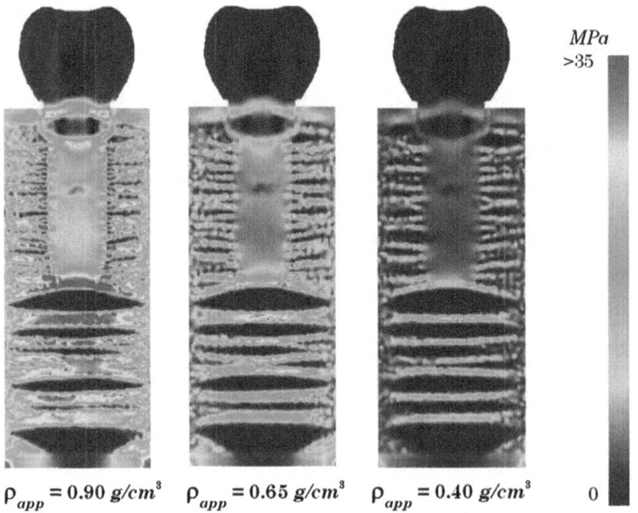

Fig. 7.62 Obtained trabecular architecture for the selected apparent density control values

In Fig. 7.61 are presented the obtained trabecular architecture for each one of the control values considering the proposed remodelling algorithm. In Fig. 7.62 are presented the correspondent von Mises effective stress distributions.

The obtained trabecular distributions show the same bone resorption below the implant and well defined horizontal trabeculae. These results, resembling real trabecular structures, are in accordance with clinical observations for the considered mandible section (plane *Oyz*).

7.3.2 Femoral Prosthesis

The long-term effect of the implant on the bone tissue can be predicted by the bone tissue remodelling simulation. Additionally, this useful numerical tool can be used to select and/or optimize the implant appropriate shape and/or material.

The aged population presents a higher incidence of fractures. Epidemiologic studies regarding osteoporotic fractures [29] concluded that the lifetime risk of an osteoporotic fracture at age 50 years in developed countries (UK, Sweden, USA and Australia) varies between 39.7–53.2 % for women and 13.1–22.4 % for men.

Hip fracture patients present high morbidity and experience a significantly loss of the quality of life due to the fracture. Generally, the hip fracture treatment requires the implantation of a structural element with the objective of securing the bone weight-bearing capability. In the particular case of the hip fracture, with the introduction of the implant, the loads initially applied in the femoral head and in the trochanters will change drastically. The new forces applied in the set implant/femur will depend on the fracture type and will change the stress field on the remaining host bone. This phenomenon is commonly known as stress shielding. The new stress distribution will catalyse the local bone tissue remodelling.

It is possible to find in the literature many clinical cases describing the avascular necrosis of the femoral head after the surgical treatment of intertrochanteric fractures [30–32]. In the work of Guimarães et al. [32] it is described a clinical case in which a femur bone with an intertrochanteric fracture was operated and a dynamic hip screws system was inserted Fig. 7.63a. However, after nine months the radiographs show clear signs of avascular necrosis of the femoral head Fig. 7.63a. The authors believe that the necrosis was probably caused by the loss of functionality of the great trochanter (due to a secondary fracture or injury during the first intertrochanteric fracture). Therefore, a new operation was performed and a stem similar with the femoral implant presented in Fig. 7.63c was inserted, Fig. 7.63b.

The numerical example proposed in this section aims to analyse the bone tissue remodelling after the second clinical intervention. Therefore, it is studied the bone tissue remodelling due to the insertion of an implant on a femur bone showing an intertrochanteric fracture combined with the loss of functionality of the great trochanter. The two-dimensional model of the fractured femur bone was constructed using a X-ray plate of a clinical case [32], Fig. 7.63a, b. Using Fig. 7.63a the contour of the complete proximal femur on the coronal plane was obtained, then based on Fig. 7.63b, the initial complete femur model was used to obtain the contour of the fractured proximal femur on the coronal plane. A similar imaging procedure allowed to construct the contour of the Exeter™ Femoral Stem V40™ represented in Fig. 7.63c. This stem was selected since the geometric properties are available on the free literature and it resembles the stem in Fig. 7.63b.

In Fig. 7.64 it is presented the geometry of the model contour. In this example the implant/femur system presented in Fig. 7.64 is studied considering four distinct analyses. For all the analyses the model domain is discretized with the same

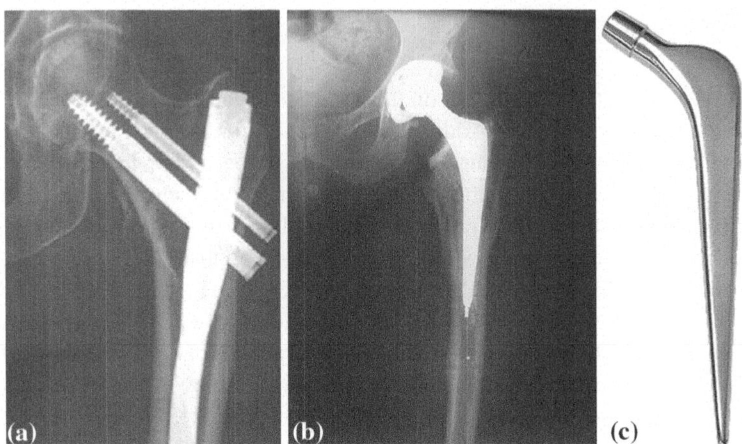

Fig. 7.63 **a** Anteroposterior postoperative radiograph, with nine months of evolution, showing consolidated fracture, with clear radiological signs of avascular necrosis of the femoral head [32]. **b** Anteroposterior postoperative radiograph, with twelve months of evolution, after a full hybrid hip arthroplasty reoperation [32]. **c** Exeter™ Femoral Stem V40™

irregular nodal distribution presented in Fig. 7.65a. Notice that the implant domain is discretized with 1,525 nodes and the support bone with 3,402 nodes.

In this section, all the studied examples consider an initial uniform density distribution $\rho_{app}^{max} = 2.1\,\text{g/cm}^3$ and a Poisson ratio $\upsilon = 0.3$. For the α and β parameters ruling the growth and the decrease of the bone tissue it is considered: $\alpha = \beta = 0.01$. Additionally, four distinct medium bone density control values were assumed: $\rho_{app}^{control} = 1.20\,\text{g/cm}^3$, $\rho_{app}^{control} = 1.10\,\text{g/cm}^3$, $\rho_{app}^{control} = 1.00\,\text{g/cm}^3$ and $\rho_{app}^{control} = 0.90\,\text{g/cm}^3$.

In the first analysis the model is submitted to the three loads suggested by Beaupré et al. [14, 15], which can be found in Fig. 7.39. Since it is being simulated an intertrochanteric fracture combined with the loss of functionality of the great trochanter, all the forces applied in the great trochanter are disregarded. Thus, only the forces applied in the femur head are considered, Fig. 7.65b. In the present analysis, the forces suggested by Beaupré and co-workers are applied directly in the implant following a uniform distribution. The magnitude, the direction and the number of cycles of each considered load case are presented in Table 7.5.

Regarding the essential conditions, the model the model is constrained in the basis, along x and y directions, Fig. 7.65b.

The mechanical properties of the bone tissue depend on the current apparent density and are obtained with the considered phenomenological law. The mechanical properties of the titanium implant stem are: $E = 110$ GPa and $\upsilon = 0.32$. It was not considered in the analysis a cemented interface between the implant and the bone tissue.

In order to simulate the permanent natural cortical layer that can be found in the femur bone, whose function is to provide protection and additional rigidity to the

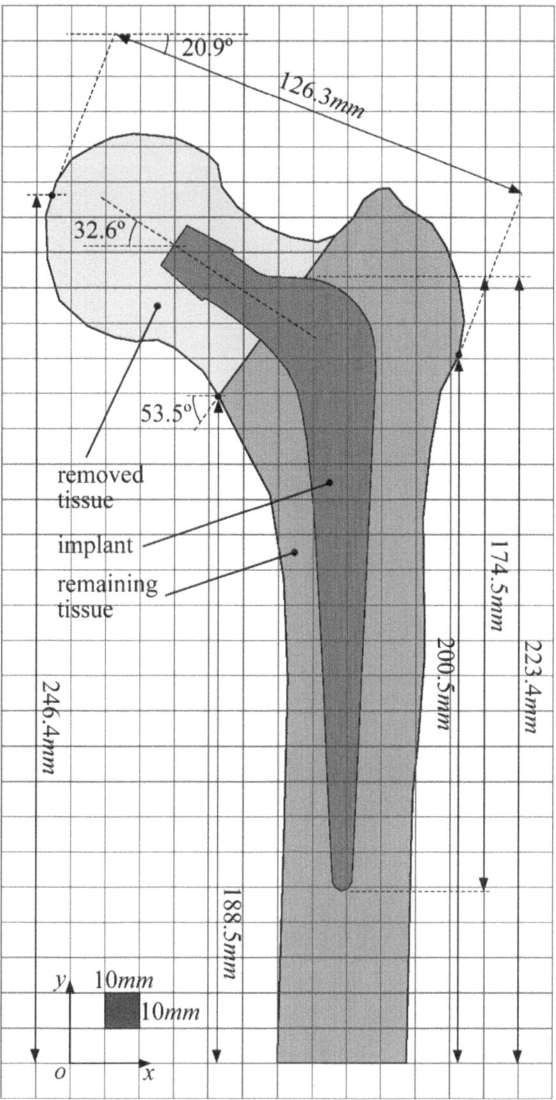

Fig. 7.64 Geometry of the analysed model

bone structure, a permanent cortical bone perimeter with 0.5 mm thickness was considered in the analysis.

The obtained trabecular distributions for each one of the control values are presented in Fig. 7.66. Notice that all the load cases, individually applied, predict the drastic reduction of bone tissue on the femur head.

In order to obtain a closer result to the real implant/femur X-ray plate it is necessary to simultaneously apply all load cases, Table 7.5. The results of the three load cases simultaneously applied are presented in Fig. 7.67.

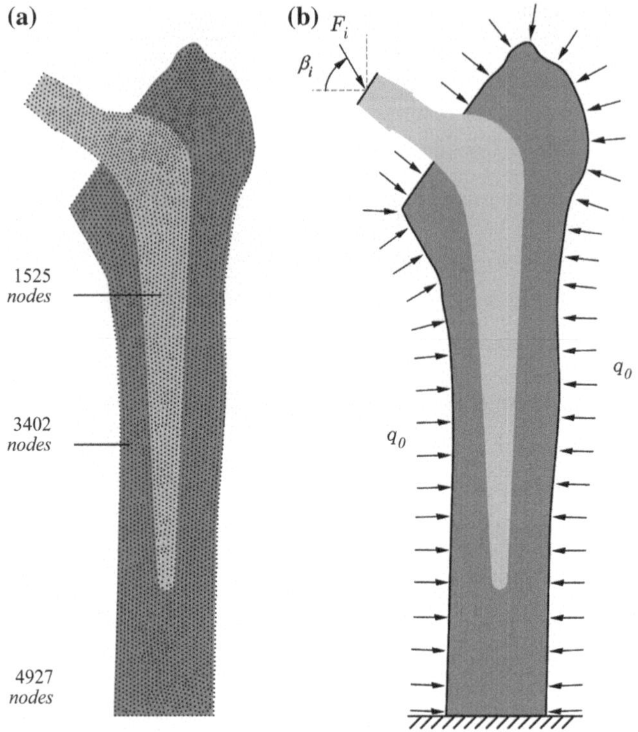

Fig. 7.65 **a** Nodal discretization of the problem domain. **b** Essential and natural boundaries of the numerical model

Table 7.5 Considered load cases

Load case	F_i (N)	β_i (°)	Load cycles
LC1	2,317	66	6,000
LC2	1,158	105	2,000
LC3	1,548	34	2,000

It is possible to confirm in Fig. 7.67 the loss of bone tissue on the femur head, in accordance with X-ray plate in Fig. 7.63b.

The obtained von Mises effective stress distributions for each load case and for the combination of all load cases are presented in Fig. 7.68 for $\rho_{app}^{control} = 1.20\,\text{g/cm}^3$.

Despite the similarity between the obtained numerical solution and clinical observation, a second study as performed. This second analysis aims to examine the importance of considering a constant hydrostatic pressure on the bone surface.

In the literature it is possible to find research works [33] indicating the existence of significant fluid pressures compressing the bone tissue up to 20 kPa. The magnitude of such internal fluid pressures can eventually lead to bone necrosis and the formation of fibrous tissues similar to that observed surrounding loose hip or knee prosthesis [33].

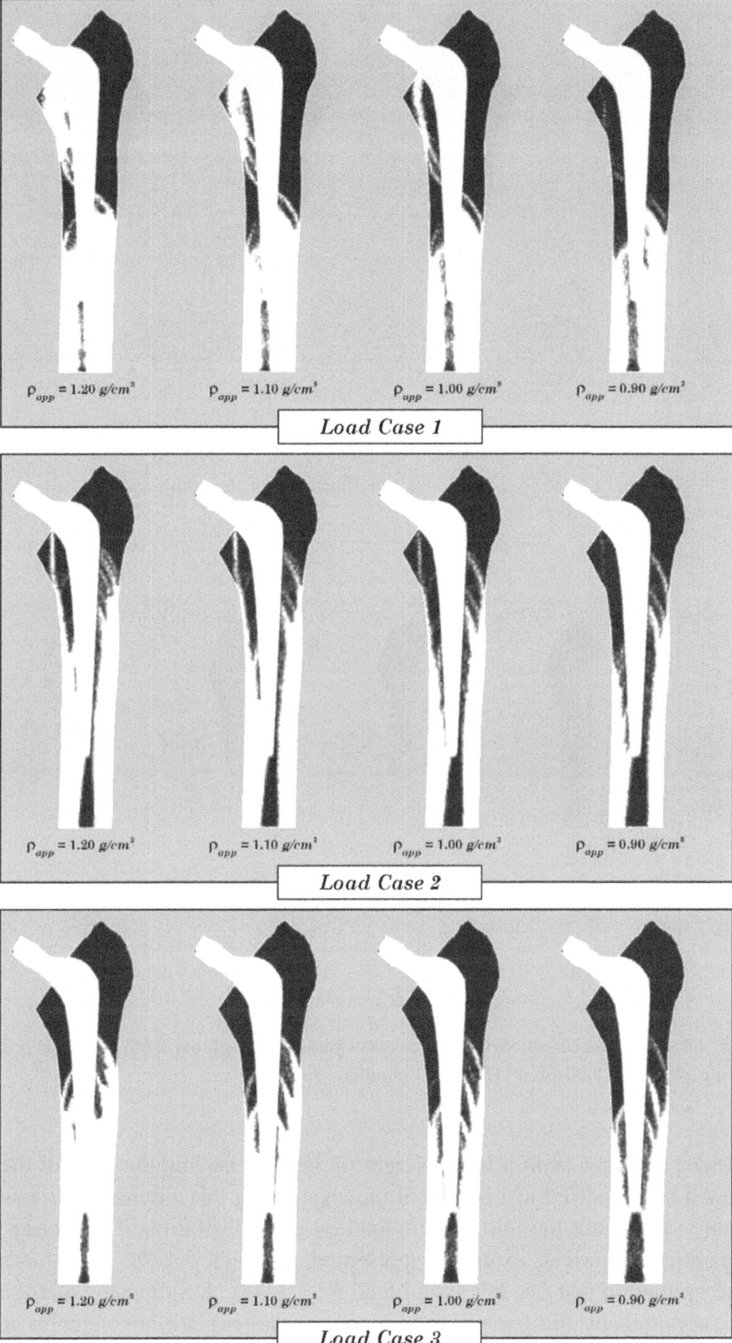

Fig. 7.66 Evolution of bone tissue trabecular architecture for each one of the studied load cases (Analysis number 1)

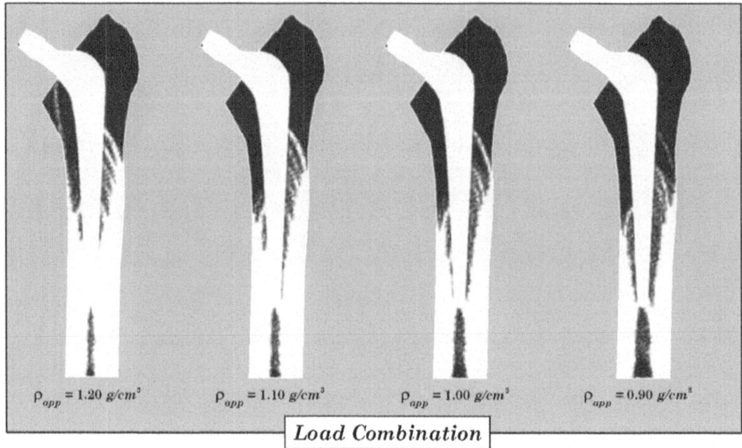

Fig. 7.67 Evolution of bone tissue trabecular architecture for the combination of all the studied load cases (Analysis number 1)

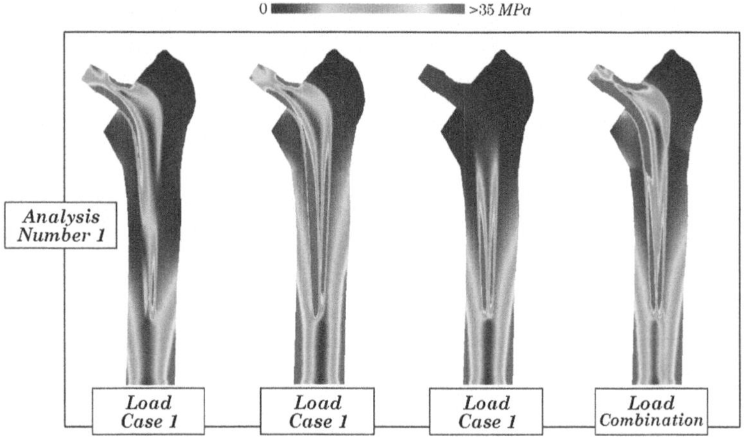

Fig. 7.68 Obtained von Mises effective stress distribution for each one of the analysis performed considering $\rho_{app}^{control} = 1.20\,\text{g/cm}^3$ (Analysis number 1)

Consider a human with a body weight of 100 kg. During the gait all the body weight will be supported just by one of the legs. Being this a dynamic process, it is acceptable to increase the initial static load by a factor of 2 (as it was observed in the dynamic numerical examples presented in Sect. 5.4.4). Therefore, it is assumed applied in just one leg a total load $P = 2{,}000$ N. Supposing that this total load is supported by the leg muscles, a simple and expeditious calculus using a hypothetic leg perimeter $p = 0.50$ m allows to obtain a hydrostatic pressure $q =$ 50 kPa on the muscle tissue, which is not far from the value suggested in the

Table 7.6 Considered load cases

Load case	F_i (N)	β_i (°)	q_0 (kPa)	Load cycles
LC1	2,317	66	50	6,000
LC2	1,158	105	50	2,000
LC3	1,548	34	50	2,000

literature [33]. Thus, a muscle tissue hydrostatic pressure, $q_0 = 50$ kPa, will be applied on the bone surface, Fig. 7.65b. With the inclusion of a permanent uniform hydrostatic pressure three new load cases are established, Table 7.6.

In this second study all the previous considerations regarding the bone phenomenological law and the implant material properties are maintained. Also the interface between the femur bone and the stem are the same as in the first example.

The results of the numerical analysis, considering individually the load cases, are presented in Fig. 7.69 for each one of the considered apparent density control values. Comparing the solutions of Fig. 7.69 with the solutions presented in Fig. 7.66 it is possible to conclude that both studies produced similar results. Nevertheless, the apparent density distributions presented in Fig. 7.69 are more smooth, indicating that the peripheric hydrostatic pressure leads to a more uniform stress distribution along the bone domain.

Afterwards, the three load cases were simultaneously applied considering the load cycles presented in Table 7.6. The results for each apparent density control value are presented in Fig. 7.70. In Fig. 7.71 are shown the von Mises effective stress distributions for each load case and for the combination of all load cases considering $\rho_{app}^{control} = 1.20\,\text{g/cm}^3$.

Comparing Fig. 7.70 with Fig. 7.67 it is possible to observe some slight differences in the obtained trabecular distribution. Nevertheless, these results do not permit to conclude about the relevance of considering a permanent hydrostatic pressure on the bone domain boundary. The comparison of Figs. 7.71 and 7.68 show that both analysis produce very similar results.

In a third study the permanent peripheral cortical layer is removed from the model. The objective is to determine the relevance of the inclusion of the peripheral cortical coat on the evolution of the bone tissue remodelling. All the previous geometric and material considerations are maintained. The essential and natural boundary conditions presented in Fig. 7.65b are imposed and the load cases described in Table 7.6 are considered.

The obtained trabecular distributions are presented in Fig. 7.72 for each one of the considered control values. As in the two previous studies, the results presented in Fig. 7.72 predict the same drastic reduction of the bone tissue apparent density on the femur head. The results regarding the combination of the three load cases, respecting the load cycles presented in Table 7.6, are presented in Fig. 7.73.

Once more, in accordance with X-ray plate in Fig. 7.63b, it is possible to confirm in Fig. 7.73 the loss of bone tissue on the femur head. In Fig. 7.74 are shown the von Mises effective stress distributions for each load case and for the combination of all load cases considering $\rho_{app}^{control} = 1.20\,\text{g/cm}^3$.

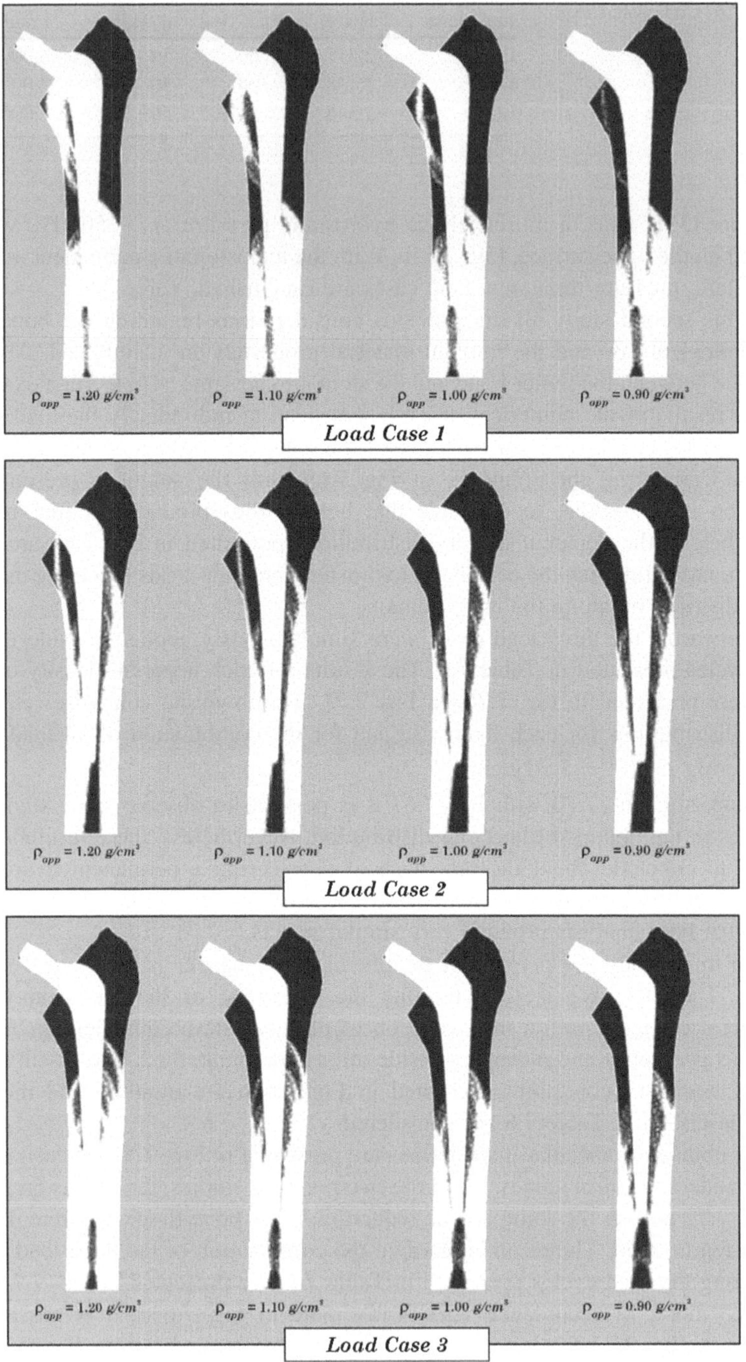

Fig. 7.69 Evolution of bone tissue trabecular architecture for each one of the studied load cases (Analysis number 2)

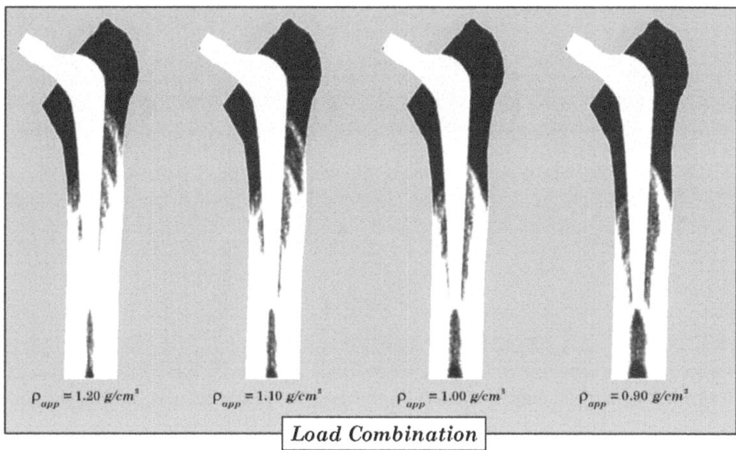

Fig. 7.70 Evolution of bone tissue trabecular architecture for the combination of all the studied load cases (Analysis number 2)

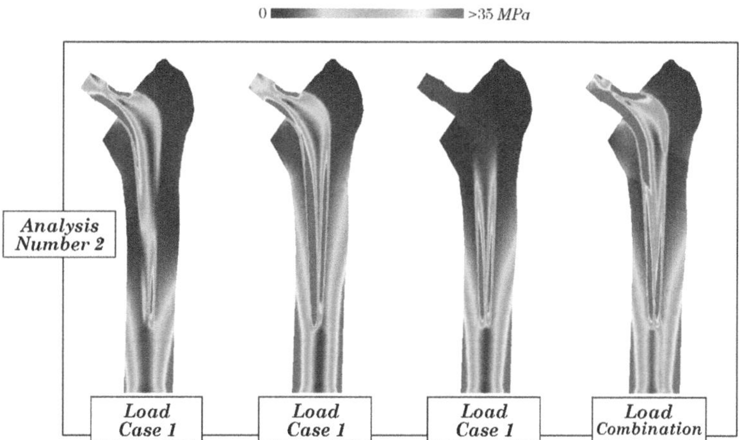

Fig. 7.71 Obtained von Mises effective stress distribution for each one of the analysis performed considering $\rho_{app}^{control} = 1.20\,\text{g/cm}^3$ (Analysis number 2)

Comparing the results regarding the control value $\rho_{app}^{control} = 1.20\,\text{g/cm}^3$ presented in Figs. 7.67, 7.70 and 7.73, it is possible to visualize that the results on Figs. 7.67 and 7.73 show a more close resemblance to the X-ray plate presented in Fig. 7.63b. The left side vertical trabecular is clearly formed and the right side trabeculae fan oriented also appear. Nevertheless, all the results are very similar between each other. To facilitate the comparison Fig. 7.75 is displayed.

The fourth study aims to investigate the influence of the material properties of the femoral implant. In this analysis the implant/bone system does not change, the

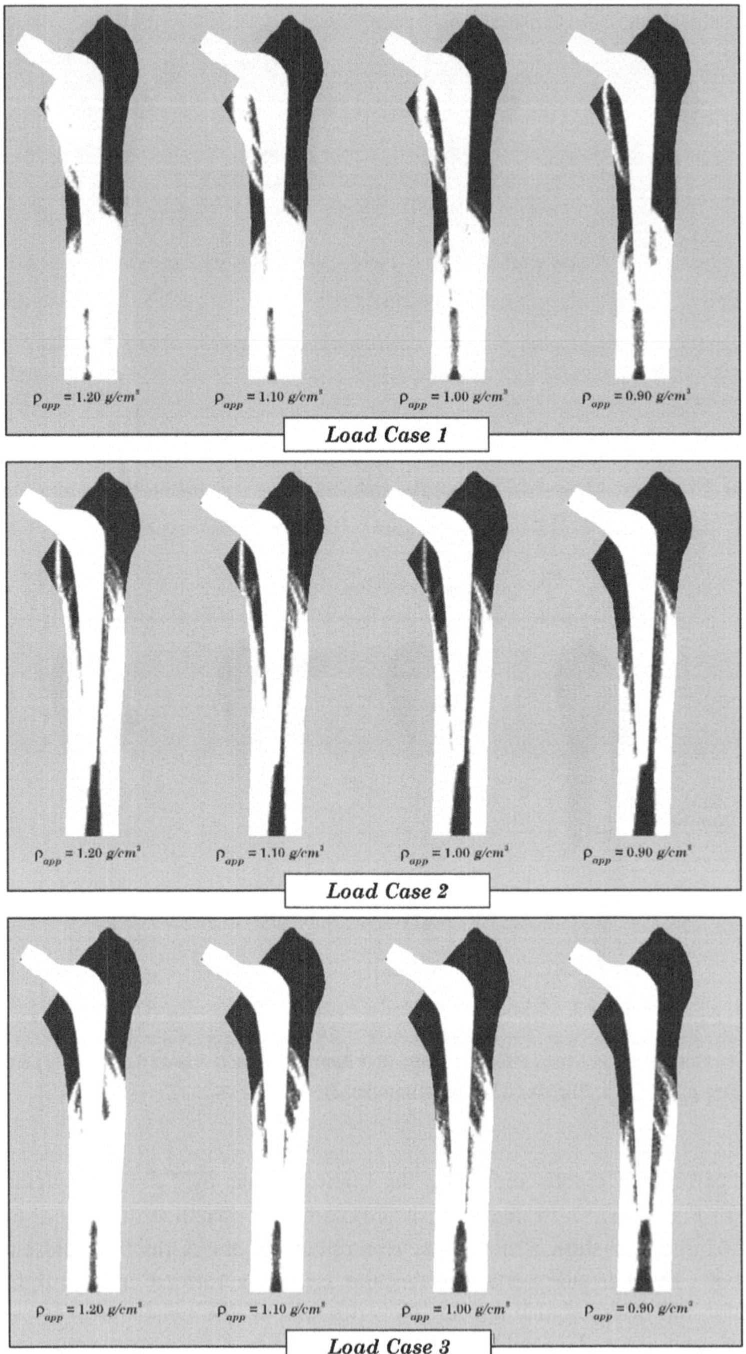

Fig. 7.72 Evolution of bone tissue trabecular architecture for each one of the studied load cases (Analysis number 3)

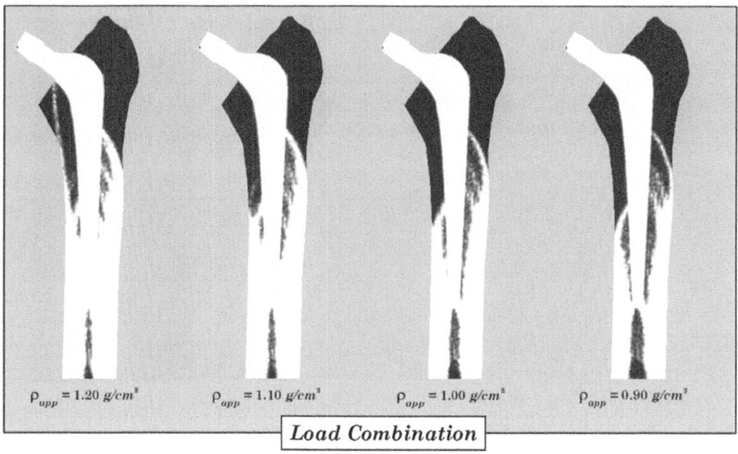

Fig. 7.73 Evolution of bone tissue trabecular architecture for the combination of all the studied load cases (Analysis number 3)

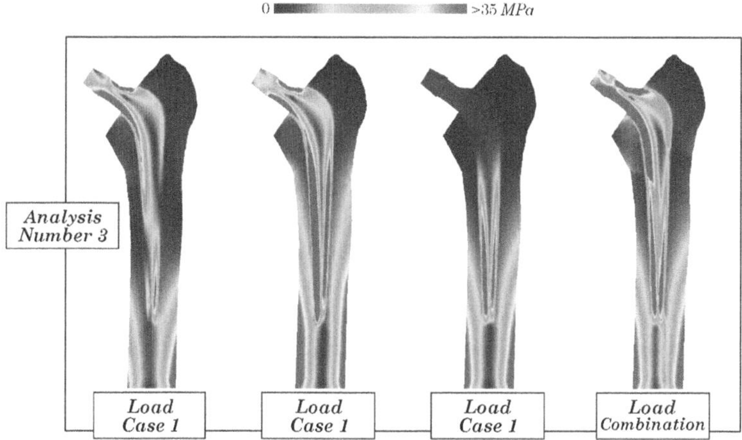

Fig. 7.74 Obtained von Mises effective stress distribution for each one of the analysis performed considering $\rho_{app}^{control} = 1.20\,\mathrm{g/cm^3}$ (Analysis number 3)

model assumes the same geometry presented in Fig. 7.64 and the same irregular nodal distribution presented in Fig. 7.65a. The natural and essential boundary conditions presented in Fig. 7.65b are assumed and the load cases described in Table 7.6 are considered again.

In this fourth study the peripheral cortical layer was disregard. Therefore, the present study is similar with the third study previously described. The only difference regards the mechanical properties of the implant stem. In this example the

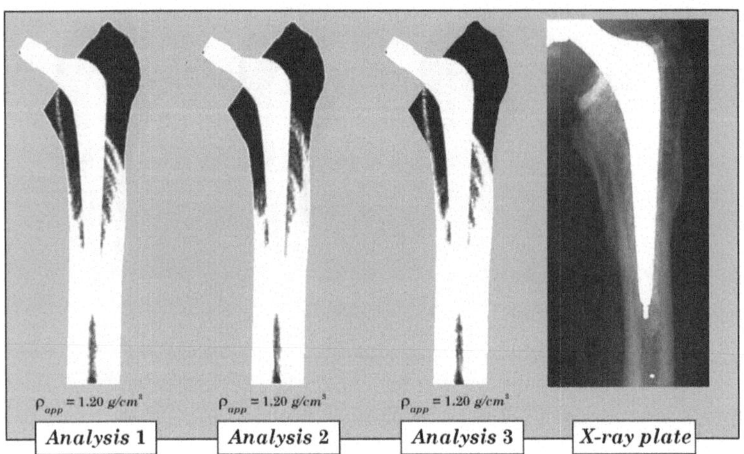

Fig. 7.75 Comparison between the three distinct analysis and the X-ray plate from [32]

femoral implant is made of a homogeneous and isotropic bonelike material with the following mechanical properties: $E = 13.7$ GPa and $\upsilon = 0.30$.

The mechanical properties of the bone tissue depend on the current apparent density and are obtained with the considered phenomenological law. Again, it was not considered in the analysis a cemented interface between the implant and the bone tissue.

The obtained trabecular distributions for each one of the control values are presented in Fig. 7.76. It is evident a notorious reduction of the apparent density of the bone tissue on the femur head. Nevertheless, comparing the results in Fig. 7.76 with the results of the three previous analyses it is clear that varying the material properties of the femoral implant leads to important changes in the trabecular arrangement of the surrounding bone tissue. Combining the three load cases, respecting the load cycles of Table 7.6, it is possible to obtain the results shown in Fig. 7.77.

Comparing Fig. 7.77 with Fig. 7.75 it is evident the modification of the bone tissue apparent density field. The femur head is now sustained by a denser trabecular column on the left side and on the right side of the femur the trabecular architecture has clearly changed.

In Fig. 7.78 are presented the von Mises effective stress distributions for each one of the considered load cases and for the load combination assuming the control value $\rho_{app}^{control} = 1.20\,\text{g/cm}^3$. In opposition to previous solutions, the results presented in Fig. 7.78 show a stress concentration just on the implant neck, near the intertrochanteric section.

Based on the presented results, it seems that using femoral implants, with mechanical properties similar with the bone tissue mechanical properties, can help decreasing the necrosis of the femur head after the surgical treatment of intertrochanteric fractures combined with the loss of functionality of the great trochanter.

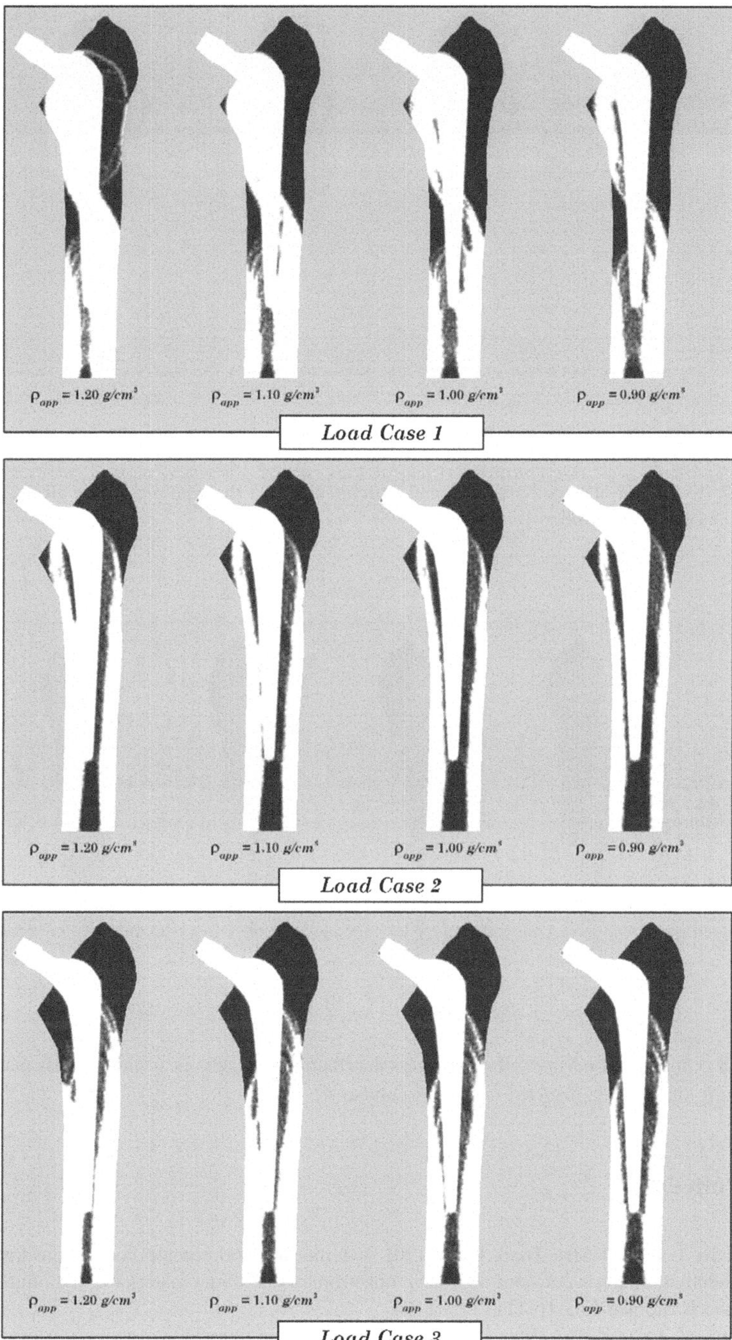

Fig. 7.76 Evolution of bone tissue trabecular architecture for each one of the studied load cases (Analysis number 4)

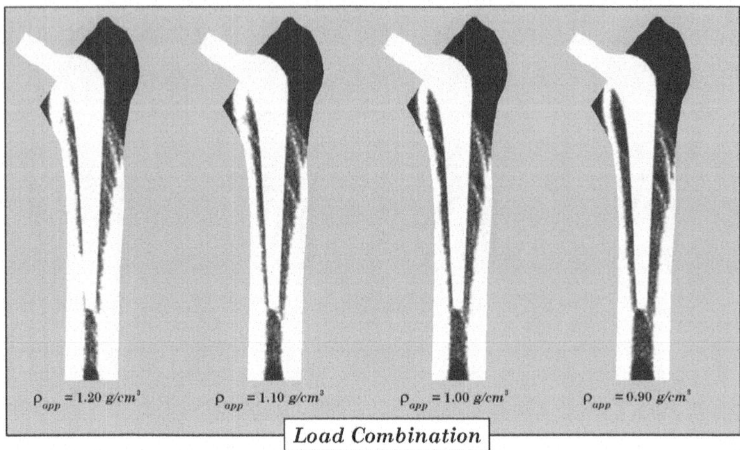

Fig. 7.77 Evolution of bone tissue trabecular architecture for the combination of all the studied load cases (Analysis number 4)

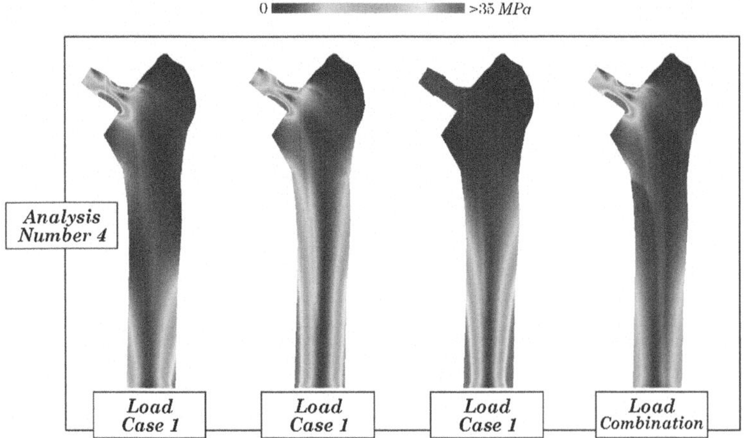

Fig. 7.78 Obtained von Mises effective stress distribution for each one of the analysis performed considering $\rho_{app}^{control} = 1.20\,\text{g/cm}^3$ (Analysis number 4)

References

1. Belinha J, Jorge RMN, Dinis LMJS (2013) A meshless microscale bone tissue trabecular remodelling analysis considering a new anisotropic bone tissue material law. Comput Meth Biomech Biomed Eng 16(11):1170–1184
2. Belinha J, Jorge RMN, Dinis LMJS (2012) Bone tissue remodelling analysis considering a radial point interpolator meshless method. Eng Anal Boundary Elem 36(11):1660–1670
3. Mullender MG, Huiskes R, Weinans H (1994) A physiological approach to the simulation of bone remodeling as a selforganizational control process. J Biomech 27(11):1389–1394

4. Xinghua Z, He G, Dong Z, Bingzhao G (2002) A study of the effect of non-linearities in the equation of bone remodeling. J Biomech 35:951–960
5. Chen G, Pettet G, Pearcy M, McElwain DLS (2007) Comparison of two numerical approaches for bone remodelling. Med Eng Phys 29:134–139
6. Shefelbine SJ, Augat P, Claes L, Simon U (2005) Trabecular bone fracture healing simulation with finite element analysis andfuzzy logic. J Biomech 38:2440–2450
7. Poiate I, Vasconcellos A, Mori M, Poiate E (2011) 2D and 3D finite element analysis of central incisor generated by computerized tomography. Comput Methods Programs Biomed 104(2):292–299
8. Cheng R, Zhou X, Liu Z, Hu T (2007) Development of a finite element analysis model with curved canal and stress analysis. J Endod 33(6):727–731
9. Scarfe W, Levin M, Gane G, Farman A (2009) Use of cone beam computed tomography in endodontics. Int J Dent (ID634567):20
10. Kim W, Voloshin AS (1995) Role of plantar fascia in the load bearing capacity of the human foot. J Biomech 28(9):1025–1033
11. Gefen A (2002) Stress analysis of the standing foot following surgical plantar fascia release. J Biomech 35(5):629–637
12. Cheung JTM, Zhang M, An KN (2006) Effect of Achilles tendon loading on plantar fascia tension in the standing foot. Clin Biomech 21:194–203
13. Cheng HYK, Lin CL, Wang HW, Chou SW (2008) Finite element analysis of plantar fascia under stretch—The relative contribution of windlass mechanism and Achilles tendon force. J Biomech 41(9):1937–1944
14. Beaupré GS, Orr TE, Carter DR (1990) An approach for time dependent bone modelling and remodelling. Theoretical development. J Orthop Res 8(5):651–661
15. Beaupré GS, Orr TE, Carter DR (1990) An approach for time dependent bone modelling and remodelling. A preliminary remodelling simulation. J Orthop Res 8(5):662–670
16. Jacobs CR, Levenston ME, Beaupre GS, Simo JC, Carter DR (1995) Numerical instabilities in bone remodelling simulations: the advantages of a node-based finite element approach. J Biomech 28(4):449–459
17. Jacobs CR, Simo JC, Beaupré GS, Carter DR (1997) Adaptive bone remodeling incorporating simultaneous density and anisotropy considerations. J Biomech 30(6):603–613
18. Pettermann H, Reiter T, Rammerstorfer FG (1997) Computational simulation of internal bone remodeling. Arch Comput Meth Eng 4(4):295–323
19. Doblaré M, García JM (2002) Anisotropic bone remodelling model based on a continuum damage-repair theory. J Biomech 35(1):1–17
20. Rossi JM, Wendling-Mansuy S (2007) A topology optimization based model of bone adaptation. Comput Meth Biomech Biomed Eng 10(6):419–427
21. Coelho PG, Fernandes PR, Rodrigues HC, Cardoso JB, Guedes JM (2009) Numerical modeling of bone tissue adaptation—A hierarchical approach for bone apparent density and trabecular structure. J Biomech 42(7):830–837
22. Jang IG, Kim IY (2010) Computational simulation of simultaneous cortical and trabecular bone change in human proximal femur during bone remodeling. J Biomech 43:294–301
23. Doblare M, Garcia JM (2001) Application of an anisotropic bone-remodelling model based on a damage-repair theory to the analysis of the proximal femur before and after total hip replacement. J Biomech 34:1157–1170
24. Doblaré M, Cueto E, Calvo B, Martínez MA, Garcia JM, Cegoñino J (2005) On the employ of meshless methods in biomechanics. Comput Methods Appl Mech Eng 194:801–821
25. Martínez RJ, García JM, Domínguez J, Doblaré M (2009) A bone remodelling model including the directional activity of BMUs. Biomech Model Mechanobiol 8:111–127
26. Jang IG, Kim IY (2010) Application of design space optimization to bone remodeling simulation of trabecular architecture in human proximal femur for higher computational efficiency. Finite Elem Anal Des 46(4):311–319

27. Lian Z, Guan H, Ivanovski S, Loo YC, Johnson NW, Zhang H (2010) Effect of bone to implant contact percentage on bone remodelling surrounding a dental implant. Int J Oral Maxillofac Surg 39:690–698
28. Chou HY, Jagodnik JJ, Muftu S (2008) Predictions of bone remodeling around dental implant systems. J Biomech 41:1365–1373
29. Johnell O, Kanis J (2005) Epidemiology of osteoporotic fractures. Osteoporos Int 16(2):S3–S7
30. Chen CM, Chiu FY, Lo WE (2001) Avascular necrosis of femoral head after gamma-nailing for unstable intertrochanteric fractures. Arch Orthop Trauma Surg 121:505–507
31. Vicario C, Marco F, Ortega L, Alcobendas M, Dominguez I, López-Durán L (2003) Necrosis of the femoral head after fixation of trochanteric fractures with Gamma Locking Nail—A cause of late mechanical failure. Int J Care Injured 34:129–134
32. Guimarães JAM, Guimarães ACA, Franco JS (2008) Evaluating the use of a proximal femoral nail in unstable trochanteric fracture of the femur. Rev Bras Ortop 43(9):406–417
33. Van der Vis HM, Aspenberg P, Tigchelaar W, Van Noorden CJ (1999) Mechanical compression of a fibrous membrane surrounding bone causes bone resorption. Acta Histochem 101(2):203-212

Index

J. Belinha, *Meshless Methods in Biomechanics*, Lecture Notes in Computational Vision and Biomechanics 16, DOI: 10.1007/978-3-319-06400-0,

N

O

P

Q

R

S

T

U

V

W

X

Y

Z

The manufacturer's authorised representative in the EU is Springer Nature Customer Service Centre GmbH, Europaplatz 3, 69115 Heidelberg, Germany. If you have any concerns regarding our products, please contact ProductSafety@springernature.com

Printed and bound by CPI Group (UK) Ltd, Croydon, CR0 4YY
15/07/2026
02167619-0005